S0-CAH-160

CANDID SCIENCE III

More Conversations with Famous Chemists

Also by the same author

The Road to Stockholm: Nobel Prizes, Science, and Scientists, Oxford University Press, Oxford, 2002.

Candid Science II: Conversations with Famous Biomedical Scientists, Imperial College Press, London, 2002.

Candid Science: Conversation with Famous Chemists, Imperial College Press, London, 2000.

In Our Own Image: Personal Symmetry in Discovery (with M. Hargittai), Kluwer/Plenum, New York, 2000.

Upptäck Symmetri! (Discover Symmetry!, in Swedish, with M. Hargittai), Natur och Kultur, Stockholm, 1998.

Symmetry through the Eyes of a Chemist (with M. Hargittai), Second edition, Plenum, New York, 1995.

Symmetry: A Unifying Concept (with M. Hargittai), Shelter Publications, Bolinas, CA, 1994.

The VSEPR Model of Molecular Geometry (with R.J. Gillespie), Allyn & Bacon, Boston, 1991.

The Structure of Volatile Sulphur Compounds, Kluwer/Reidel, Dordrecht, 1985.

The Molecular Geometries of Coordination Compounds in the Vapour Phase (with M. Hargittai), Elsevier, Amsterdam, 1977.

Forthcoming

Candid Science IV: Conversations with Famous Physicists (with M. Hargittai), Imperial College Press, London.

Edited books

Symmetry 2000 (with T.C. Laurent), Vols. I, II, Portland Press, London, 2002.

Strength from Weakness: Structural Consequences of Weak Interactions in Molecules, Supermolecules, and Crystals (with A. Domenicano), Kluwer, Dordrecht, 2002.

Advances in Molecular Structure Research (with M. Hargittai), Vols. 1–6, JAI Press, Greenwich, CT, 1995–2000.

Combustion Efficiency and Air Quality (with T. Vidóczy), Plenum, New York, 1995.

Spiral Symmetry (with C.A. Pickover), World Scientific, Singapore, 1992.

Fivefold Symmetry, World Scientific, Singapore, 1992.

Accurate Molecular Structures (with A. Domenicano), Oxford University Press, Oxford, 1992.

Quasicrystals, Networks, and Molecules of Fivefold Symmetry, VCH, New York, 1990.

Symmetry 2: Unifying Human Understanding, Pergamon Press, Oxford, 1989.

Stereochemical Applications of Gas-Phase Electron Diffraction (with M. Hargittai), Vols. A, B, VCH, New York, 1988.

Crystal Symmetries, Shubnikov Centennial Papers (with B.K. Vainshtein), Pergamon Press, Oxford, 1988.

Symmetry: Unifying Human Understanding, Pergamon Press, Oxford, 1986.

Diffraction Studies on Non-Crystalline Substances (with W. J. Orville-Thomas), Elsevier, Amsterdam, 1981.

CANDID SCIENCE III

More Conversations with Famous Chemists

István Hargittai

Edited by Magdolna Hargittai

ICP

Imperial College Press

Published by

Imperial College Press
57 Shelton Street
Covent Garden
London WC2H 9HE

Distributed by

World Scientific Publishing Co. Pte. Ltd.
5 Toh Tuck Link, Singapore 596224
USA office: Suite 202, 1060 Main Street, River Edge, NJ 07661
UK office: 57 Shelton Street, Covent Garden, London WC2H 9HE

István Hargittai
Budapest University of Technology and Economics
Eötvös University and Hungarian Academy of Sciences
H-1521 Budapest, Hungary

Magdolna Hargittai
Eötvös University and Hungarian Academy of Sciences
H-1518 Budapest, Pf. 32, Hungary

British Library Cataloguing-in-Publication Data
A catalogue record for this book is available from the British Library.

CANDID SCIENCE III
More Conversations with Famous Chemists

ISBN 1-86094-336-5
ISBN 1-86094-337-3 (pbk)

Printed in Singapore by World Scientific Printers

Foreword

It has been a good many years since I read my first book on the history of science. Although I no longer remember the author or title of that book, I do remember the impression it made on me. First, it awakened in me a deep and abiding interest in science, an interest which has endured and become ever more intense with the passage of time. Then too, it has inspired me to read other books on the history of science and these in turn, as in a chain reaction, impelled me to read far more of science history than I could possibly absorb. Admittedly then, much of what I read in those early years was beyond my comprehension; however, over time, understanding eventually did follow.

There was one facet of this early activity which I found particularly fascinating and rewarding. This was the aspect of science history which was concerned with the lives of the great scientists themselves, the challenges they met, the struggles they faced, and the obstacles they overcame — when they were successful. Not only are these stories of great intrinsic interest, but I believe there is no better way to stimulate the creative impulse than to learn at first hand from the words of the masters, those whose work has survived the test of time.

Who, for example, can fail to be inspired by reading, in his own words, Fermat's statement of his famous "Last Theorem" and his claim to have found a truly marvelous proof of this magnificent theorem? An additional bonus is the insight one gains into the process which guided Fermat to his remarkable and surprising conclusion. The fact that the problem itself eluded all attempts at solution for some 350 years, finally yielding in the most unexpected way, adds an appropriate air of mystery to this extraordinary tale. The proof itself must be regarded as a major mathematical triumph, if not the supreme mathematical gem, of the twentieth century.

I remember, too, with what pleasure I read, for example, Archimedes' account of his discovery of the volume and surface area of the sphere and of Gauss's proof, in his own words, of the constructibility, with straight edge and compass, of the regular polygon of 17 sides, totally unanticipated at the time. In both these cases I felt as if I were sharing in the joy of their discoveries, almost as if I were participating in the discovery process itself, a feeling which I think can be conveyed in no other way. I believe in fact that no third person account can possibly generate the same sensation of pleasure one derives by sharing with the author himself his own feelings of excitement as he recounts the circumstances of his discovery.

These examples then provide, in my opinion, convincing evidence that an important purpose is served by those who take upon themselves the task of scientific biography which attempts to throw light on the origins of discovery, as Professor Hargittai attempts to do in his book. I believe, too, that much of the earlier biographies which already exist were written after the death of the scientist involved, often long after, so that no first hand account is available. Then, too, these earlier biographies were often written by those with little or inadequate scientific background so that the circumstances and the full significance of the scientist's contribution were not always made clear.

Professor Hargittai's contribution, on the other hand, serves the unique and important purpose that it presents in a timely way, while the scientist himself is still alive, and in the first person, the significance of his contribution and the context in which it was made, as he himself has seen it. For this reason the reader himself feels a closer connection with the scientist than would otherwise be possible and develops a deeper appreciation of the author's contribution than he otherwise might. In any event this has been my own experience and I believe that Professor Hargittai's book will serve the same function, the importance of which cannot be exaggerated, since I believe it will stimulate the reader to think in new directions.

Buffalo, New York, September, 2002 — Herbert A. Hauptman

PREFACE

This is the third volume of the *Candid Science* series and I have been gratified by the warm reception for the first two volumes. In this volume, I am returning to chemistry, but the classification in this collection is as free as in the previous two volumes. This collection contains again interviews with a diverse group of scientists. It covers a broad spectrum of research areas in chemistry, including the discovery of new elements and compounds, the VSEPR model, computational chemistry, organic synthesis, natural products, polysaccharides, supramolecular chemistry, peptide synthesis, combinatorial chemistry, X-ray crystallography, reaction mechanism and kinetics, electron transfer in small and large systems, non-equilibrium systems, oscillating reactions, atmospheric chemistry, chirality, and the history of chemistry. The topics are discussed first-hand by, we think, the most appropriate persons.

Sometimes it happens that when one produces a book, it may stay on one's shelf after it has appeared in print, but I do not find this with the *Candid Science* volumes. I freely admit that I often return to them for information or just for reading. I consulted my interviews extensively when I was working on my later book about the Nobel Prize (*The Road to Stockholm*, Oxford University Press) and as I was preparing lectures for my course at the Budapest University of Technology and Economics, entitled "Great Discoveries in 20th Century Science." This academic year, I am presenting the course for the fourth time and this provides an additional stimulus for continuing with my interviews. I can now see how our students absorb the stories about the discoveries and the human backgrounds of the great scientists of earlier times, which give life to the actual scientific papers.

My fascination with interviewing has not waned over the years. I never thought that one could get addicted to such activity, but apparently I did. I still prepare with the same curiosity and anticipation for my next interview as I did for the first. We are working on the fourth volume of the series (M. Hargittai, I. Hargittai, *Candid Science IV: Conversations with Famous Physicists*) and a fifth volume now seems to be emerging as well.

This volume contains more unpublished interviews than the previous volumes. For those published earlier, I express my appreciation to the now defunct magazine *The Chemical Intelligencer* (Springer Verlag, New York) and to the magazine, *Chemical Heritage* (Chemical Heritage Foundation, Philadelphia) for their kind cooperation.

I thank the enthusiastic staff of Imperial College Press (London) and of World Scientific (Singapore), and especially my editorial contact Mr. Suwarno, for their dedicated efforts in bringing out these volumes. I also thank Mr. István Fábri and Ms. Judit Szücs (both in Budapest) for untiring technical assistance. I am grateful for the generous support from the Hungarian Academy of Sciences and the Budapest University of Technology and Economics as well as for the support of our molecular structure research by the Hungarian National Scientific Research Foundation and the Ministry of Education of Hungary. Our scientific research brings us to meetings and laboratory visits whose byproducts are often the interviews presented here and elsewhere. Our family vacations provide additional opportunities in this respect.

Interactions with our children, Balazs, an Assistant Professor of Chemistry at St. Francis University in Loretto, Pennsylvania, and Eszter, a Ph.D. student in the Sociology of the Internet at Princeton University, have brought encouragement for these projects. My wife and fellow scientist, Magdi, Editor of these volumes, continues to be an inspiration.

Budapest, Fall of 2002 István Hargittai

CONTENTS

Glenn T. Seaborg, 1995 (photograph by I. Hargittai).

1

Glenn T. Seaborg

Glenn T. Seaborg (1912–1999) was University Professor of Chemistry, Associate Director-at-Large of the Lawrence Berkeley Laboratory, and Chairman of the Lawrence Hall of Science at the University of California, Berkeley. From 1937–1939, following receipt of his Ph.D. in chemistry from the University of California at Berkeley, Seaborg served as the Personal Research Assistant to the great physical chemist Gilbert N. Lewis, Dean of the College of Chemistry at the University of California at Berkeley. Seaborg shared the Nobel Prize in Chemistry in 1951 with Edwin M. McMillan, "for their discoveries in the chemistry of the transuranium elements." Seaborg was one of the co-discoverers of plutonium (element 94). During World War II, he headed the group at the University of Chicago's Metallurgical Laboratory that devised the chemical extraction processes used in the production of plutonium for the Manhattan Project. He and his coworkers subsequently discovered nine more transuranium elements. Not long before he passed away, Seaborg was honored by having the element 106 named after him to be "seaborgium," with the chemical symbol Sg. In 1944, Dr. Seaborg formulated the actinide concept of heavy-element electronic structure, which accurately predicted that the heaviest naturally occurring elements, together with synthetic transuranium elements, would form a transition series of actinide elements, analogous to the rare-earth series of lanthanide elements. This concept is one of the most significant changes in the periodic table since Dmitri I. Mendeleev's 19th century design. Seaborg's discoveries also include many isotopes that have practical applications in research (such as iodine-131, technetium-99m, cobalt-57, cobalt-60, iron-

55, iron-59, zinc-65, cesium-137, manganese-54, antimony-124, californium-252, americium-241, and plutonium-238), as well as the fissile isotopes plutonium-239 and uranium-233.

I recorded a conversation with Glenn Seaborg on April 2, 1995, in Anaheim, during a chemistry meeting. This conversation follows first.*

Is nuclear chemistry losing importance nowadays?

It's not as central a field as it used to be, but there is still an awful lot of interesting work going on. There are also many interested young people in the field though perhaps not as many as used to be in the past. Environmental chemistry is more popular these days. However, there is no contradiction between nuclear chemistry and environmental chemistry. Also, I don't think that people's fear of nuclear power plants is justified, but the media paint a bad picture and connect it with the waste disposal problem.

Glenn Seaborg and István Hargittai in Anaheim, 1995 (by unknown photographer).

*This part was originally published in *The Chemical Intelligencer* **1995**, *1*(4), 1–7

How do you feel about the controversy in naming new elements?

I'm disappointed and it's so unjustified. The reasons they give just don't make any sense. They say that they don't want to name an element after a living person. However, that has been done throughout history. Gallium was named after a living person in 1875. They also say that they want a perspective of history. Well, my discoveries were half a century ago — plutonium and the actinide concept for placing the heavy elements in the periodic table and so forth. You can't get much more perspective of history than that, and I don't believe they are going to decrease in importance. The actinide concept is going to be part of the periodic table forever.

Do you consider placing the actinides in the periodic table your most important achievement?

Yes. It is the most important and even that was 50 years ago. Most of the people for whom elements have been named have had a much shorter period of evaluation between their discovery and when the element was named than 50 years.

You have had a large time span to observe the development of chemistry.

It has increasing importance especially in biological chemistry. The general advances in the chemistry of disease are extremely important. The role of chemistry in the biological sciences may determine our future.

Why, do you think, students find chemistry difficult?

It just reflects back. They read the newspapers also. However, if you have teachers who are motivated by science, that helps a lot. We often have chemistry teachers who are not trained in chemistry and, therefore, are not particularly inspiring to their students. The whole precollege education system just needs more money — more money for buildings, for equipment, to pay the teachers competitive salaries. We are not going to solve the problem of precollege education in the United States until we put more money in it.

You have recently edited a book, which is a compilation of your publications. It shows not only your interest in nuclear chemistry but a much broader interest and a desire to communicate to a broader circle of people. What is your most important message to this broader readership?

It is that scientists pay attention to the sociological aspects of their work and the impact their work has on society. When I was a young man, it was frowned upon to be concerned at all with the impact of your work. You were just supposed to carry out your research for research's sake. That has changed very much.

How should scientists pick a problem?

I don't say that they should change from basic research to applied research; they still may be carrying on the basic research, but they need to explain to the public the importance of it. So many things are discovered. For example, in the case of my discovery of all these radioactive isotopes — technetium-99m, iodine-131, cobalt-60, and cesium-137, all turned out to have tremendous practical applications in nuclear medicine. There are millions of applications per year now.

Did you anticipate those applications?

No, not at all.

Did you write research proposals for support?

No, we didn't. Some of us were paid as research associates or faculty members. We had a little bit of a budget to buy equipment and so forth. This was before the era of research proposals. The discovery of plutonium, for example, was personal research. It was not supported by the government or any other research foundation. Support is easier today. However, it is more difficult for young scientists today to find a position. There is a little oversupply of scientists in both chemistry and physics. However, we have been through that before, and it's the law of supply and demand sort of adjustment and takes care of itself.

You wrote a beautiful article about G. N. Lewis. Although he has made one of the most important discoveries in chemistry, the two-electron covalent bond, he never received the Nobel Prize. However, his fame didn't suffer from this.

No, his fame didn't suffer from this, but just the same, really it's an injustice. If I may say so, it is nice to have a Nobel Prize.

In what way did the Nobel Prize change your life?

It did not change as much as most people's because I was so young — I was 39 years old — and I just kept going. I do read about recent Nobel Prize winners who claim that it adversely affected their lives; it did not in my case. I was very active after and continued.

How do you feel about the gap between what C. P. Snow called the two cultures.

We need to educate the general public about science. That's an area where I put a lot of my efforts for the last 40 years. The general public is not sufficiently conversant with the science that affects their lives so much. That's the way I view the two-culture problem. It all goes back to not having a sufficient number of teachers trained in the subject matter who are embued for the subject. For example, I was completely uninterested in science until my junior year in high school. Then I took a course in chemistry that I really had to take in order to be eligible to go to the state university, UCLA. My high school was in Watts, which was the Harlem of Los Angeles. We lived in a neighboring town, but we didn't have a high school so I was bused to Watts. So I took chemistry there and I was absolutely turned on by an inspiring high school teacher. He just told us about interesting work that was going on, controversies, and the big discoveries that were made. He had us on the edge of our seats, and I said, that's for me. He had a Master's degree in chemistry, and he knew chemistry. Somebody with a degree in education couldn't have done that.

Did he live to see your success?

Yes, and I told him all about it. His name was Dwight Logan Reid. He determined my career. Then next year, my senior year, he taught physics. Ours was a small high school and we just alternated, and I liked his physics course even better. He was knowledgeable in physics too. In those days, of course, nobody knew what a physicist was, so I majored in chemistry but I took a lot of physics at UCLA. Then when I went to Berkeley to do my graduate work, I did it in nuclear physics. However, I consider myself a chemist. That's been the key to my discoveries, plutonium and everything else.

There was then another teacher very important to me, a teacher of physical chemistry at UCLA, James Blaine Ramsey, and then G. N. Lewis when I went to Berkeley. In the first two years after I'd got my Ph.D. I was his personal assistant and I could not have had a better start, even though it was not in nuclear physics.

Do you have opportunities to tell about all this to students?

Yes, I do.

I noticed that you still have a position in the Lawrence Berkeley Laboratory.

I am a University Professor of Chemistry and I am Associate Director-at-Large of the Lawrence Berkeley Laboratory. I am also Chairman of the Lawrence Hall of Science. It's a public science teaching facility for precollege science, named after Earnest Lawrence. The Hall develops precollege curricula for high school and even more for grammar school and trains teachers in sciences. It also has interactive exhibits. It was my idea. Ernest Lawrence died in 1958 when I was Chancellor of the University of California at Berkeley. The idea for a memorial arose, and I made the suggestion that there a Hall of Science be created, and then it was given the name Lawrence Hall of Science. Thousands of teachers get educated there yearly.

Is it very expensive running such a facility?

We have a problem. It has never been completely supported by the University. I left after I'd created it. I became Chairman of the Atomic Energy Commission, so I was not there to see it take root, and so we never got the support we should from the University itself. We work on a budget that comes from grants and public donations.

Do you have family?

Six kids.

Did any of them follow you in their choice of profession?

One did a little bit — went into zoology — but most of them avoided it because they had a disadvantage. They had my name. When they went into a class, they figured, if I take chemistry or physics, they are going to think that "I am a dumb person" because I can't live up to my father's reputation. It makes it difficult for children if they have a very prominent parent. You'll find time after time that it affects the children.

How about your parents?

They were working people. My mother was born in Sweden, and she just had a sixth-grade education. My father was a machinist, a high-school graduate in a little mining town in Northern Michigan, Ishpeming. My father's father was a machinist, and his father was a machinist. I have a feeling that they were doing something as close to science as they could in their environment. They couldn't be scientists, nobody even knew what the word meant. My father was born in Ishpeming but both of his parents came from Sweden. I am 100% Swedish. I spoke Swedish as my first language. My mother communicated to me in Swedish. When I go to Sweden, which I have not done recently, in many cases my relatives can't speak English so we communicate in Swedish. I still have many relatives there. In fact, I responded to the toast at the dinner following the Nobel ceremony by speaking Swedish.

Did your parents live to witness your success?

Yes.

Could they appreciate it?

Absolutely.

> In 1998, I asked Dr. Seaborg to up-date our 1995 interview. I always felt that our original conversation was unduly short, besides, there had been a pleasing development in the Seaborgium story.†

Seaborgium (106) The element with the atomic number 106 was synthesized and identified over 20 years ago (1974) but was not officially given a name until last year. The investigators who have now been officially sanctioned as the discovery team were a group from the Lawrence Berkeley Laboratory (LBL) — Albert Ghiorso, J. Michael Nitschke, Jose R. Alonso,

---

†This part was originally published in *The Chemical Intelligencer* **1998**, *4*(4), 37–40

Carol T. Alonso, Matti Nurmia, and I — and from the Lawrence Livermore National Laboratory (LLNL) — E. Kenneth Hulet and Ronald W. Lougheed. The experiment was performed at LBL's Heavy Ion Linear Accelerator (HILAC) by bombarding ^{249}Cf with ^{18}O to produce the isotope $^{263}106$. The new nuclei were shown to decay by the emission of alpha particles with a half-life of 0.9 second and a principal alpha energy of 9.06 MeV to the previously known $^{259}_{104}Rf$, which in turn was shown to decay to the known $^{255}_{102}No$. Thus the atomic number of the new nucleus was firmly established by a genetic relationship to its daughter and granddaughter.

At about the same time, another claim to the discovery of element 106 was made by a Russian group working at the Laboratory of Nuclear Reactions at the Joint Institute for Nuclear Reactions in Dubna, Russia — Georgiy N. Flerov, Yuri Ts. Oganessian, Yu. P. Tretyakov, A. S. Iljinov, A. G. Demin, A. A. Pleve, S. P. Tret'yakova, V. M. Plotko, M. P. Ivanov, N. A. Danilov, and Yu. S. Korotkin. They reported the observation of a spontaneous fission activity with a half-life of 4–10 ms, produced by bombarding $^{207}_{82}Pb$ with $^{54}_{24}Cr$, and which they assigned to $^{259}106$ on the basis of reaction systematics.

Because of the competing claims, the two groups agreed not to propose a name for element 106 until it could be determined which group had priority for the discovery. In 1976, an international group of scientists proposed criteria for the discovery of new chemical elements and suggested that one of these should be that new elements not have a name proposed by the discoverers until the observation is confirmed.

In 1984, another group of investigators at the Dubna laboratory showed that the spontaneous fission activity attributed in 1974 to element 106 by the Flerov group was actually due primarily to the daughter of element 106 and not to element 106, thus effectively invalidating the Dubna claim to the discovery of element 106. In 1992, the Transfermium Working Group credited the discovery of element 106 to the Berkeley-Livermore group.

In 1993, another LBL group of investigators, using the 88-inch cyclotron, repeated and confirmed the results obtained by Ghiorso *et al.* on the discovery of element 106 in 1974. Thus, the way was cleared for the Ghiorso group to suggest a name for element 106.

The eight members of the Ghiorso group each suggested a name or names with very disparate results. Some of the names suggested to be honored were the famous and extraordinarily versatile nuclear physicist Luis

Alvarez; the eminent French nuclear scientist Frédéric Joliot (a transfer of the Dubna suggestion for naming element 102); the scientific giant Sir Isaac Newton; the famous inventor Thomas Edison; the famous scholar and inventor Leonardo da Vinci; early explorers such as Christopher Columbus, Ferdinand Magellan, and the mythical Ulysses; the great American statesman George Washington; a Russian scientist such as Peter Kapitza or Andrei Sakharov; and the native land of a member of the discovery team, Finland.

Then Ghiorso took the initiative to suggest another approach. He met with (or contacted) the other six members of the discovery group (i.e., without my participation) and suggested that element 106 be named "seaborgium" (symbol Sg). He received unanimous agreement on this suggestion. Ghiorso then met with me and made this proposal for the naming of element 106. Caught completely by surprise, I asked for time to consider it. After discussing it with my wife Helen and a number of others, all of whom were enthusiastically in favor of this suggestion, I assented to this action.

It was decided to have the announcement made by Kenneth Hulet of the Lawrence Livermore National Laboratory at the meeting of the American Chemical Society (ACS) in San Diego to be held in March 1994. The occasion would be the opening session on Sunday, March 13, of the symposium in honor of Hulet in recognition of his receipt of the prestigious ACS Award for Nuclear Chemistry. This was done by Hulet in a dramatic session, well attended, that morning in San Diego.

The name seaborgium for element 106 has been endorsed by the ACS Committee on Nomenclature and Board of Directors.

Then followed a period of trouble extending over a period of a couple of years. The Commission on Nomenclature of Inorganic Chemistry (CNIC) of the International Union of Pure and Applied Chemistry (IUPAC), which should approve the names for newly discovered elements, rejected the name "seaborgium" on the unreasonable basis that I was still alive (as I mentioned in my 1995 interview.) However, the CNIC and IUPAC had not followed its own rules, which required that the provisional names for new elements should go out to the world scientific community for comments. When this was done they received an overwhelming endorsement for the name "seaborgium" for element 106. The IUPAC, at its meeting in Geneva, Switzerland on August 30, 1997, adopted this name, and it is now official for use throughout the future.

The chemical properties of seaborgium have been studied and found to resemble those of tungsten, its homologue in the periodic table. I am looking forward to obtaining evidence for seaborgous chloride, seaborgic sulfate, sodium seaborgate, etc.

Element 110 In August-September 1991, an experiment to synthesize element 110 by the ${}^{59}_{27}Co + {}^{209}_{83}Bi \rightarrow {}^{267}110 + n$ reaction was performed at the Lawrence Berkeley Laboratory by a group under the leadership of Albert Ghiorso. One event with many of the expected characteristics of a successful synthesis of ${}^{267}110$ was observed. The basic idea of the experiment was to bombard ${}^{209}Bi$ with ${}^{59}Co$ ions, furnished by the Super-HILAC, at an energy near the barrier to observe the production of a very short-lived alpha particle emitter (predicted energy 11–12 MeV) that would be connected genetically to known nuclei of lower atomic number. (This is the method of establishing the atomic number of a new element through its genetic relationship with known descendants of decay, such as was used for the discovery of seaborgium, element 106, as described in the previous section.)

A new gas-filled magnetic separator and detector was especially constructed for this experiment. One unique event was observed in the entire 41-day interval, the element-110 candidate, with the observed magnetic rigidity, energy, and energy loss expected for element 110. This atom of element 110, presumed to be ${}^{267}110$, decayed with the emission of an 11.6-MeV alpha particle after 4 μs (reasonably consistent with predictions). Due to a failure of the electronics, the alpha decay of the daughter ${}^{263}_{108}Hs$ to ${}^{259}_{106}Sg$ could not be detected. Presumably, the ${}^{259}Sg$ then underwent undetected electron capture decay to ${}^{259}_{95}Ha$, which in turn decayed in 6.0 seconds to ${}^{255}_{103}Lr$ by the emission of an alpha particle which escaped in the backward direction. Finally, the known 22-seconds ${}^{255}Lr$ decayed in 19.7 seconds with the observed full energy of its known 8.3-MeV alpha particle.

Alternative scenarios, such as might be devised from transfer-product decays or random events, were found to be unlikely by orders of magnitude and, on balance, the association of this event with the formation of element 110 appears to be the simplest and most probable explanation of the observation. Unfortunately, the Super-HILAC has been shut down so that it is impossible to repeat the experiment in the very near future.

In November 1994, an international group working at the GSI laboratory in Germany announced the synthesis and identification of an isotope of element 110, $^{269}110$ (half-life, 270 μs, 11.13-MeV alpha particles), produced in the reaction $^{208}_{82}\mathrm{Pb} + ^{62}_{28}\mathrm{Ni} \rightarrow ^{269}110 + n$. The identification was made by the observation of four subsequent alpha-particle decays.

A Russian-American collaboration led by Yuri A. Lazerev of Dubna and Ronald W. Lougheed of the Lawrence Livermore National Laboratory announced in January 1995, the synthesis of element 110 by the reaction $^{244}_{94}\mathrm{Pu} + ^{34}_{16}\mathrm{S} \rightarrow ^{273}110 + 5n$ at Dubna.

The work at the GSI laboratory is the most definitive, and our work at the Lawrence Berkeley Laboratory and the work at Dubna will have to be confirmed in order to allow these two groups to be recognized as participants in the discovery and naming of element 110.

Precollege Education Standards for Science I have been involved for more than 40 years, since before the days of the wake-up call of the Soviet Sputnik of October 4, 1957, in advocating and working toward the improvement of precollege education in science. A notable contribution was my participation as a Member of the National Commission on Excellence in Education in the preparation and issuance of the influential report "A Nation at Risk."

Most recently, I have been involved with the group "Associated Scientists," which has vied for the task of writing the Academic Content Standards for precollege science in the state of California. As a result of this effort, I was appointed in January 1998 by California Governor Pete Wilson to the Commission for the Establishment of Academic Content and Performance Standards and to serve as Chairman of its Science Committee. In this capacity, I am striving to come up with stringent standards emphasizing basic concepts and a definite curriculum for each grade, kindergarten through the 12th grade.

Predictions for Chemistry in the Upcoming Millennium Much progress will be made in the chemistry of life processes — in biochemistry, molecular biology, and related areas concerning the study of proteins, enzymes, nucleic acids, and other macromolecules. Chemical and biological investigations at the molecular and cellular levels, aided by enormously efficient computers, will elucidate the origin of life and perhaps lead to the artificial creation of life. Biochemical genetics will give us a great deal of control over the genetic code and, beneficently applied (which will pose a real challenge), should result in a reduction or elimination of genetic defects.

Immunochemistry, computer-aided molecular medicine, and chemotherapy should lead to the alleviation, treatment, cure or prevention of our major ailments, including "mental" illness, and to a slowing of the aging process. We will understand the structure, mechanism, and functioning of nerve and brain, leading to control of our memory, through investigations in neurochemistry and the related fields of neuroanatomy, neurophysiology, statistical biology, and experimental psychology. Biochemical engineering should make available implantable (microcomputer-assisted) artificial hearts, kidneys, eyes (instruments to permit the blind to "see"), ears (instruments to permit the deaf to "hear"), and other bodily parts and organs.

Medicine underwent a revolution in the middle of this century when it became possible to cure some infectious diseases with antibiotics. Now medicine is undergoing another revolution based on the applications of molecular biology — the mapping, cloning, and study of human genes to understand the body's normal functions at the molecular level, made possible by the unraveling of the role of DNA by James Watson and Francis Crick in the 1950s. A consequence of this will be gene therapy, one of the most exciting ramifications of molecular biology. The first successful gene therapy in a mammal was reported in 1984, when researchers injected a growth hormone gene into a fertilized mouse egg to overcome a genetic deficiency of growth hormone. The power of recombinant DNA technology lies in the opportunity it affords, for the first time, to produce virtually unlimited quantities of practically any protein and the promise it gives for the production of new drugs.

We are now engaged in a multi-year, multi-disciplinary, technological undertaking to order and sequence the human genome, the complete set of instructions guiding the development of a living organism. The actual planned ordering and sequencing involves coordinated processing of some 3 billion bases from a reference human genome, and the consequent more complete understanding of human genes should lead to great advancement in the diagnosis and treatment of human diseases.

Another biological advance lies in monoclonal antibodies based on proteins produced by white blood cells in response to a foreign substance, such as a virus. Monoclonal antibodies are ideal for diagnosing infectious diseases. This technique, along with sophisticated computers, should make possible much more rapid diagnoses and, consequently, earlier selection of therapy.

Chemistry, properly supported, will help solve our energy, food, and mineral resource problems, even with our expanding worldwide population (hopefully at a diminished rate). Our endless supply of solar energy will be put to practical use through processes yet to be discovered — direct catalytic conversion to electricity, splitting of water to produce hydrogen fuel, or widespread bioconversion of vegetation and waste products. Chemists will increase the efficiency of extraction of new sources of minerals, and materials scientists will synthesize substitute materials from more abundant supplies.

Chemists will synthesize millions of new compounds tailored for a wide spectrum of practical uses. Nuclear chemists will be involved in the synthesis of additional chemical elements, hopefully in the region of the superheavy elements predicted to exist in the "island of stability."

These are but a few of the areas in which chemical progress will be made, and I have placed emphasis on the practical applications. The advances in theoretical chemistry will be tremendous in scope, and I shall not even attempt to make any predictions.

Some Thoughts for the Future An important factor in the future, transcending the science of chemistry, will be the new public attitudes toward basic science and science in general — that is, the growing attitude toward ethical and human value considerations. The focus of this concern often is not on the question of whether the work is worth doing but instead on whether its potential harmful impact may outweigh any good it could do — that is, whether the research or project should be initiated at all. This attitude is affecting work on energy resources and technologies, biological research, aircraft development, and advances in the social sciences and education. This is going to have an increasing effect on the support and conduct of science, and I think most scientists are recognizing this.

As in the other cases of new influence, it is going to have its good and bad effects. Essentially, it is vital that science does serve the highest interest of society and contribute to the fulfillment of human values. And I believe that the science community for the most part is acting very responsibly and responsively in this direction. In many areas of research, such as genetic experimentation, atmospheric work, and the effects of chemicals on human health and the environment, it has taken the lead in placing human concerns above all.

But it should be realized that while there are certain values and ethical codes of a universal nature, there are also values that are more closely associated with the tastes, likes and dislikes, habits, and culturally induced beliefs of various individuals and groups attuned to certain so-called lifestyles. In a democratic society — and particularly one of growing advocacy and activism — there are bound to be many conflicts over these. And science and technology, with their increasing influence on life in general, certainly will be caught up in many of these. If this is the case, it may be essential that we find a way to establish some broad codes of conduct and values by which we can use science and technology to maximize human benefits within a framework of some type of consensus value scale. It seems to me that we must do this in order to avoid being paralyzed by a kind of case-by-case value judgment of all that we do. This does not mean that technology assessments and risk/benefit studies of individual concepts should not be conducted. Nor does it mean that science should not maintain a most profound sense of responsibility toward safeguarding society from possible errors on its part or misapplications of its work. It does mean, however, that we must find a way to avoid having a "tyranny of parochial interests" when it comes to the possibility of advancing the general good through scientific progress.

Perhaps I can summarize by suggesting that future directions of chemistry, and science and technology in general, may be influenced by two broad goals: more fully establishing the boundaries — physical, environmental and social — in which we can operate; and providing the knowledge capital that will allow us to operate within them. That knowledge capital — a product of basic research — upon which we have drawn so heavily in the recent past and which we must replenish with new ideas might also allow us to compensate somewhat for declining physical capital and higher cost resources.

Finally, a few general thoughts. Our success in chemistry, and science in general, over the past century, and especially the last few decades, has brought us to a high level of material affluence, but this success also has fostered many new problems for the world. It also has given many people the notion that science should move us toward a utopian, problemless, riskless society. But this is a false notion. We live and always will live in a dynamic situation, amid problems whose solutions will breed other kinds of problems, and in a society where the leaps of progress will be proportionate to the risks taken. Even within the bounds of a "steady-state society," a "no-growth society," or any other scheme of population-resource-energy equilibrium we might achieve, there always will be change and creative growth that will challenge the human intellect. There always will be dangers, risks, and increasing responsibilities that will drive us toward a new level of excellence in all we do or try to achieve. This is the process of human evolution at work, a process that started with man's ascendancy and will continue for some time.

William N. Lipscomb, 1995 (photograph by I. Hargittai).

2

William N. Lipscomb

William N. Lipscomb (b. 1919) received the chemistry Nobel Prize in 1976 "for his studies on the structure of boranes illuminating problems of chemical bonding." Our conversation took place in his office at the Gibbs Building on the campus of Harvard University. Dr. Lipscomb is currently Abbot and James Lawrence Professor Emeritus of Chemistry at the Department of Chemistry, Harvard University, Cambridge, Massachusetts. I spent a pleasant morning with Professor Lipscomb, on Monday, July 24, 1995. He instructed his secretary to bar all but the most urgent calls. Three got through. One was about some chemical experiments and the other two were in reference to Dr. Lipscomb's rehearsals and forthcoming performances as a clarinetist. I thought to mention this here because in spite of my original intention, I forgot to ask him about his activities as a performing musician, and I had known this to be an important part of his life.*

Please tell us about your early days and your background.

I grew up in Kentucky. My father was a physician, an MD, and my mother a music teacher. She taught voice. When I was 11 years old, she gave me a chemistry set, one of those things you buy in a store, and I started experimenting with it. That's not unusual. Many boys get these sets and play with them for two weeks. Only I didn't stop playing. Then I discovered that I could buy additional apparatus and chemicals using my father's privilege

*This interview was originally published in *The Chemical Intelligencer* **1996**, *2*(3), 6–11 © 1996, Springer-Verlag, New York, Inc.

at the drug store. He had a special rate because he was a physician. So I began building up a home laboratory. I did both chemical experiments and electrical experiments. I also put a wire across the street to some friends who were also interested in science. In fact, within about four houses either way there were several boys about my age. Among this group of boys, there were eventually two Ph.D.s in physics, two MDs, one civil engineer and me. We had quite a group interested in science. This is very important.

Would this be typical Kentucky?

They have three Nobel laureates now. If all the 50 states had three laureates, we would have 150 in the Union. But it's very unusual, and you're right, it's not Massachusetts, it's not California, where there are probably more, or New York. The high school that I went to actually had very poor facilities. It was in the county, not the city, on the edge of Lexington. It was Picadome High School and it eventually became Lafayette High School. They had physics and they had biology courses but they didn't have a chemistry course. As I learned later, my father went to the Superintendent of the county schools and said, you'd better have a chemistry class there or I'll take my son out and put him downtown in the big high school. So they made a chemistry course, and I helped make it actually. When I came to the course, the teacher said, well you already know all the chemistry we are going to have, why don't you just show up for examinations. So I stayed in the back of the room and did some experiments, a little piece of original research, but I didn't publish it. I still have the manuscript.

What was it about?

The teacher, Fred Jones, gave me his book on organic chemistry, and I saw this reaction: sodium acetate plus sodium hydroxide gives sodium carbonate plus methane. This is a way of preparing organic compounds, and then there was an extension of the series to produce C_2H_6. This was in the organic book. And I said to myself, why don't I go the other way and start with sodium formate and sodium hydroxide to make hydrogen. That's not going to be in the organic book. It's inorganic chemistry. So I did that and it worked fine. I made hydrogen and analyzed it to make sure it was pure. This I did when I was in high school.

I had a very good teacher although he didn't know much chemistry. The second year of the chemistry course, when I was a senior, I also helped teach the laboratory part of the course.

Why did your father give so much importance to having a chemistry course?

Because he saw that I was going that way, and he was very supportive. He didn't particularly want me to become a physician, although if I had, I would have been the fourth in a row in our family, but he let me do what I wanted. Then I went to the University of Kentucky, which was not a very good university by the standards of the Big Ten. As I went through college on a music scholarship, I had a separate program of my own of reading and trying to work out things that were not taught. For example, nobody in the chemistry department knew anything about quantum mechanics, and I studied it on my own. One of the physics professors helped me a little bit. I also studied mathematics with a man named Fritz John who went on to the Courant Institute in New York and eventually became a member of the National Academy of Sciences. He was a very great mathematician, but they didn't know that then. He had just come over from Germany because of Hitler's deportation and killing of the Jews. He escaped and was at an early stage of his research. I learned a lot of new mathematics from him, group theory, matrices, vectors, and so on. And I taught him about Maxwell's equations when I was an undergraduate because he wanted to know something about physics. He was a very important individual in my life.

I had a pretty rigorous training from my school. Then, after graduation, I had some offers, one to go to Michigan at $150 a month without any duties except to go to graduate school, and another one from Caltech at $20 a month to assist in their physics lectures. I went to Caltech to study physics, and I took the lowpaying offer because I wanted to be in a good school. During my first year, I heard some lectures by Linus Pauling and switched to chemistry. I did my thesis with him and some people who worked with him. That was a big part of my beginning.

You dedicated your Nobel lecture to your sister.

I had two sisters. I have one now. She lives in Ohio and she is a housewife. The other one was a composer and a very fine musician, and she was hampered because she had polio, and she could not walk. Since the age of 16 she had to spend most of her life in a wheelchair. I helped her as much as I could when I was around. She was the fine musician of the family. She wrote music, and she taught music; she died at the age 51, about 20 years ago.

William Lipscomb displaying the Nobel Diploma, 1995 (photograph by I. Hargittai).

Then you had your own family.

I have two children who are close to 50. My second wife is a designer, and we have adopted a little girl from Thailand, who is now six and a half.

You've mentioned that you did some service for the war effort.

I was at Caltech a total of five years, and for three and a half years, I was employed on war projects. The major one was nitro-derivative propellants for rockets, and I was weighing nitroglycerin with my hands. So I really had only one year and a half for graduate work.

You were very lucky with your teachers.

Oh, yes.

How about your pupils?

I have a long list of first-rate pupils, but the most outstanding is probably Roald Hoffmann. But I had some other very outstanding ones. Don Wiley, who worked on the immune system, is really first class. Douglas Rees is another one, who, by the way, comes from Kentucky, and worked on the oxidation-reduction processes in proteins. I've had almost a hundred graduate students.

Is it correct to say that you're best known for your borane studies?

I suppose so, although for the last 37 years I have been working also on the structure of proteins. I think that that work is very well known. It depends on whom you're talking to.

Were you the first who thought of the two-electron three-center bond?

No. Hirschfelder, Taylor, and Eyring did it for the molecule H_3^+ back in the 1930s. Then a year later Coulson did some work on it. Actually, the first proposal came from a man named Dilthey in the 1920s, before quantum mechanics. He didn't understand it, but he proposed the three-center bond, and nobody believed him and particularly Mulliken didn't believe him. Then it was Longuet-Higgins who first proposed the three-center bond with a bridging hydrogen, in about 1946. So I was not the first to invent the three-center bond, but we were the first to show the generality of the three-center bond involving boron. The Nobel Prize is not always given to somebody who made the first discovery, but it's given to the person who changed chemistry with it, and that's what we did.

Do you think this is why G. N. Lewis didn't get the Nobel Prize?

I think the problem was that nobody understood the two-electron bond until Heitler and London's research in 1926. Then Hund and Mulliken developed the molecular orbital method. It's true that Lewis lived much beyond 1926, but Lewis himself did not understand the nature of the electron-pair bond in his early research. Failing that, I think his time sort of passed. It may have been a good idea to give him the prize but I can understand why they didn't, because Lewis himself certainly didn't formulate the physical basis of the bond.

There's a lot of polyhedral chemistry in boranes, including icosahedral geometry, which is now even more important because of buckminster-

fullerene. Somewhere you used the expression, "pleasing polyhedral shapes." The word "pleasing" is something that I'd like to ask you about.

People who make discoveries do not necessarily use the aesthetic side until afterward. Then they recognize it and say, yeah, that's beautiful. Of course, we were not the first to find the boron icosahedron either. That was found by Sevastyanov and Zhdanov in C_3B_{12}, boron carbide, in 1941. That was a missed opportunity because nobody thought then that you could make fragments of this polyhedron and arrive at a structure of decaborane. That was not an idea that was around. It's an idea we missed.

Who missed?

Everybody. Pauling missed it, I missed it. The idea that boron hydrides should be fragments of icosahedron came much later. It came from the structure of decaborane, $B_{10}H_{14}$, by Harker *et al.*, around 1950.

Who were the other important players in boron chemistry?

H. C. Longuet-Higgins and M. de V. Roberts. They did the structure of $B_{12}H_{12}^{2-}$. Then M. F. Hawthorne was a very important player. He prepared many, many compounds. These people made such important contributions. When the prize was announced and I found out it was on boron chemistry, I wondered why they didn't share it with Longuet-Higgins or with Hawthorne because they made very important contributions.

At some point around 1960 you switched to enzymes. Why?

We had been doing X-ray diffraction work and I was interested in moving up the scale of molecular sizes, and I also had an interest in biological problems from Linus Pauling and from my father. So we began doing large structures of organic compounds, and eventually it became feasible to do protein structures so I said that I'd like to do that too.

Was your work on enzymes as successful as that on boranes?

Yes.

What's the most important finding there?

We've been working on the behavior of zinc in enzymes and also what are called allosteric enzymes, that is, enzymes that undergo conformational change when they have different subunits, different proteins, to make a change in shape. One form is active and the other form is almost inactive, and we would like to understand how the enzyme is made nearly inactive by this conformational change. I've been working on this problem for many years. We've also done many other structures.

We met almost 30 years ago in a crystallographic meeting in Moscow. That was in 1966, and I was working on the structures of some compounds containing sulfur-nitrogen bonds, using gas-phase electron diffraction. You had studied similar systems by X-ray crystallography, and already then, you paralleled your experiments with molecular orbital calculations. Those calculations were not as helpful then as they are now, but obviously you found it important to carry them out.

That's right. It got started with the boron chemistry, where we had structures already in 1953. We knew the structure of B_2H_6 from Price. Then there was the B_4H_{10} structure with some controversy. We also had B_5H_9, B_5H_{11}, and preliminary work on B_6H_{10}. It was these compounds that challenged the theory and that led to the 1954 paper by Eberhardt, Crawford, and Lipscomb. In that paper we recognized first that parts of these structures look like diborane with bridge hydrogens and terminal hydrogens. This was generally true throughout the boron hydrides that you could recognize diborane in their structures. The larger structures were too complicated to yield a valence theory because of the need for a very large number of possible resonance forms. However, the smaller structures were simple enough to make a consistent theory out of the boron hydrides, involving tetrahedral boron, two-center bonds, three-center bonds, and bridge hydrogens. We could make a theory out of these. First, we assigned bonds this way, but when we came to the more symmetrical ones, B_4H_4, for example, we also used molecular orbitals. Consideration of diborane-like fragments was especially useful. Later, we transformed the delocalized boron orbitals to localized bonds.

This method involves no assumptions beyond the wave function itself. It's an objective method, maximizing the repulsion of two electrons when they were in the same orbital and minimizing their repulsion when they were in different orbitals. This method didn't have to give three-center bonds, but it did. In a sense we used known theory to transform molecular orbitals to localized bonds, which immediately tested the way that we drew bonds in 1954. That was soon after my beginning in theory. We then went on to much more complicated molecules. Shortly after I came to Harvard, I proposed a computerized version of an extended Hückel theory, which I suggested to Lawrence Lohr and Roald Hoffmann. It's an approximate method. Even now, it is the only method that can be used throughout the periodic table. For most chemical applications, you don't need precise energies and you don't need precise densities. All you need to know is where the nodes are in the wave function and an approximate order of energies of molecular orbitals. This is the method that Roald Hoffmann used to test the Woodward-Hoffmann rules, to test the fragment idea in transition metals. The next step was to produce methods that gave more accurate results in my group, providing better descriptions of the bond and also providing a logical basis for extended Hückel theory. So I've been interested in theory for a long time.

It seems to me that the big challenge for chemistry in this century was to find out the nature of the chemical bond.

That's right.

What remains?

I can tell you what remains. It's intermetallic compounds. There are 20,000 compounds, 5000 structures, and no progress. We don't understand them at all. That's very interesting.

Would you name a simple example?

KHg_{13} or Cu_5Cd_8 or Sb_2Tl_7. What do you do about valency in these cases? There are 5000 different three-dimensional structures that we don't understand, and I don't understand them either. I'd thought that I would spend a few years on boron chemistry and then move to the intermetallic compounds, but I never did.

Another important area is the theory to understand bonding of the transition states. This is important in order to understand the rates of reactions. We studied the transition state for the reaction of BH_3 with B_2H_6 to give $B_3H_7 + H_2$. This is a very interesting transition state. The question is whether there is an intermediate B_3H_9, and we think there is not.

The level at which theory is possible is getting much better all the time thanks to both the theory and computers. I don't have any students working on this kind of problem. Since I turned seventy, I cannot take graduate students, although I now work with postdoctoral fellows.

But you are terribly busy.

Yes, I work on protein structures, and so on.

By the way, do you know what a dual polyhedron is? If you take a boron octahedron and I put a carbon in the center of each of the faces and throw away the borons, then I get cubane. Now, here is a proposal I made in 1978, before buckminsterfullerene was discovered; it has 32 borons and 60 faces. If I put a carbon in the center of each of its faces and throw away the borons, then I get buckminsterfullerene. This $B_{32}H_{32}$ is the dual polyhedron of buckminsterfullerene. When the discovery of buckminsterfullerene happened in 1985, I remembered that I'd done something along those lines. Then later Lou Massa and I looked at these analogues in general so we could go on to propose other examples, all related by the Descartes-Euler formula, which can be found in Coxeter's book.

Do you think the discoverers of buckminsterfullerene should get the Nobel Prize?

I'm not sure. Probably, but I'm not sure. It has had a big impact on chemistry, but I don't know if it's had a bigger impact than some other discoveries. I just don't know. I wouldn't be surprised.

Harry Kroto, István Hargittai, and William Lipscomb (all three are Kentucky Colonels) at the reception of the Royal Swedish Academy of Sciences during the centennial Nobel celebration in Stockholm, December 2001 (photograph by M. Hargittai).

Neil Bartlett, 1994 (photograph by I. Hargittai).

3

NEIL BARTLETT

Neil Bartlett (b. 1932 in Newcastle-upon-Tyne, England) is Emeritus Professor of Chemistry of the University of California at Berkeley. He is best known for his synthesis of the first compound of a noble gas. He received his B.Sc. degree in 1954 and his Ph.D. in 1958, both from King's College, University of Durham at Newcastle, which is today the University of Newcastle. He was a faculty member of the University of British Columbia (UBC) in Vancouver, Canada, from 1958 to 1966, was Professor of Chemistry at Princeton University from 1966 to 1969, and has been at Berkeley since 1969. He has also served as Principal Investigator at the Lawrence Berkeley Laboratory.

Dr. Bartlett is a Fellow of the Royal Society (London) and Foreign Associate of the National Academy of Sciences of the U.S.A. and of the French Academy of Sciences. His many other honors include the Robert A. Welch Award (1976).

Our conversation was recorded in Neil Bartlett's office at Berkeley on May 11, 1999.*

Today is the last day of operation for your laboratory at the University of California at Berkeley, but this place gives the impression of frantic activity rather than of being closed down. What is your latest manuscript about?

I have completed a comprehensive review of my fluorine chemistry[1] and I've also just written a piece for Ron Gillespie's 75th birthday.[2] In this

*This interview was originally published in *The Chemical Intelligencer* **2000**, *6*(2), 7–15

paper for Ron, I've given my final statement on $XePtF_6$, "Concerning the Nature of $XePtF_6$." It's an unfortunate thing in some ways, fortunate in others, that I've never been able to get $XePtF_6$ crystalline. It's a solid but it doesn't crystallize, and when you heat it, it changes into something else. At the very beginning, this was very frustrating. Once I had set up to do my PtF_6 preparations very cleanly, and it took some time because I had been used to working in glass, I was able to handle it with very little decomposition. Nevertheless, I could never get really clean 1:1 stoichiometries with xenon. In my very first experiment, I had used a large excess of xenon. I attributed the fact that the combining ratio wasn't 1:1 to some attack on the glass, which one could see. When, in my later experiments, I did get close to 1:1, there was no X-ray powder pattern. When I had excess PtF_6, there was always a faint pattern. However, the pattern was always the same, regardless of the stoichiometry. Eventually, I realized that we could drive the composition to $Xe(PtF_6)_2$ by using excess PtF_6, and its X-ray pattern was the same as when the ratio of PtF_6 to xenon was 1.3:1. When my student, N. K. Jha, warmed this material of composition close to 2:1, we saw then a new pattern. When we pyrolyzed it at 160°C, and we did this at the end of October, 1962, we obtained a colorless material. By then, the Argonne National Laboratory (Argonne) had made and reported XeF_4 but, wouldn't you know, the material we got was XeF_4 also. That was very puzzling because XeF_4 forms a complex with XeF_2, a semi-ionic 1:1 adduct, yet our xenon fluoride had the powder pattern of pure XeF_4. If $Xe(PtF_6)_2$ was a xenon(II) compound, heating it, we thought, should produce XeF_2. But we didn't see XeF_2, only pure XeF_4! The answer to all of this came only years later, but it stimulated us to do some research that led us to some very nice findings. One of these findings was that we could very easily fluorinate xenon to XeF_6. Then we made $XeF_5^+PtF_6^-$. That crystallized beautifully. That was in 1966. We could never find complexes of XeF_4. Joe Berkowitz at Argonne had done some nice ion-molecule reactions in which he assessed the ionization enthalpy for the process

$$XeF_x \rightarrow XeF_{x-1}^+ + F^-$$

and he showed that it's easier to take a fluoride ion from XeF_6 than from XeF_2 and that the most unfavorable ionization process was for XeF_4. The reason is that XeF_6 is a crowded molecule, the nonbonding electron pair being sterically active. This helps drive off a fluoride ion. So XeF_6 is the best fluoride ion donor, XeF_2 is next, and XeF_4 is by far the worst.

When did you realize that the lone pair contributes to the overcrowdedness of XeF_6?

We realized that when we did the crystal structure of $XeF_5^+PtF_6^-$.

But hadn't Gillespie predicted as early as 1962, using valence shell electron pair repulsion (VSEPR) arguments, that XeF_6 should have a distorted octahedral geometry?

Yes, of course, but it's a subtle matter. It's true though that the proceedings volume of the first conference on noble gas compounds[3] has a paper by Ron Gillespie in which he correctly predicts the geometries of the xenon fluorides and oxyfluorides. Although the really good electron diffraction people, like Ken Hedberg and Larry Bartell, gave experimental evidence for the distorted geometry of XeF_6, some other people were misled by $XeOF_4$ impurity. XeF_6 is a tremendous scavenger for oxygen, and it makes $XeOF_4$ easily. Ron Gillespie, again, predicted that $XeOF_4$ would be a nice square-based pyramid with the O–Xe–F angle approximately 90°. That impurity therefore perturbed the radial distribution curve of XeF_6. It began to look like that of a regular octahedron. That was one of the reasons why the early studies were a bit fouled up.

At about the same time, we were looking at XeF_5^+. Ours was the first XeF_5^+ salt to be established as such. That came about when Don Stewart arrived from Australia. He was an experienced fluorine chemist when he joined me in January 1966. I suggested to him an approach that I hoped might give crystalline $XePtF_6$. First, we took O_2PtF_6 with a large pressure of xenon to drive out oxygen. Elsewhere in the apparatus, we put sodium metal in a container to combine with the oxygen. But all we got was a red, sticky mess, never anything crystalline. Then we took a simpler approach, starting with PtF_5, which I had made earlier, a lot of xenon, and a little bit of fluorine. Don took rather modest amounts of fluorine, less than 1 atm in his first experiment, and a lot of xenon, and right away he got some crystals. We took some Raman spectra, and Jim Trotter mounted a crystal on the diffractometer, and we quickly got a unit cell. We could tell from the unit cell that this material had to have a rather large formula weight. It indeed proved to be $XeF_5^+PtF_6^-$. The XeF_5^+ was a beautiful square pyramid with a sterically active nonbonding electron pair in complete agreement with the VSEPR model. It also convinced me that the nonbonding electron pair must be sterically active in XeF_6. When the crystal structure of XeF_6 was solved by Robinson Burbank at Bell

Labs, he found tetrahedral and octahedral clusters of fluoride with XeF_5^+ and F^-.

We just jumped into the discussion of xenon-fluorine compounds, but I had planned to start our conversation with a quote from Primo Levi's book The Periodic Table. *On the first page of the first chapter he mentions your discovery. He says: "As late as 1962 a diligent chemist after long and ingenious efforts succeeded in forcing the Alien (xenon) to combine fleetingly with extremely avid and lively fluorine, and the feat seemed so extraordinary that he was given a Nobel prize."*

I greatly admire Primo Levi, and his suicide perplexes me. I suppose he was a typical survivor and felt guilty in surviving. The person who drew my attention to Primo Levi's book was Emilio Segrè. The remark didn't upset me.

Although you did not receive the Nobel Prize, and you could very well have, receiving this mention by Primo Levi is unique. Of course, your discovery of the preparation of $Xe^+[PtF_6]^-$, which you published in June 1962,[4] makes you unambiguously the pioneer of the field. The paper is very brief, less than half a page long and cites only three references. You did not mention Linus Pauling's prediction of the possibility of noble gas compounds from the early 1930s.

In my reviews I mention that. The beginnings of attempts to make noble gas compounds go back to Henri Moissan and William Ramsay. Ramsay was a friend of Moissan. The first inert gas to be discovered was argon because it's so plentiful. Moissan had made fluorine about nine years before. When Ramsay discovered that this monoatomic gas just wouldn't do anything with anything, he sent some argon to Moissan. This was in the mid-1890s. Moissan passed argon mixed with fluorine through an electric discharge. He looked very carefully at the effluent, but there was no sign of any combination. He communicated back to Ramsay that argon didn't react with fluorine. That indication of chemical inertness undoubtedly had impact. There were a number of other attempts, in particular by von Andropoff in Germany, who erroneously believed that he had made a bromide of krypton.

If you look at the classic paper by Walther Kossel on electronic theory of valence, what you see is a remarkable foresight; he lists the ionization

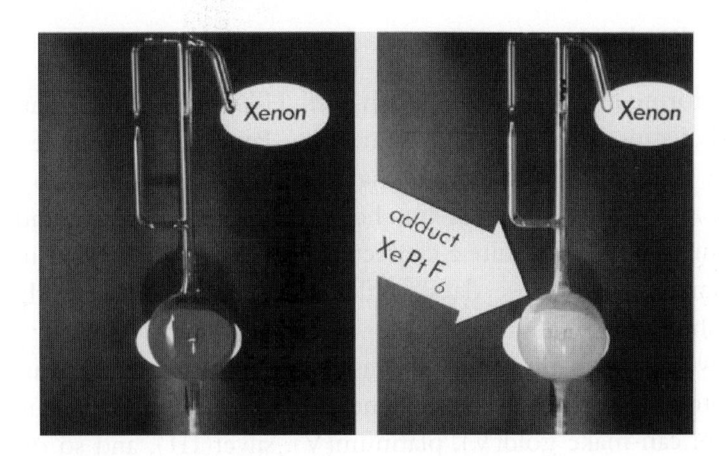

The historic experiment producing the first noble gas compound was recently recreated (Photographic & Digital Imaging Services, Ernest Orlando Lawrence Berkeley National Laboratory, courtesy of Neil Bartlett).

potentials of the noble gases as they were then. They weren't the present values but the relative ionization potentials are right. Kossel says right there, if there's going to be any chemistry, look at krypton and xenon. Many years later, in 1932, Linus Pauling, on the basis of looking at periodic trends, remarked that xenon should make an octafluoride, krypton a hexafluoride, and so on. The one thing that was certainly right in Linus's prediction was that he said that one may expect to find XeO_6^{4-} compounds. Eventually, they were made, in 1963.

Linus was so convinced that there should be some xenon chemistry that he wrote his old teacher Fred Allen, at Purdue, and begged from him a sample of xenon. It was difficult to get samples of xenon in those days. Linus then took xenon to his Caltech colleague Don Yost, who had a student, Kaye, and Yost and Kaye put xenon and fluorine in a quartz bulb. Then they made a mistake. They should have gone out into the Caltech sunlight. Instead, they passed an electric discharge through the bulb and all they got was attack on the quartz container. Quartz is not very resistant to fluorine, you make silicon tetrafluoride easily and liberate oxygen. They reported this in the *Journal of the American Chemical Society* in 1933 in a paper on the nonreactivity of xenon with fluorine. That, I suppose, persuaded Pauling that he had been a little too optimistic. The xenon fluorides are close to the limit of thermodynamic stability after all, and accessible xenon chemistry depends upon that slight stability.

How did Pauling react, 30 years later, to your discovery?

He was informed very early on by myself. His reaction was that Yost and Kaye had distinctly missed the boat. He, Pauling, had made the right prediction. Of course, Yost could have suggested to Linus to try it for himself. I never met Don Yost but he writes in an engaging and amusing way. He must have been quite a character. He writes somewhere, in referring to their failure, that those days of fluorine chemistry "were the days of wooden ships and iron men."

What we do with fluorine today is done at room temperature by just using photons to dissociate the fluorine. Everything is done in fluorocarbon plastic. You can make gold(V), platinum(V), silver(III), and so on in liquid anhydrous HF at room temperature, with fluorine and sunlight in such an apparatus. Such containers did not exist in Don Yost's days. He was working in quartz because he knew well that fluorine is extremely difficult to handle in ordinary glass. You could use Pyrex if got it really dry. Yost probably didn't know that and he probably had some HF in his fluorine.

What were you using?

We were using Pyrex glass in the very beginning. But we had learned important things, largely as a consequence of the Manhattan Project. People involved in making UF_6 learned that you could work with fluorine more easily if you scrubbed HF out of it. You could do that by passing the fluorine over sodium fluoride, which made sodium bifluoride. Then one had fluorine, which was a good deal easier to contain in dry glass. If there is any water in the glass or quartz apparatus, however, then you make HF and a cycle of reactions consumes the fluorine and destroys the container.

There seems to have been a lot of activity in the area of noble gas compounds right around and following your discovery.

My response requires some historical background. I had been very lucky to be doing my early chemistry in glass. I saw the chemical reactions taking place. My most important finding was made at the end of my graduate student work. I discovered that platinum formed a reactive volatile, red compound when heated in fluorine. I thought it was a platinum(VI) oxyfluoride, $PtOF_4$, because when I observed this material forming in the glass apparatus, the fluorine was attacking the glass and, in the process, liberating oxygen.

Neil Bartlett at the time of the production of the first noble gas compound (courtesy of Neil Bartlett).

In the meantime — and these were the days of conscription and I didn't fancy going into the military, so I taught school for a year in lieu of that — you can imagine my consternation, while I was teaching school, when a paper appeared on platinum hexafluoride. This was the work of Bernard Weinstock and his group at the Argonne National Laboratory. Weinstock was an extraordinarily good fluorine chemist. When I got to Vancouver the following year, in 1958, I decided to press on and find out more about the compound I had made back in England. Initially, my student Derek Lohmann and I thought we had evidence that it was $PtOF_4$. This conclusion, and our separate evidence for PtF_5, were given in a paper that we published in 1960. $PtOF_4$ was wrong. Even by the time our paper came out, I realized that something was wrong. The reason was that the red solid we had been making was extremely difficult to analyze. It was probably the most difficult piece of chemistry I've ever done. When we put this red solid into water, which was then our normal way of carrying out an analysis (we used aqueous alkali to catch all the fluorine as fluoride), we certainly generated ozone. One could smell the ozone. Some of the platinum was precipitated as the metal and some went into solution as PtF_6^{2-}. What we had was a major analytical problem and it took us some time to solve it. I still have apparatus here on the shelves which we used to measure the oxygen released from the oxygen salt.

Eventually, it dawned on me that the way to analyze this material was to add bromine trifluoride to it and so measure the oxygen content directly as oxygen gas. We were afraid that we might be losing some fluorine as the gas F_2O, so we devised a way of putting the material into a gelatin capsule with sodium on the outside, in a Parr bomb. We fused the mixture, and so combined all the fluorine. All the platinum came down as the metal. Then we got consistent analyses and, my God, I couldn't understand this, the composition was O_2PtF_6, "What did that mean? Was the platinum 10-valent?" Here I was, a 29-year old neophyte, looking at this material and thinking, "Is this in fact *a trans* PtO_2F_6 molecule with oxygen sitting on a threefold axis with six fluorines?" The compound had a simple cubic X-ray pattern. It was also paramagnetic and this ruled out Pt(VIII). This convinced me that what I had to have here was a PtF_6-containing entity. That was confirmed also by the fact that when we slowly hydrolyzed this material with moist air, and then took it up with alkali, we had a lot of PtF_6^{2-} in solution. PtF_6^{2-} exists in aqueous solution, although it's thermodynamically unstable with respect to hydrolysis. So there had to be six fluorines attached to the platinum in order to generate that. I thought, "This has to have a PtF_6-containing entity, $PtF_6^+O_2^-$ or $O_2^+PtF_6^-$." The former is energetically absurd if you think through the cycle. So it had to be $O_2^+PtF_6^-$." That formulation posed the biggest intellectual struggle.

When I sat down and looked at the steps in the Born-Haber cycle, I realized very quickly that the lattice energy that one would get for $O_2^+PtF_6^-$ would be, at the most 135 kcal/mol. Now one is expending 280 kcal/mol to take away the electron from oxygen. Of course, one is also fighting the fact of the entropy loss in going from O_2 gas and PtF_6 gas to O_2PtF_6 crystalline solid, beautifully organized in a cesium chloride-type arrangement. One probably needs 20 to 30 kcal at room temperature for that, of course. This compound was subliming at 100°C in a vacuum. In the end, I came to the conclusion that the electron affinity of PtF_6 had to be around 180 kcal for $O_2^+PtF_6^-$ to be a stable salt at ambient temperatures. That's more than twice the electron affinity of the fluorine atom and it seemed inconceivable. I remember that we all used to go for coffee about 10 a.m. each morning.

Neil Bartlett explaining an experiment to Magdolna Hargittai, 1999 (photograph by I. Hargittai)

One morning I said to Jack Halpern, who was then at UBC, "Jack, I'm sure this salt has to be $O_2^+PtF_6^-$ even though the electron affinity of the PtF_6 has to be enormous." Jack was skeptical. Indeed, that was a reasonable response from an excellent scientist. He was the person whose judgment I most respected. His response caused me to go over each piece of the experimental evidence with great care. I came to the same conclusion! (Incidentally, Jack Halpern is now Professor Emeritus at the University of Chicago and is a Vice President of the National Academy of Sciences.) Eventually, I presented a departmental seminar on $O_2^+PtF_6^-$. I thought it was a convincing story, but afterward a colleague said, "Maybe something has been overlooked." That stimulated me to think about other ways to demonstrate what I was now convinced of, that PtF_6 was the most potent oxidizer ever discovered. It was at that point that I started thinking about things that would be more difficult to oxidize than molecular oxygen.

You were at UBC and so was Jack Halpern; was it a strong school then?

It had a number of very good people. Robin Hochstrasser, who is now at the University of Pennsylvania, was also a colleague at UBC. I used to ask him to explain quantum-mechanical difficulties and spectroscopy because I wasn't then *au fait* with group theory. Robin was then an Instructor at UBC. Ian Scott, who is now at Texas A&M, came as an Associate Professor in the middle of all this. Howard Clark was an inorganic chemist colleague. He eventually became the President of Dalhousie University. It was an excellent department.

Gobind Khorana was also at UBC at that time.

Yes, but he wasn't in the Chemistry Department — an absurdity. I knew about Khorana before I knew about UBC. He was a member of the research staff of the British Columbia Research Council. It was a ridiculous situation. Why he was not invited to join the Chemistry Department, I have no idea. He was by far the most distinguished organic chemist in the whole of British Columbia. He was well known for his work on the synthesis of coenzyme A. I myself had started off in organic chemistry.

Once you were certain that PtF_6 was such an extremely strong oxidizing agent, what was your next target for oxidation?

Early on, I thought of one-electron oxidation of nitrogen trifluoride. We tried it and it didn't work. The ionization potential of nitrogen trifluoride to go to NF_3^+ is about 13.5 eV. We weren't anywhere near. I knew, of course, that N_2^+ was out of the question because I'd been always doing my chemistry in the presence of nitrogen. I also happened to remember that helium had an enormous ionization potential. In late January or early February of 1962, I was preparing a lecture for an inorganic class and as I was leafing through the text looking for something, the familiar plot of ionization potentials against the atomic number flicked by and I noticed that the ionization potential of xenon looked awfully like that of molecular oxygen. I estimated that the lattice energy would be about the same for $Xe^+PtF_6^-$ and $O_2^+PtF_6^-$. So the oxidation of Xe by PtF_6 ought to go. So I ordered some xenon. I only had money for 250 cc. At that point, although I had two graduate students, they were newly arrived, and they didn't know how to glassblow.

Once the xenon was received, it took me about a week to get everything set up. I had to use a very sensitive quartz sickle-gauge to measure volume and the pressure of the gases. It wasn't until 7 o'clock on a Friday evening, March 23, that I was all ready for the experiment. When I broke the seal between the PtF_6 and xenon, there was an immediate reaction. My two graduate student coworkers (N. K. Jha and P. R. Rao) had had to leave to go to dinner because they were staying in the residencies and dinner was served at 6:30 p.m. So I went out into the corridor to find somebody to come and see. There wasn't a soul in the building. So I went back to the lab. Suddenly, some doubts occurred to me, "Maybe the xenon was impure, maybe there was some oxygen present, maybe I'm just fooling myself." Then I sublimed the product, by heating it under vacuum. Then I condensed water onto the sample and allowed it to react with the solid, the evolved gases being contained in that part of the apparatus. I gave this section of the apparatus to David Frost, a mass spectrometrist colleague at UBC. I asked him, "Please, look for xenon, and there will also be some oxygen because the platinum will oxidize the water." He looked at the relative abundances and found at least as much xenon as oxygen. That proved it.

By the time I got the mass spectrum back from David, it was Tuesday or Wednesday. I knew that I had got to write this up. I was sure by then that the Argonne people, having seen my O_2PtF_6 paper, were going to open their stopcocks on their supplies of PtF_6, and not only that, they also had rhodium and ruthenium hexafluoride, which they had made. I was sure that rhodium hexafluoride would be the most powerful of all. That was the thing I kept thinking about after this, "They're going to carry out the reaction chemistry with all of these hexafluorides that they have bottled up. They must surely be embarrassed by the fact that for five years they had PtF_6 and hadn't discovered that it would oxidize oxygen. They're going to turn the tables on me, so I had better get this thing out." I wrote it up for *Nature* and sent it off on April 2, which was a Monday. I sent it as a letter and heard nothing for a month. Then I found out that *Nature* no longer put the date of receipt on letters, and I wrote them, "Please, assure me that this paper that I sent three and a half weeks ago will be published within a month. Otherwise, I will withdraw it." I heard nothing. In early June, I did receive, by sea mail, a card acknowledging receipt of my manuscript. By then, I wrote back and said, "Please, consider my paper withdrawn." I then sent it to the *Proceedings of the Chemical Society*,[4] the forerunner of *Chemical Communications.*

So the Argonne National Laboratory was a big competitor.

Absolutely. Once the $XePtF_6$ paper appeared, they realized that, indeed, their hexafluorides should be capable of doing this. They had immediately a publication with a very large number of authors, in which they looked at the reactions of not only platinum hexafluoride but ruthenium and even plutonium hexafluoride. When they did the reaction with ruthenium hexafluoride, they saw a colorless product and a green product. When you see a green fluoride from ruthenium hexafluoride, you know that it is ruthenium pentafluoride. Then you know that you must be transferring fluorine to xenon. They quite correctly (they being Claasen, Selig, and Malm) decided at that point just to take xenon and fluorine. They had a mixture of xenon and fluorine in a can and heated it to something like 450°C and they pumped off the fluorine at liquid nitrogen-temperature, at which temperature xenon is retained. But there wasn't any xenon there. This is how they made xenon tetrafluoride, a colorless solid. By chance, they had hit on the conditions that favor mainly XeF_4, the XeF_2 or XeF_6 impurity being small. (Later we showed how to purify xenon tetrafluoride.)

The Argonne people sent off their paper in August 1962, and while they were doing this, I was trying to make rhodium hexafluoride in order to try to oxidize krypton. I expected rhodium hexafluoride to be the strongest oxidizing agent. If we had been left to our own devices, we would have found xenon tetrafluoride and we would have found pure xenon tetrafluoride. Clearly, Argonne had been unlucky and I'd been tremendously lucky. For five years they had lots of platinum hexafluoride.

What may have been the reason for your being lucky and their being unlucky?

Two things. One is that I was working in glass so I always saw what was going on. Every chemical change that occurred, I could see. The Argonne people were always working in metal lines with metal valves and taking the greatest care not to expose their very reactive fluorides to glass or any other thing that would react with them. They were doing beautiful spectroscopy and beautiful physical studies. In collaboration with Moffitt of Harvard University, Weinstock and Goodman of Argonne did an exquisite study on the optical properties of these hexafluorides. They looked at the Jahn-Teller features of the hexafluorides. They showed a second-order Jahn-Teller effect for platinum and osmium hexafluoride. They were looking at the spectroscopy of these things across the board but not looking at the chemistry. Their strong orientation towards the physics of the materials was the other difference between us, while I was very much oriented toward the chemistry. Later, John Malm said that they did see this O_2PtF_6 but they thought it was a hydrolysis product. They were shoveling it out of the apparatus some times. They thought they had a moisture leak. If the Argonne people had been working in glass, they would have seen the formation of O_2PtF_6, they couldn't have missed it.

Why did you then leave UBC?

It was very complicated and I can't tell you all of it on the record. First of all, there was the problem that I had shot up from being a lecturer, the lowest rank, to full professor in five years, and people didn't always look at it charitably. Secondly, I had a very serious accident in January of 1963. I was very lucky again. I could have lost more than some vision in my right eye, which is all that I lost. I had decided after visiting Argonne in October 1962 that there was no way I could compete with them on the fluorine chemistry of xenon. What I did perceive was that they were unlikely to make an oxide. I thought that it should be possible to make an oxide of xenon. I recognized also that it would be thermodynamically unstable. So I had one of my students, P. R. Rao, make a xenon oxide. I thought that the best way to do it was to derive it from xenon tetrafluoride. We didn't want much oxide because it was going to be thermodynamically unstable. My idea was to put xenon tetrafluoride into a quartz vessel and then distill water in and hydrolyze it. Then take off the HF as quickly as possible, because the HF would attack the quartz, and see what we could get. In our first experiment, we had only about 0.2 g of XeF_4, and Rao got a small amount of colorless solid from the aqueous solution and we analyzed that for xenon. The analysis indicated that the composition was such that it had to be $XeO_2.H_2O$ or $XeO_2.(2H_2O)$. The precision was not very high and we didn't do an iodine titration to see what the oxidation state was.

In any case, we decided to take a bigger sample, about half a gram, for the next experiment. Rao did the same hydrolysis as before. This was on January 27, 1963. After dinner, I went to the lab to see how he was getting along. Everything was going well, he had a colorless solution, he pumped off the water and HF, and he called me to show his nice crystals. Sure enough, they were beautiful long crystals coming out of the aqueous solution. We each had one of these plastic visors on. Mine was somewhat scratched, and in my head was the notion, "Oh, yes, this is a hydrate, maybe." So I said to him, "I wonder if those crystals will fall to a powder under vacuum," that is, whether they would lose the water of crystallization.

I said, "Be careful, take off all the water and let me know when you got them dry." And Rao told me when they were dry and I came over to look at them. Although I had very good eyesight, and I still do in my left eye, I couldn't see the crystal surface clearly because of the visor so I put it up and just at that moment the sample, in its quartz tube, blew up. When I put my visor up, Rao put his visor up too. We were both injured in one eye, he less so than myself. I immediately lost all capability to see with my right eye because it was sliced. We were carted off to hospital. We were in hospital for four and a bit weeks. They thought I had a detached retina. I was sandbagged for two weeks. When I got out of hospital, I received a paper on the crystal structure of xenon trioxide from Allan Zalkin at Berkeley. Things were moving so quickly. We probably wouldn't have succeeded in characterizing it, but it's hard to know. If I had been more patient and had taken out the crystal and had taken it to Jim Trotter, we might have had the structure, but it's by no means certain.

What I didn't know was that there was still some glass in my eye. Every two or three weeks, this damn eye used to give me hell. The glass was moving around inside of my eye. I put up with that for 27 years. In 1990, a piece of glass started coming out of my eye, and my wife sent me off to an excellent ophthalmologist. He took out all of the troublesome glass.

The accident in 1963 had its impact. There was also the problem of having risen rapidly in the ranks but, worst of all, there was a bad political situation. Suddenly, the university was short of money and the politicians decided to increase the number of universities by adding another three universities. At the same time I was approached by Princeton and asked if I'd go there. So, unwisely, I left. I should never have done so.

The sort of work I do, the way I work — I'm very much a loner — I didn't need to have better facilities or a bigger department. I think I could have achieved all that I have accomplished at UBC. After all, I'm a European, and Vancouver is politically and culturally closer to my British background. I was "at home" there.

Later, you left Princeton too.

I recognized right away that I was already a West Coast person because of living for eight years in Vancouver. The abrasive nature of New York and the awful climate in the summer and the somewhat isolated nature of Princeton itself. Although Princeton and the surrounding countryside were attractive, the nearest cities were not. When I had an invitation from Berkeley, I accepted it with relief.

Are there any applications for the noble gas compounds?

Very little yet. XeF_2 is a clean fluorinator since Xe itself is of such low reactivity that it is usually evolved when XeF_2 is used as a reagent. I'm more hopeful for our thermodynamically unstable metal fluorides like nickel trifluoride as fluorinators. We have an application for a patent for them. The process that is used in industry, by companies like 3M, to fluorinate organic materials involves electrochemical fluorination at a nickel anode. It has been postulated for many years that that process probably involves a higher fluoride of nickel. I'm sure that's right. Now we have a higher fluoride of nickel and we have its crystal structure. It's sufficiently stable kinetically once you get it. It's a black solid, $Ni^{2+}NiF_6^{2-}$. One can take an adamantane-like molecule and simply treat it with nickel trifluoride at ice temperature, and it is quantitatively converted to the perfluoro compound. It's very convenient and there is no carving up of the carbon skeleton, no cleaving of any C–C bonds. It's an extremely effective fluorinating agent.

I would like to ask you about ONF_3.

I'd been intrigued by the fact that CO_2 interacts so easily with OH^- to make $CO_2(OH^-)$, the bicarbonate ion, and thought that one should similarly be able to make CO_2F^-. I had an undergraduate student at UBC, Miss Tingle, trying to make CO_2F^-. At one point she got material with CsF that behaved like that. It was a white solid, and when we put it into water, it liberated CO_2. The problem with this white solid was that it had a very complicated X-ray powder diagram. She had been using acetonitrile as a solvent and it turned out that she wasn't drying it nearly well enough. After drying it, she couldn't make this material, and I suggested to her to take a powder diagram of cesium bicarbonate. The picture turned out to be very much the same as the previously recorded pattern from her white solid which reacted with water. Her white solid was evidently a mixture of cesium bicarbonate and cesium bifluoride, generated from the water in the acetonitrile. It was obviously very difficult to make CO_2F^-. Just at that point, a person called Chris Willis, who had been a postdoctoral fellow at UBC, published a paper describing the synthesis of COF_3^- salts. I found that very interesting since I realized that making CO_2F^- should be harder than making the known molecule NO_2F because the nuclear charge of nitrogen is greater than that of carbon. For similar reasons, if OCF_3^- did exist, it should surely be easier to make ONF_3. I gave this to a student called Passmore, who is now a professor at the University of New Brunswick. He started by fluorinating NOF in nickel bombs. He got some interesting infrared spectra. Simultaneously, another of my students, Steven Beaton, who was looking at the reactions of ONF with PtF_6 and IrF_6, saw the same signals in his spectra. I thought that the new infrared features might be due to ONF_3. I could also rationalize that structure with a tight tetrahedron around nitrogen in ONF_3. Finally, we got Tedd Wells at Simon Fraser University to run an NMR of this gas for us and he proved it was ONF_3.

Then I got a nasty surprise at the 1965 IUPAC meeting in Moscow. A colleague from Utah came up to me after my talk in which I had mentioned ONF_3 and said, "Nice talk but the new molecule you mentioned is known." It turned out that it was known in the classified literature. It had been prepared by William B. Fox at Allied Chemical. He was a regular visitor to UBC as a recruiter and had given us some money for research, though not for this work. I did tell him straightforwardly, "This is embarrassing because the people in your company will think that you have told me." And he said, "That's possible." What he didn't tell me was that there was yet another group who had independently made ONF_3. I suggested that I send him our paper as part of our next report. Then he could contact me and say that it ought to be published. I sent him the manuscript and he wrote back that we should delay publication because they couldn't get their paper released in time. I agreed to the delay but asked that they also submit their paper to *Chemical Communications.* Alas, he sent it to the *Journal of the American Chemical Society* whereas we had sent ours to *Chemical Communications.* Fritz Sladky, who was a postdoc with me at Princeton, told me of gossip in Europe that I was supposed to have known that ONF_3 existed and that I'd stolen the idea. Horrible. That left a really bad taste in my mouth. Allied Chemical published only what we had in our paper. No more. I had to say to Jack Passmore, "It would be a waste of time to do more than complete the aspects that we have briefly described in our note." By the way, the third place where ONF_3 had been made was Rocketdyne, or a company that would later become Rocketdyne. Since these things are no longer classified, I should go back and see what was in fact known before our publication appeared. When I think about it, it still upsets me. I've always avoided working in areas where I knew that other people were working actively. That's why I never deliberately made PtF_6 prior to discovering the $O_2^+PtF_6^-$ formulation of our "oxyfluoride." Once I knew it was $O_2^+PtF_6^-$, I then made some PtF_6. I then deliberately mixed it with oxygen.

Is there anything you would like to add to the story of your fluorine chemistry?

In the beginning when we were making PtF_6, we were using Weinstock's method, burning platinum wire in fluorine. Now, Weinstock himself, who was a delightful person, would be very pleased to see that you can make it at room temperature. It's easy to make a high oxidation state in an anion because an anion is electron-rich. The electronegativity is lower for a given oxidation state in an anion than it is in a neutral molecule. That, in turn, is lower than it is in a cation. If I take silver and I expose it to fluorine in the presence of fluoride ion, in HF, and expose it to light to break up F_2 to atoms, I convert the silver to silver(III), AgF_4^-. This is easy because the Ag(III) is in an anion. I can then pass in boron trifluoride and precipitate silver trifluoride, which is now a much more potent oxidizer than AgF_4^- because the electronegativity in the neutral AgF_3 is much higher than it is in the anion. If I can now take away a fluoride ion, and make a cation, I drive the electronegativity even further up. With such a cation, for example, AgF_2^+, I can steal the electron from PtF_6^- and make PtF_6. This can be done at room temperature with a better than 70% yield of PtF_6.

What originally turned you to chemistry?

What attracted me was the fact that one could make such beautiful materials, both in color and in crystallinity. As a child, I grew crystals, cuprammonium salts. One could get these beautiful, royal blue crystals and make things, playing God, create things. That was at the age of 11 in my first chemistry lab at the grammar school.

Would you care to say something about your parents?

My mother was widowed very early. She was a remarkable woman. She never had anything other than an elementary school education. Her father was born in Heligoland with the name of Vock, a German, and he died when she was 2. So my mother had a hard upbringing. She became a saleslady in a shoe-selling shop. She was so good at it that she quickly became a top sales lady for the countrywide company and she made sufficient money to buy a little grocery store. My father was a shipwright. They started their married life in the Great Depression, and they both ran this corner shop. My father had served in World War I and was severely gassed. He was never in robust health after that. Eventually he died of cancer. My mother was such a good businesswoman that she made a small fortune in this little corner shop. She had an excellent memory and was able to carry all manner of diverse information in her head. There were three children, but we were never hard up. She was also extremely tolerant and let all my chemical experiments go on with bangs and nasty smells. She encouraged me to do more. My brother and I used to make ice cream on weekends as a little side business. I bought all my chemicals and books with money from selling ice cream. Mother just smiled benignly on all of this. She had a difficult time during World War I with the name Vock, "What are these damn Krauts doing here?" So mother changed her name to Voak to make it sound more English. The original name is from a Scandinavian area of Europe. I've done some checking. My grandfather had the name Friedrich Wilhelm Heinrich Johannes Vock. Evidently, his family were admirers of the Prussians. This Vock appears to have been an adventurer and migrated to Newcastle. At that time Heligoland was British, and he was a British subject. Queen Victoria and the Kaiser exchanged Heligoland and Zanzibar in 1890. This switching made my maternal grandfather German, and his birth certificate was sent from Somerset House to Berlin.

You are closing your lab now. Are you also winding up your activities?

I'm going off to do a little bit of lab work in Bordeaux, France, but I'll end my activities this year.

What are you planning to do?

There are many things that I've postponed in my life. My wife certainly deserves more attention than I've given her. One of the things I've postponed is working with my hands. I used to do watercolor painting but gave it up in the 1970s. I enjoy doing woodwork. I make bits and pieces of furniture. I also developed an interest in silver smithing, making round things from a flat sheet. Whether or not I now have the appropriate skills remains to be seen. I have a number of hobbies, including my garden.

Is there any compound that you would have liked to make and never succeeded?

Gold hexafluoride is a molecule that I have long wanted to make because it should have an electron affinity one electron volt higher than that of PtF_6. That's what we were trying to make when we made gold(V) compounds way back in the early 1970s. We have recently tried to make it by oxidizing AuF_6^- through nickel(IV) and silver(III) cation chemistry. We found that we could make platinum hexafluoride, ruthenium hexafluoride, and rhodium hexafluoride at room temperature, very easily. However, not gold hexafluoride. Somebody else will now have to find the effective route to it. It should exist, if made at low temperatures and kept cold.

References

1. Bartlett, N. "Forty Years of Fluorine Chemistry: King's College, Newcastle, The University of British Columbia, Princeton University, and The University of California at Berkeley." Edited by Eric Banks (University of Manchester Institute of Science and Technology), 2000.
2. Graham, L., Graudejus, O., Jha, N. K., Bartlett, N. "Concerning the Nature of $XePtF_6$." *Coord. Chem. Rev.* **2000**, *197*, 321–334.
3. Hyman, H. H. (ed.) *Noble Gas Compounds.* University of Chicago Press, 1963.
4. Bartlett, N. "Xenon Hexafluoroplatinate(V) $Xe^+[PtF_6]^-$." *Proc. Chem. Soc.* **1962**, 218.

Ronald J. Gillespie, 1998 (photograph by I. Hargittai).

4

Ronald J. Gillespie

Ronald J. Gillespie (b. 1924 in London, England) is Emeritus Professor at McMaster University in Hamilton, Ontario, Canada. He received his B.Sc., Ph.D., and D.Sc. degrees from London University in 1945, 1949, and 1957, respectively. He then taught at the Department of Chemistry of University College, London. Since 1958, he has been at McMaster University. He was elected Fellow of the Royal Society of Canada in 1965 and Fellow of the Royal Society (London) in 1977. I have known Ron Gillespie since 1972 when we first exchanged correspondence about the VSEPR model of molecular geometry. In 1991, we co-authored a book, *The VSEPR Model of Molecular Geometry* (Allyn & Bacon, Boston). We recorded this conversation during a molecular structure meeting in Austin, Texas, in March 1998, and Ron then augmented it with some "afterthoughts" in 1998.*

What started it all?

I was born in 1924 in London. When I was 11, I was awarded a scholarship to the local grammar school. Although my parents did not have to pay fees, they did have to pay for school uniforms, sports clothing, and train fares, which was difficult for them. I enjoyed school and did fairly well. During the term, my performance was only average because I spent a lot of time playing sports, but I did well on the final exams, particularly in

*This interview was originally published in *The Chemical Intelligencer* **1999**, 5(3), 6–10

science. During the final two years, we had to specialize in preparation for university. I chose to study chemistry, physics, pure math, and applied math. I was never good at math, but by working hard I managed to do reasonably well. I did much better in both physics and chemistry, but the physics class was deadly boring because the teacher simply read the textbook. The chemistry teacher used the lessons to make chemistry much more exciting than the textbook and gave us a lot of interesting lab work. It was the chemistry teacher, George Cast, who inspired me to go into chemistry. He also inspired a classmate, William Moffitt, who was really brilliant at math in school, to also go into chemistry. He went to Oxford, worked with Coulson, and eventually became Professor of Theoretical Chemistry at Harvard, but tragically died at an early age.

We were well into the war in 1942 when I passed the final examination at school and did well enough to be awarded a bursary to do a special two-year wartime degree in chemistry at University College, London. We were evacuated to the University of Wales in Aberystwyth, and that was where I spent my two undergraduate years. Toward the end of the second year, Ingold, the head of the department, said to me, "You are going to stay and do research, aren't you?" That summer (1944), the department moved back to London; it seemed that the war was coming to an end. Ingold had a team working on aromatic nitration. He put me on cryoscopy of nitric acid in sulfuric acid to look for the nitronium ion. Someone else (Jim Millen) was working on the Raman spectra of these solutions, and another student (Dan Goddard) was attempting to prepare stable salts of the nitronium ion, and there were others working on the kinetics of aromatic nitration. That's how I got started on acids. After the nitronium ion, I got intrigued with what else you could do in sulfuric acid.

Did you like Ingold?

I never really got to know him. Although he could be friendly in a formal way, he was a very remote person. For several years I was in awe of him. He would come around the lab about once a month, usually with Ted Hughes, for a short chat about what you were doing. That was almost the only time that you saw him except at the department research colloquia. He always had a good knowledge of what you were doing or should have been doing. We never talked about anything outside my research project. After I found the cryoscopic evidence for the nitronium ion, I made suggestions about other things that I might do. He always said, "OK, go ahead." He let me do more or less what I wanted. I began to study the ionization of other solutes in sulfuric acid. That was the beginning of my studies of sulfuric acid and later of other superacids.

Ingold published seven papers on my work in the *Journal of the Chemical Society* in 1950. All were written up by Ingold, but several of them carried my name only. He wrote the papers and gave the final version to me to read, but I didn't dare to suggest any changes. He didn't consider my area of research to be his main interest, which was in kinetics and reaction mechanisms. I suspect that if I had been working in that area, I might have had less freedom. But Ingold nevertheless had very wide interests. He was an organic chemist and a physical chemist and knew some inorganic chemistry.

I got my Ph.D. in 1949, but before that, and it was typical Ingold, he came to me one day and said, "I'd like you to give a few lectures. We need a course on molecular properties." That was it. A few weeks later, I got a letter from the university saying that I had been appointed an Assistant Lecturer. I received no suggestions about the content of the course but prepared a series of lectures on molecular geometry, dipole moments, the parachor, magnetism, etc. Thus, I was already on the faculty before I got my Ph.D. I knew almost nothing of what was going on outside University College. I never considered any other university.

I continued with my research on sulfuric acid, now quite independently from Ingold. From time to time, he sent a student to me. Ingold chose his students and allocated others to the faculty. I started to branch out, looking at the conductivity of solutions of other substances in sulfuric acid and measuring acidities. That's how I got interested in superacids.

In 1953, I got a Commonwealth Fund Fellowship to work in the U.S. I had the idea that sulfuric acid had a very high dielectric constant because the solutions seemed to behave in an almost ideal way. There was an expert (Robert Cole) in dielectric constants at Brown University, so I went there for a year. It was a difficult measurement because 100% sulfuric acid is very highly conducting because of its extensive self-dissociation. However, everything went well though the dielectric constant wasn't quite as high as I had thought, turning out to be 120.

The Commonwealth Fund Fellowship gave me sufficient money to buy a second-hand car to tour the U.S. and then write up my impressions of the States. My wife had stayed behind in Britain, but she came over for the summer and we toured and camped all around the U.S. and into

Ronald Gillespie in 1981 (courtesy of Ronald Gillespie).

R. J. Gillespie, R. F. W. Bader and P. J. MacDougall in 1988 (courtesy of Ronald Gillespie).

Canada. It was then that I began to think that it might be a good idea to move to North America.

When I returned to University College, I tried to further expand my research, but it was not easy. I wanted to use Raman spectroscopy, for example, but that was Jim Millen's field. It was frustrating because everything had to be obtained through Ingold. I had no research money. If I wanted anything, I always had to go and ask him.

Eventually, I was fully into research and teaching in inorganic chemistry. Ron Nyholm, a newly appointed faculty member from Australia, also taught inorganic chemistry. I got to know him well and had many discussions with him. When I was teaching bonding, I was never satisfied with the textbook explanations of bonding in terms of sp^3 hybrid orbitals. It seemed to me to be a circular argument. Then I came across a paper by Sidgwick and Powell, who showed that geometry could be related to the number of electron pairs. I also found another paper by Lennard-Jones on the importance of the Pauli principle in determining geometry. By then I was also teaching quantum mechanics. I talked to Nyholm quite a lot and we decided to write a review on inorganic stereochemistry. I would do the main groups and he would do the transition metals. Writing this review, I formulated the rules that later became the VSEPR model. Even in that article, I pointed out that it was not just an electrostatic model but is based on the Pauli principle. But people did not think much about that; they just took the rules and used them widely, although it took 15 years to get VSEPR into the U.S. textbooks.

Soon after, I decided it was time to move. There were a lot of us at University College who were quite dissatisfied at the time. People were moving. I was interested in Australia but my wife thought that it was too far away. I was also interested in Canada, possibly Vancouver or McMaster. Nyholm and I had decided that we needed an NMR spectrometer, and we built a machine. It was tiny and not very useful so we decided to try to buy one. Nyholm was trying to get some money while I went to the States to talk to Varian. On the way back, I visited McMaster. I knew Art Bourns, who was the Dean of Graduate Studies, from a period he had spent at University College. I knew rather little about McMaster, except that it was growing rapidly from a small Baptist college under Harry Thode, who envisioned it as a major research university. I soon had an offer from McMaster, which doubled my salary, and we decided to move. I went on the condition that I got a Raman spectrometer and an NMR spectrometer.

I went to Canada at a very good time as the National Research Council of Canada was awarding what seemed to me to be large grants. This was the first time I had my own research money. Peter Robinson came over with me as a postdoc to help get research going.

Are superacids and VSEPR your most important areas of research?

My research has been mainly in superacids, sulfuric acid, fluorosulfuric acid and hydrogen fluoride, which led me naturally into fluorine chemistry. I didn't do a lot more on VSEPR except for trying to promote it. I was excited by the discovery of the noble-gas compounds, and it was an obvious opportunity to show that it was easy to use VSEPR to predict their structures. At one of the early conferences on noble-gas chemistry, Larry Bartell asked me to predict the structure of XeF_6. I believe he had already determined the structure by electron diffraction but not yet published it. I told him that it could not be octahedral, which the MO theorists believed they had proved, but was probably a distorted monocapped octahedron, which turned out to be correct. Since then, he has been a great fan of VSEPR and has done much to promote it as a useful theory.

At times, you almost seem to have been embarrassed by the simplicity of the VSEPR model ...

I would not say that I was embarrassed by its simplicity but I was trying to improve it mainly to try to account for the exceptions, which have always challenged me. One of my colleagues, Richard Bader, originally did not believe in the VSEPR model because he had shown that electrons are not as localized as the model claims. It was not until he began to study the Laplacian of the electron density in the mid-1980s that he was converted to VSEPR. He had a student, Preston MacDougall, who found that the Laplacian of the electron density provides strong support for the model. This is how I got interested in Bader's work. About this time, in 1989, I officially retired. I planned to continue my research but my funding from NSERC (the Natural Sciences and Engineering Research Council of Canada) for my experimental program stopped. However, I had already decided that I wanted a change so this was the opportunity to finally do some real research on VSEPR. In collaboration with Richard Bader and Peter Robinson, I began to look at the apparent exceptions and more deeply at the physical basis of the model. Now I have a small grant from NSERC to work on transition-metal molecules.

Getting back to superacids, I remember George Olah, in writing a preface to one of his books, expressing his admiration for your seminal and fundamental work on superacids. Eventually he got the Nobel Prize but you didn't.

I can understand why. He exploited superacids in organic chemistry to an enormous extent. I was never very interested in organic chemistry and I stuck to inorganic chemistry. But going back in history, I should tell you that I first met George when he spent a very short time at University College after he left Hungary as a refugee. Then I moved to Canada and he did also to work at Dow in Sarnia. At that time I had the only NMR spectrometer in Canada, and George sent his technician from Sarnia to obtain spectra of his samples on our machine. I must say that these samples often looked like black gunk. If my students got anything that sort of color, I sent them back to the bench. In those black solutions, George found the trimethylcarbonium ion. I was too insistent on purity, and perhaps I shouldn't have been. One could say I missed an opportunity, but I always had in the back of my mind similar-looking solutions in sulfuric acid that we had prepared which did not give any useful results. Although I did most of the first experiments showing how to make superacids and measuring their acidity, George thoroughly exploited superacids in organic chemistry.

Afterthoughts. Giving this interview got me reflecting on what I had contributed to chemistry and how it was that I had done reasonably well. At the time I had just finished reading *Darwin, A Life in Science* by John Gribbin and Michael White. At the end of this book, there is an Appendix, which quotes what Darwin wrote about himself in his autobiography. I was struck by the fact that much of what he wrote I could have written about myself although not in the same elegant, if rather ponderous, Victorian English. It seemed to be of interest, therefore, to quote some of what Darwin wrote about himself, even if comparing oneself to Darwin might appear somewhat pretentious. The attributes that contribute to success are many and varied, but I was quite surprised to find that I shared many of these attributes with Darwin, as I am sure many other scientists do. This is not to say, of course, that quite different qualities might not be important for others, but it seems that, for a certain type of researcher, the qualities that Darwin possessed are important for success. Darwin wrote:

> I have no great quickness of apprehension or wit which is so evident in some clever men, … I am therefore a poor critic:

a paper or book when first read excites my admiration, and it is only after considerable reflection that I perceive the weak points. My power to follow a long and purely abstract train of thought is limited: I should moreover never have succeeded with ... mathematics. My memory is extensive yet hazy: it suffices to make me cautious by vaguely telling me that I have observed or read something opposed to the conclusion which I am drawing, or on the other hand is in favor of it; and after a time I can generally recollect where to search for my authority. So poor, in one sense, is my memory, that I have never been able to remember for more than few days a single date or a line of poetry [or a humorous story; RG].

... my love of natural science has been steady and ardent. This pure love however, has been much aided by the ambition to be esteemed by my fellow naturalists [chemists; RG]. From my early youth I have had the strongest desire to understand or explain whatever I observed, that is, to group all facts under some general laws. These causes combined have given me the patience to reflect or ponder for any number years over any unexplained problems. As far as I can judge I am not apt to follow blindly the lead of other men. I have steadily endeavored to keep my mind free, so as to give up any hypothesis, however much beloved, as soon as facts are shown to be opposed to it. On the other hand I am not very sceptical, — a frame of mind, which I believe to be injurious to the progress of science.

Therefore, my success, as a man of science, whatever this may have amounted to, has been determined, as far as I can judge, by complex and diversified mental qualities and conditions. Of these the most important have been — the love of science — unbounded patience in long reflecting over any subject — ... and a fair share of invention as well as of common sense. With such moderate abilities as I posses it is truly surprising that I should have influenced ... the beliefs of scientific men on some important points.

Darwin also noted that he had "... had ample leisure from not having to earn my own bread." By this I think he meant he had plenty of time to pursue what interested him, because for Darwin, as for many other scientists, it is often difficult to distinguish work from leisure. The "gentleman scientist" of the Victorian era has disappeared today but as a privileged academic, and even before as a graduate student, I have been lucky enough almost my whole life to have been able to do, very largely, just what interests me — clearly another important factor in contributing to success in science as Darwin recognized.

Lawrence S. Bartell, 1996 (photograph by I. Hargittai)

5

Lawrence S. Bartell

Lawrence S. Bartell (b. 1923 in Ann Arbor, MI) is Philip J. Elving Professor Emeritus at the University of Michigan. He has spent his research career in structural chemistry, working on a variety of topics, in several of which he made pioneering contributions. Thus he was the first as early as in 1955 to call attention to the importance of the physical meaning of the geometrical parameters determined by different physical techniques. He studied the effects of temperature on the molecular geometry extensively. His invention that elicited the most popular attention was electron holography, in 1972. It even yielded a direct way to measure interatomic distances in molecules. In greatly modified form it is having an impact in the field of electron microscopy and also made it to the *Guinness Book of World Records* as the World's Most Powerful Microscope. His latest interest is nucleation. Dr. Bartell was named "Michigan Scientist of the Year" by the Science Museum, Lansing, Michigan, in 1986 and the most influential member in the past century by the Michigan chapter of the Alpha Chi Sigma professional chemistry fraternity. He received a "Creativity Award" from the National Science Foundation in 1982. Our conversation was recorded in his office at the Department of Chemistry, University of Michigan, Ann Arbor, on November 18, 1998 and completed in Wilmington, North Carolina on April 10, 1999.*

*This interview was prepared by Magdolna Hargittai.

Would you care to say something about your family background?

My father was the first in his family ever to go to college. He was born on a farm and was a very strong fellow who excelled in sports as he worked his way through the small Michigan college, Albion. After he got a degree in chemistry he went to Iowa to teach chemistry at Simpson College. He was also asked to coach all of the athletic teams of the college. He coached so well that his women's basketball team won the state championship in Iowa. After several years, he was offered the position of Professor of Chemistry, of Director of Athletics, or both if he wished. By this time, he decided he should get more education, so he came to Ann Arbor, got a Ph.D. in chemistry, joined the faculty, and stayed for the rest of his life.

My mother's father was a physician in Virginia. When mother came to Michigan, she was one of the rare female students in chemistry. After her bachelor's degree she stayed on for a Master's degree working under the direction of my father. Being very smart she got all A's (this *before* she started to date my father, I have been led to understand!). There was a nepotism rule forbidding the university from hiring two members of a family, but she helped my father in many ways. My father and my mother — whose name, curiously, was Lawrence S. Bartell, the same as mine — published a paper together when I was about nine years old. Later, people supposed it was me and that I must have been a child prodigy.

Mother and dad never talked shop at home. I never heard any enthusiastic discussions about science. It seemed to me then, as now, that my parents lacked a romantic attachment to science, or sense of adventure in research, or sense of fantasy that can lead to creativity.

My enthusiasm for science started from my model airplanes and the experiments I did with them. Then physics opened my eyes to other things that one could experiment with. Later, I even found organic chemistry interesting, especially the laboratory except for the dishwashing. In physical chemistry, experiments were even more challenging. Then, of course, there was the war. I was deferred from military service as long as I remained in my undergraduate chemical curriculum. Because of the war, we had accelerated training. I was only 20 when I left the university to be interviewed by Glenn Seaborg at the University of Chicago. Seaborg offered me the position of Research Assistant on the Manhattan project. It was supposed to be secret but I had a pretty good idea before I went that the project was about the atomic bomb. When I was a student I had become convinced that the most fascinating area of chemistry was radiochemistry, so I accepted Seaborg's offer with alacrity. I worked for one year on the project. It was so secret that the draft-boards did not know what it was. Therefore, a lot of research assistants working on the project, including my roommate, got drafted into the army and sent back to the project as members of the Corps of Engineers. This created interesting situations. Because we research assistants all had bachelor's degrees but not doctor's degrees, those who went into the Corps of Engineers went as enlisted men (i.e., not officers). When you have enlisted men, of course, you must have officers to tell them what to do. But these officers weren't scientists, weren't cleared by security, and therefore didn't know what was going on. To find out what was going on, they sometimes tried to "pull rank" and got rebuffed. This made them very angry but they could do nothing about it. It created some delightful friction.

What was your work in the Manhattan project?

My first assignment was to take a gram of plutonium, probably plutonium nitrate (which I was told was worth over ten million [in today's] dollars), and do experiments on one tenth of that amount. My assignment was to test Seaborg's suggested procedure for separating plutonium produced in a reactor from the huge excess of uranium and from the exceedingly active fission products. The procedure I worked on was that which was actually used to extract the plutonium for the Nagasaki bomb. We were desperately afraid that the Germans might get there first, because it was known that Heisenberg was running the program and he was a world-class physicist. Recently it was remarked that one reason we made enormously greater progress than the Germans in harnessing nuclear energy was that while Heisenberg was indisputably a *great* physicist, he was not a particularly *good* physicist. He was quick and dirty about details and his project missed important points. In our efforts to beat Germany we worked very hard. As a matter of fact, on the night of D-day I happened to be working all night in the laboratory and kept hearing reports on the radio making it clear that some huge operation was beginning.

We were contaminated perpetually. Today people say that back in those days the dangers of radiation weren't recognized. That is not true. Even Dr. Markham told us in freshman chemistry about the terrible radiation disease occurring in women who painted radium on phosphorescent hands of watches and, in order to get a fine point on the brush they used to paint with, licked it. To keep us fit we had a health physics department and were checked regularly.

Actually we were made to take many precautions. Each time we left "New Chem," the temporary building where we worked, we had to put our hands into a counter about 50 feet from the door. If the count was above some critical amount lights would flash and bells would ring and then we had to decontaminate our hands. We did this by repeated oxidation-reduction cycles. We would wash our hands with potassium permanganate until they became black, then with sodium bisulphite until they were bleached back to whiteness. After enough cycles the counts decreased sufficiently for us to pass. It is really quite amazing how much abuse one's skin can take. My hands always rang those confounded bells, whatever precautions I took with the intensely radioactive solutions. We wore rubber gloves but there was so much radioactivity to contend with that it was impossible to avoid getting contaminated.

You may remember the famous west stands of the Chicago football stadium where the world's first nuclear reactor was constructed by Fermi and his group. After I had been in Chicago for awhile, a "hot lab" was set up in the west stands to handle particularly active material. It wasn't quite ready when I had the "distinction" of being the first person told to try it out by centrifuging some hot material. When I returned from the hot lab and entered New Chem, I set off the counter 50 feet from the door. Health physicists took all my clothes and swabbed every orifice. I never was compensated for the clothes I lost and don't remember how I got home that day.

After working in the field of radiochemistry every day for a year, I got my fill of radiochemistry — enough for a lifetime. So it was almost a relief when I received a "Greeting" from the President of the United States informing me that I was going to be inducted into the armed services. Actually, I had to be persuaded by my boss at Chicago not to volunteer for the military service some time before I was drafted. When you hear about your best friends getting shot at while you are enjoying civilian life in Chicago you feel guilty. As it turned out, I was the last person to be drafted from the project in Chicago and the only one drafted not to be taken immediately into the Corps of Engineers. I never understood fully why this honor befell me. Anyway, if I were headed into the regular armed services I had no interest in muddy trenches. Therefore, I enlisted in the navy. I have to say in retrospect that I contributed to the victory of the allies far more as a civilian than as a sailor in the Navy because the atomic bomb, of which I was a tiny part, ended the war. In the Navy, I was pretty much a detriment.

Would you care to comment on how you feel about your participation in the bomb project, in view of the terrible consequences to innocent civilians who happened to be in Hiroshima or Nagasaki when the bombs were dropped?

Naturally, I am sorry about the suffering inflicted but never regretted for a moment my part in the project. After all, the bomb DID end the war, saving perhaps millions of lives, Japanese as well as American. While we were working on the project we were alarmed at the prospect that Hitler might get the bomb before we did and the consequences would by unimaginably terrible. Many of my friends who were scheduled to be in the invasion of Japan are grateful that they were spared the horrors that the invasion would have involved. And the only reason Hiroshima and Nagasaki were more or less intact by the time of the bombing instead of having been gutted by the fire bombing that had already laid waste most large centers in Japan was that they had been saved as targets for atomic bombs. Bombing an already devastated area would not have produced the political and psychological effects that the bombing did achieve. More people died by fire bombs than by atomic bombs. What was different was that the "normal" bombing took thousands of bombers while an atomic bomb, only one. And don't forget that the Japanese soldiers, during the infamous "Rape of Nanking," wantonly slaughtered far more innocent civilians in that one place alone than were killed by both atomic bombs. Most Monday morning quarterbacks who harangue about the monstrosity of dropping atomic bombs on Japan do not factor into the picture that the military leaders who controlled Japan's policies were intransigently opposed to surrender. They submitted a plan to defeat the invasion of Japan that they estimated might cost the lives of 20 million people, making the tragedy of Hiroshima pale by comparison. It is not well known that Japan had its atomic bomb project, too, run by exceptionally gifted physicists. It was not well supported by the military, however, so its progress was minimal. But after the bombs fell on Japan, the physicists were told to produce the bomb within three months. An absurd order! Finally, to round out the response to your question, it is a great misfortune that nature does not rule out the atomic bomb. It was just a matter of time before bombs would be designed and made. If they had not been used to end World War II, illustrating the horrors of their usage not only to philosophers but, more importantly, to politicians, you know very well that some political leader would have used them in anger by now, now that they are thousands of times more powerful than in World War II. So they have been a deterrent to further global warfare. Through the balance of terror, we have experienced the longest period of relative peace in recent history.

In the navy, I spent nine months in the hospital with rheumatic fever acquired from complications of scarlet fever. During my recovery I read about astronomy and became so fascinated with the field that I spent my naval mustering out pay to buy a 10-inch mirror and built a powerful telescope. Since I was going to go to graduate school I knew I wouldn't have the time to grind my own mirror. It was a thrill to observe the heavens with my telescope, an instrument larger than most amateurs possess. The diffraction limit to resolving power of telescopes got me interested in physical optics and the interference of waves. Astronomy taught me most of what I needed to learn, later, when I began to earn a living probing matter with electron waves.

You never considered becoming a professional in astronomy?

No. I loved the subject but didn't think it was what I wanted to make a living doing. In graduate school I took courses that very few chemists did. Probably the most brilliant professor I ever had was the theoretical physicist Otto Laporte, the man who discovered what is now known as the Laporte selection rule in spectroscopy. He discovered the rule when he was a student in Germany. He was a very colorful character, a superb lecturer who drew marvelous diagrams on the blackboard with colored chalks that he expected us to copy into our notes in colored ink. Laporte's genius was not restricted to theoretical physics. He was uncommonly steeped in knowledge of all cultural disciplines. For example, when we students heard he was interested in ancient Chinese history, some of the guys in our class went to the library and looked up details about one of the most obscure Ming dynasties they could find. In a conversation with Laporte one of the students casually mentioned the name of the dynasty whereupon Laporte immediately expounded upon that dynasty in interesting detail. He was unreal!

Any other teachers who influenced your future interest?

I am sure that Kasimir Fajans did. In his rambling, disorganized way, he had been more stimulating to me than any other chemist. He was a brilliant but contentious man. He had worked his way up to become a Professor and Head of an Institute of Physical Chemistry in Munich. Because he was Jewish, he had to leave Germany in the 1930s and accepted an appointment at the University of Michigan. Fajans was legendary. Socially, he was the model of the perfect European gentleman. At his parties he would be the most gracious host you could imagine. Moreover, from his exceptional grasp of chemistry he was able from an empirical approach to construct a theory of electronic structure of atoms and molecules that was closely parallel to molecular orbital theory. He called it his "quanticule theory." Fajans thought mathematical theory was out of place in chemistry. He argued against Pauling's ideas again and again and with too little tact or diplomacy. Although he was a perfect gentleman in purely social circumstances, his ability to get along with competitors in science was too poorly developed for him to succeed. I have to say in retrospect that Fajans was probably right at least half of the time in his arguments with Pauling. But he argued in such an obnoxious, clumsy way that he began to be perceived by referees and editors as a crank and eventually found himself unable to publish in reputable journals.

Fajans himself independently developed the rules for radioactive decay that were the same as the Rutherford-Soddy rules. The winners of the Nobel Prize in Chemistry, however, were Rutherford and Soddy, not Fajans. Another Nobel story involves Fajans. He told me once how one of his assistants came to him one day describing his confusing experiment in which he precipitated some silver halide in the presence of a dye. He had observed a color change he couldn't explain. This assistant was a rather mediocre student according to Fajans. Although the assistant failed to comprehend what he had done, Fajans told me that he, himself, understood it immediately. This was the origination of the famous Fajans indicator, which used to be used in analytical titrations of silver or halogens. This "mediocre" assistant turned out to be Odd Hassel who later won the Nobel Prize in 1969 for the discovery (via electron diffraction) of conformational equilibria in certain cyclohexane derivatives. So twice Fajans perceived that others, no more deserving than he, got the prize for. I never heard him complain about it directly, however.

Of course, there was Lawrence Brockway who was one of our better known physical chemists and a protégé of Pauling. Brockway had won the prestigious American Chemical Society Award in Pure Chemistry and was undeniably an extremely talented man. He even became the father-figure of vapor phase electron diffraction around the world, although it was Mark and Wierl who had originated the field. But somehow Fajans had a profound distaste for Pauling and Pauling's theories, and therefore constantly criticized Brockway who promoted Pauling's ideas. These challenges were often launched in a most intemperate manner in public and Brockway was not strong enough to withstand the barrage. So Brockway never achieved the stature in chemistry that he might have under different circumstances.

Weren't you Brockway's doctoral student?

Yes. After my discharge from the Navy and my recuperation in the Southwest Desert, I came back to Michigan. At Michigan, the only person whose work seemed interesting to me was Brockway, because electron waves and wave interference seemed fascinating, and being able to see the wave nature of electrons visually any time I wanted to, just by diffracting them from some diffraction grating, molecular or crystalline, was compelling. So I decided to work for Brockway. I have never regretted that choice. Unfortunately, Brockway's heart was no longer in research because of the acrimonious assaults by Fajans. So I didn't see him around very much. He was often away consulting in industry and the government. He was also a minister in the Reorganized Latter Day Saints Church. As a matter of fact, he married my wife and myself. I got more stimulation from Pauling's junior colleague Verner Schomaker than from Brockway when I was a graduate student. As it happened, Verner passed through Ann Arbor a number of times, at first to see Brockway but later he even stopped when Brockway was out of town. Verner was celebrated for his imaginative ideas about chemistry, and was great fun to talk to. He inspired me to persevere. Brockway gave me great freedom to design equipment and make it or have it made, and I enjoyed the building and testing. In the process I taught myself to use standard tools in the machine shop like drill presses, lathes, and milling machines and this turned out to be very important when I began my academic career at Iowa State University where I had to make my own research instruments. Knowing how to machine is also a great help in designing equipment to be made by others.

Wedding picture of Lawrence and Joy Bartell in Ann Arbor, 1952. From left to right: Lawrence Brockway (Bartell's thesis advisor and minister), Connie and Will Bigelow, Lawrence and Joy Bartell, Bartell's mother and father (courtesy of Lawrence Bartell).

Lawrence Bartell and Verner Schomaker with a guide in Moscow, 1959 (courtesy of Lawrence Bartell).

As an illustration of Brockway's degree of guidance I mention my design of the diffraction unit to replace Brockways's old unit. When he saw my drawings for the first time, the night before the deadline the machine shop imposed for submission of the plans, he was appalled because the plan looked nothing like his design but it was far too late for him to change anything. Happily, when the diffraction unit was completed, it worked so much better than anything he had seen that he was quite pleased. When he asked me what I was going to do with the apparatus, I told him I'd measure electron distributions in atoms. He asked me "How are you going to do that?" I just said, "Don't worry, it will work" — and it did work. Incidentally, this is what eventually started me off in electron holography.

Considering your whole career, which of your results do you think of as the most important?

Which of your children you love the most?

Both of them, of course!

So there you are. I find it difficult to chose between my projects. Since you are an electron diffractionist what you probably think of first is my early treatment of the effects of anharmonicity of molecular vibrations on analyses of molecular structures. This was reviewed in detail in the prefatory chapter you and your husband solicited.[1] Surface chemists might think my invention of an ellipsometric technique to measure absorption spectra of films as thin as monomolecular might qualify as the most important. Since I was too poor to do electron diffraction work, I constructed some ellipsometers during my early years at Iowa State and discovered their unique capabilities.

How I got into surface chemistry is a curious story, for I never expected to do it. What happened was the result of a decision by the Simonize Company, a company that made wax products including what used to be the most famous car polish. Nowadays automobile finishes are so good that they don't need regular waxing. That was not true when I was young. When the Simonize directors decided they wanted to learn more about what their wax films were like, they went to my father, who was a distinguished surface chemist. It turned out that the standard techniques of surface chemistry were not up to the job of measuring the properties of these films. I had just finished my Ph.D. and my father asked me if I had any ideas about how such films might be investigated, so I was called in as a sort of assistant consultant. I tried several interference techniques in an attempt to measure film thicknesses but found they were quite inadequate, implying that the films were very thin indeed. After searching the literature I found that a technique called ellipsometry had been applied to biological films in 1945 but not in a way generally applicable to films such as Simonize waxes. Still, since the technique claimed to have a sensitivity of angstrom units, a claim I found difficult to believe, I kept reading and found in articles written in 1885 that Drude had developed the theory of reflection of elliptically polarized light. The ellipsometric technique was genuine.

Therefore, I devised a way to carry out ellipsometric studies generally applicable to surface films and I discovered that well buffed Simonize films were only about a molecule thick. Simonize films were a dramatic illustration of the effectiveness of boundary lubrication! This finding encouraged me to investigate the films by electron diffraction and the diffraction patterns showed that Simonize films were remarkably well-organized semicrystalline aggregates as opposed to some of the short-lived cheap competitors which gave diffraction patterns indicative of chaotic heaps of molecules.

In Simonize films deposited on metal surfaces by evaporation of very dilute solutions, the molecules were found to stand up in well-aligned arrays nearly perpendicular to the surface. But when I stroked the molecules with Kleenex, I discovered I could make them lie down flat. To make them respond to my bidding gave me a feeling of intimacy with molecules that was quite exhilarating. It also gave me a brand new tool with which to study surface chemistry. It was with this that I began my research program at Iowa State University. I found interesting things about the wetting of films by various liquids. Results depended upon the structure and degree of depletion of the films and upon the way the liquid molecules fit among the film molecules. Since we had this extraordinary sensitivity (fractions of Angstroms) to the variation of the film thickness we were able to do interesting things with it. It was not the *mechanical* thickness of the film that was measured but the *optical* thickness. So, while pondering ways to vary the optical thickness, I realized that one way to do it would be to vary the wavelength. If you sweep through the peak of an absorption spectrum you pass through a region of anomalous dispersion where the index of refraction varies enormously. It occurred to me that instead of varying the actual thickness of the film, we could vary the optical thickness by varying the wavelength and we ought to be able to see this anomalous dispersion. This was in about 1956. It worked right away. Then Don Churchill and I developed the method into a quantitative technique for studying optical characteristics of very thin films. Today a number of companies make ellipsometers for measuring surface spectra. The technique can be the method of choice in certain circumstances, and is a commercial success. So how can I rule that out as a child that I don't care about?

Why did you leave surface science?

Because of Russ Bonham, my very talented Ph.D. student. Actually, I remained in surface science for a number of years after Bonham joined my group but Russ wasn't interested in it. On the other hand, electron waves and structures of molecules did pique his curiosity and I could not afford to lose him. Therefore, I arranged for us to come to Michigan to use my old apparatus. That got me back into molecular structure. Our first molecular candidates to study were hydrocarbon molecules because they were readily available and enabled me to get funds from the American Petroleum Institute to support Bonham's research. Even though hydrocarbons might sound rather pedestrian, they turned out to be an extremely good choice. Particularly provocative was isobutylene because its structure forced me to formulate a radically new way of looking at bond angles and molecular structure. The ideas developed from the work led to a paper that became a "Citation Classic." Gillespie is currently expanding the ideas into new areas of the periodic table.

So Russ Bonham dragged me back into structural chemistry, and we obtained very worthwhile results right away. After working with surface chemistry for four or five years, and after Russ began to get results with data from the Michigan unit, I was told that the Ames Laboratory at Iowa State was willing to support some of my research and build an electron diffraction unit if I wished. So I designed and built one that was much better than the apparatus I designed when I was a student at Michigan. When I returned to the University of Michigan I was allowed to buy it and take it with me. We are still using it.

What else has my research group accomplished? Probably the most cerebral thing I ever did was to figure out how to treat intramolecular dynamical scattering of electrons in vapor-phase molecules — not a result that will shake the world but one which does aid certain structure determinations and solve a long-standing problem. When we studied ReF_6, ReF_7, and IOF_5, it was clear that the standard diffraction theory was insufficient to account for our experimental intensities. I had such confidence in Jean Jacob's diffraction intensities that I couldn't accept sweeping the discrepancies under the rug as others in the field did. Eventually, after a long struggle, I found a magic transformation that solved the problem. The result is probably the cleverest thing I ever did. But, of course, it is such a specialized, technical thing that it impacts very few people.

Electron correlation is another problem I attacked that did affect quite a few research laboratories. I am not a theorist but I recognized that there is something strangely unsymmetrical between gas-phase electron diffraction and X-ray crystallography. The two fields use fundamentally different scattering formulae. X-ray crystallographers average scattered amplitudes over the quantum motions of the electrons and then square to get the intensity. By contrast, gas-phase electron diffractionists square the amplitudes calculated from instantaneous configurations of atoms to get instantaneous intensities, then average intensities over the quantum motions of the atoms. So whether first to average, then square or to square, then average? As an experimentalist I found this disparity very strange. Once I even ran into Debye and asked him about it, but he just laughed.

Debye was probably the most brilliant person I ever knew. He was incredible. But he had this maddening habit of deflecting questions and making you work out the solutions for yourself. Well, after a struggle I finally realized what was going on and how it worked. And it became clear that Debye understood all of this perfectly in 1915 when he published his treatment of the scattering of X-rays. Another thing became clear. By playing elastic against inelastic scattering it was possible to determine experimentally how electron correlation influenced the spatial distribution of electrons. This is an application Debye did not follow up because quantum mechanics had not been sufficiently developed in 1915 for him to exploit such a possibility.

For me, the wonderful thing was that just after I recognized this and Bob Gavin and I completed our first quantum computations to illustrate the magnitudes to be expected, I was invited to celebrate Debye's 80th birthday at an informal symposium held at Cornell University. That was in 1964. The auditorium was full of distinguished theorists, including Lars Onsager, Christopher Longuet-Higgins, E. Bright Wilson, Elliott Montroll, Ben Widom, and, of course, Peter Debye. Moreover, we were given the opportunity to get up and say a few words if we cared to. It was an irresistible opportunity to tell the assembly what we had learned that Debye had understood in 1915, and how one could observe electron correlation experimentally. When I finished speaking, Longuet-Higgins came up to me and thrilled me by saying, "Why don't I know you?" The realization that such measurements were possible and that a long-forgotten idea of Debye was bearing fruit at his 80th birthday symposium created quite a stir.

Afterwards a number of laboratories took up my idea and carried out programs of research on the topic. For example, when I went to Paris on a sabbatical in 1973, I found three different groups of theoretical physicists in Paris extending my original research idea and so were laboratories in South America and Asia. So that work seems to have initiated research elsewhere as much as anything else I ever did. I enjoyed this work quite as much as that on surface spectra and dynamic scattering. But, of course, again, it would be known only by a small group of people. So that's another child I love as much as my other children.

After that I got involved with molecular mechanics. That work probably had a wider impact in science than our other work because its development was relevant to many more people, namely most organic chemists. You will recall that molecular mechanics is a computer technique for deducing expected structures of molecules (mainly organic molecules) from a model force field recipe. Brad Thompson, Jean Jacob, and I carried out some of the earliest, if not the earliest truly bona fide studies in which complete molecular relaxation was allowed and only physically meaningful interactions were invoked, with proper distinctions made between contributions from nonbonded interactions and bending force constants. We got pretty good mileage out of that work. Of course, the person who later embraced the method with all his heart and effectively demonstrated to organic chemists how important a tool for research it really is, was Lou Allinger. Lou once told my wife the story, " While I was giving a lecture on molecular mechanics at Wayne State University this man in the audience got up and started to attack me." Well, I wasn't attacking him, I was simply asking him a couple of questions. We are good friends now. Anyway, Allinger published a great many papers in the field, including a detailed monograph, and he made his programs attractive to and available to organic chemists. His programs are an extremely valuable tool in research for almost all organic chemists. It's not a perfect tool but is one of the better ones.

As far as model force fields are concerned, it is crucial to build anharmonic components into the molecular mechanics program if you want to have adequate predictive power in the case of crowded molecules. That was borne out strikingly in Hans-Beat Bürgi's study of tri-*t*-butyl methane where our field turned out to give an excellent representation of the molecule when others, more or less devoid of anharmonic terms, didn't.

Once when you asked me why I jumped around from field to field to field in my research I tried to explain that it was not so much jumping as evolving. When simply studying structures of gas-phase molecules in their ground states lost its appeal, it seemed it might be interesting to study excited molecules. Electronic excitations were too short lived to be feasible for our method but vibrational excitations looked possible. Since it might be able to excite them by laser irradiation, we got an infrared laser and attempted to shine its focused beam on SF_6 vapor flowing from a micronozzle. Our very first experiments showed such a spectacular degree of excitation that I couldn't believe them. The excitation was enormously greater than the amount I estimated from the literature. I was ready to throw the plates into the trashcan and tell my students to start all over but Steve Doun, bless his heart, went ahead and analyzed the plates anyway. They told a very nice story. From his results we could determine the temperature of the molecular sample from the large increase in the amplitudes of molecular vibrations and could also see that the bond lengths increased remarkably. We finally realized that what had happened was that the laser beam had grazed the tip of the micronozzle and had heated the daylights out of it — heated it almost to its melting temperature! Therefore we were not exciting molecules by the direct pumping of photons. Molecules were getting fiercely hot by streaming past a very hot nozzle tip. It was astonishing how quickly heat was transferred to the gas during such a brief passage through a hot tube. So when we saw what we could do by accident, we began to do it on purpose and studied several gases. I worked out the theory of how structure should change on heating. These calculations could not be tested using spectroscopic force fields because no entirely adequate anharmonic force fields had been obtained but we could compare results with quantum computations. Our new method allowed us to follow the structure and vibrations of SF_6 up to almost 2000 K. This was enormously higher than physicists with their standard ovens were able to attain because their molecules were cooked apart while ours, heated for such a brief time, survived long enough to be measured, intact. Since our molecules were executing vibrations of such large amplitudes, our results provided information about anharmonic components of molecular force fields that spectroscopists were blind to.

Later we constructed a different kind of nozzle out of silver which enabled us to watch what happens when we truly pumped photons directly into the molecules. We found that even with our small 50-watt continuous-wave laboratory laser we were able to pump a half dozen photons into our molecules. As the molecules swallowed the photons we could watch them swell up, and we could again infer the degree of excitation by measuring their amplitudes of vibration. There are, of course, a huge number of vibrational and rotational states in such molecules, so only a few of them can be pumped by a laser of a given frequency. How to account for our results presented an interesting puzzle to solve.

Since only a very small fraction of the population of molecules in a warm ensemble can absorb a given laser output, we supposed that we could collapse the vibrational and rotational states into a much smaller number and thereby enhance laser pumping if we cooled the molecules to a very low temperature. The simplest way to do that would be in a supersonic expansion. Once we started to think about a supersonic nozzle and what might be done with it, we imagined that there would be a lot of things possible to do with it that would be more interesting than just pumping molecules with a laser beam. So this gave a new opportunity to jump, as you called it, to another field.

It seemed to me that a worthwhile program would be to study structures of liquids by electron diffraction. In principle, electrons offer large advantages over the standard techniques of X-ray and neutron diffraction. And, after all, liquids *are* the least understood common phase of matter. For this reason, research on liquids might be especially rewarding. The trouble was that electrons are scattered by matter enormously more intensely than are X-rays and neutrons, requiring the liquid sample to be extremely thin. Something as thin as hundreds of an Angstrom unit would be needed to obtain diffraction patterns uncorrupted by multiple scattering. Liquid clusters produced in a supersonic expansion offered the only method I could think of to generate reproducible samples of the required thickness. So we built a fairly elaborate supersonic attachment to our diffraction unit, and supersonic expansions have been at the heart of my experimental program ever since.

When we decided to embark on supersonic experiments, hoping to investigate liquid clusters, we were gambling. Up to that time, the only clusters that had been observed in supersonic experiments were *solid* clusters. But, since the scientists who studied them were physicists or gas-dynamicists, the substances they studied were mainly rare gases. Being chemists, we supposed it might be useful to investigate substances so formidably complex that no physicist would ever try them — like benzene and butane. And these yielded liquid clusters right away.

We imagined that we could use tricks of working up data that we learned in gas diffraction and apply them to advantage to analyses of liquid structure. After all, with clusters and electrons we could get a lot higher signal-to-noise ratios and more finely focused beams and much more deeply supercooled liquids than were possible in X-ray and neutron diffraction experiments. This supercooling sharpens pair correlation functions, giving a more discriminating test of the potential function used in the interpretation of results. For example, in our investigation of benzene, we could study the liquid supercooled by 75 degrees and got the most accurate structure of that well-studied fluid that had been obtained.

Because our new approach worked so well, we might still be studying liquid structure — except for the fact that we found something even more interesting in our investigation of clusters. In the study of clusters of about five dozen different substances we found that half of the substances gave liquid and half gave solid clusters. More intriguing was the fact that sometimes we would see different cluster structures (different phases) depending upon the way we squirted the vapor molecules through the nozzle. How on earth could the way molecules are squirted through a nozzle control the way the molecules decide to pack when they aggregate into clusters? It almost seemed like magic. Once we began to think about it and to analyze what was going on in the supersonic expansions, we began to see part of the answer.

We found that the initial pressure, the mole fraction of the subject gas in the carrier gas, the molecular weight of the carrier gas, and the shape of the nozzle all had an effect on the temperature at which vapor condensed into clusters. When you generate clusters very rapidly in highly nonequilibrium conditions, what you get depends more on kinetics than on thermodynamics. Our analyses of the processes going on in our supersonic nozzles suggested that some of our clusters that initially condensed into one phase must have transformed into another inside of our nozzle. I suggested to Ted Dibble to adjust conditions to bring the transition outside, into the region beyond the nozzle where we could observe the transformation directly. That worked right away. So we found something even more interesting and more important than liquid structure, something that carried us more deeply into unknown territory than liquid structure. We could monitor rates of nucleation in phase transitions by an entirely new method.

Our technique is the first new experimental method for investigating nucleation in 50 years. So we evolved into a new field. We have obtained results both for the freezing of liquids and for certain solid state transitions. It is satisfying but it is also hard work.

You once mentioned that you also had some interaction with atmospheric science.

This happened because of our investigation of the rate of nucleation of ice in deeply supercooled water. Previous laboratory studies of the freezing of water occurred in substantially warmer water and were blind to the phase of ice obtained. We studied water undergoing nucleation at roughly the temperature of nucleation in cirrus clouds, I believe. I understand that what happens in cirrus clouds has an important effect on the climate. Moreover, we showed directly that the ice first nucleated was the metastable *cubic* ice, not the ordinary hexagonal ice. Atmospheric scientists had inferred that result from indirect evidence. Previously it had not been possible to carry out experiments like ours in the laboratory, which is why our work attracted the attention of atmospheric scientists.

Let me make a remarkable comparison. When we studied freezing at around 200 K, we observed nucleation occurring about 20 orders of magnitude faster than it had been observed in the prior experiments done by conventional techniques. If you just say "20 orders of magnitude," it may go in one ear and out the other without attracting much notice. Such a number is not one we are accustomed to thinking about. But if I tell you to compare the age of the universe since its birth at the "Big Bang," some tens of billions of years ago with the tiny interval of three milliseconds, a comparison of times 20 orders of magnitude different from each other, this may give you a little better appreciation of what 20 orders of magnitude means.

Lawrence Bartell at the time of the interview in his office at the University of Michigan, 1998 (photograph by M. Hargittai).

Are you a member of the scientific establishment?

I don't really know what that means, but I don't think so. This is because I have spent almost all my scientific career working in the back alleys of science, doing things that other people were mostly unaware of. I would rather do something I think is significant or at least interesting to me than to follow the leaders working on mainstream problems like most scientists do. When you see quite attractive puzzles that other people haven't yet noticed, you get to play with them at your leisure. The downside of this is that people don't notice you or care about the work you are doing, so your satisfactions have to come more from your own sense of accomplishment than from the recognition by others. But that's fine. As I mentioned in the holography article I wrote for your *Chemical Intelligencer*,[2] I got the most recognition and won the most prizes for my holographic work than for anything else I ever did — not because the work was especially important. It was done on a whim, mainly with undergraduates, as a very minor part of my program. The recognition came because the work put me into the *Guinness Book of Records* and people are more impressed by that book than by professional journals.

What do you think about the future of experimental structure determination in general?

Well, I am so old I don't have to worry very much about the future. And if you are asking about gas-phase structures, it's getting to the point where quantum computations can give you answers more easily than experiments. On the other hand, the field of crystallography will continue to flourish for many years to come because it can treat systems of great complexity. There is a possibility that a combination of *ab initio* and semiempirical computations coupled with data bases and new experiments may let you treat considerably more complex problems with fair reliability. To treat an entire protein in its natural medium with *ab initio* calculations alone at a high level of accuracy will remain beyond the capabilities of computers for years. Nevertheless, if you mix together reliable information about force fields and structures from whatever sources of information you can acquire, it may become possible to bootstrap yourself up and fashion a very powerful technique. So the future may be very bright for imaginative and aggressive people who combine various techniques for the study of really complex systems, systems that would be intractable to direct treatment by any single technique.

Any hobbies?

Listening to music is one of my enjoyable pastimes, and I love to sail and go walking in the woods. My first major hobby was building model airplanes. As mentioned earlier, when I was 10 years old, I won a model airplane contest for children under 12. Later I acquired a model large enough to fly with me in it. Such a machine is known as an ultralight aircraft, an aircraft light enough for a man to carry that can be knocked down and carried on an ordinary automobile. It was exhilarating to fly! Building models and working with my hands was always important to me.

In my early years in academia I had to build my own instruments and I enjoyed the experience. You once asked if I were an experimentalist or theorist. Clearly, I'm an experimentalist. I work with my hands.

References

1. Bartell, L. S. "Reminiscenes about Electron Waves," In *Advances in Molecular Structure Research.* Edited by Hargittai, M.; Hargittai, I., Vol. 5, JAI Press, Stamford, CT, 1999, pp. 1–23.
2. Bartell, L. S. *Chem. Int.* **1998**, *4*(4), 53–56.

Paul von Ragué Schleyer, 1995 (photograph by I. Hargittai).

6

Paul von Ragué Schleyer

Paul von Ragué Schleyer (b. 1930) was Professor of Organic Chemistry and Director of the Computer Chemistry Center of the University of Erlangen-Nürnberg in Germany at the time of the interview. He has been Graham Perdue Professor of Chemistry at the University of Georgia in Athens, Georgia, since 1998. He is a Fellow of the Bavarian Academy of Sciences and many other learned societies. His many distinctions include awards in six countries, in inorganic and theoretical as well as organic chemistry, among them the Ingold Medal of the Royal Society of Chemistry (London) and the James Flack Norris Award of the American Chemical Society. He is best known, as a pioneer computational chemist, for the discovery of basically new molecular structures, particularly those involving lithium and electron-deficient systems. Publication analyses have identified him as the 12th most cited chemist during 1965-78 and the 8th most cited during 1984–1991.

We recorded our conversation at the Third Conference on Current Trends in Computational Chemistry in Vicksburg, Mississippi, on November 3, 1995, and finalized the text in March 1997.*

*This interview was originally published in *The Chemical Intelligencer* **1998**, *4*(1), 18–25

I recently heard you say that experiments are no longer necessary in chemistry. We can compute everything.

Everything? You take my remarks out of context! Neither statement is true. However, compounds of all the chemical elements (even "super heavy" ones which haven't been detected experimentally yet) can now be computed at good *ab initio* levels, as can large molecules with a thousand atoms. Dynamic simulations of medium-sized proteins in water environments can be performed. Two dozen new drugs developed by "molecular modelling" and "rational design" in the pharmaceutical industry are on, or are coming onto, the market. The scope of computational chemistry is widening with explosive rapidity. "Everything" is not amenable to computation, but chemistry no longer is an exclusively experimental science.

I've tried to point out that many experiments carried out today yield information that could be obtained more easily, more rapidly, more efficiently, far more cheaply, and even more accurately by computations. Areas are now well established where the confidence in the reliability of good calculations is so high that experiments no longer seem necessary. Fortunately so. While experimental checks are desirable, the production of high level theoretical data far exceeds what can be verified experimentally. Computational "screens" provide leads, guide research, and replace experiments in university as well as chemical and pharmaceutical industry laboratories.

Most quantum chemists believe (with good reason) that the structures, energies, and other properties of molecules comprised of at least the first 18 elements can now be computed with high accuracy by routine procedures. Your fine 1992 book with Aldo Domenicano, *Accurate Molecular Structures*, stressed the importance of the information content of refined geometries. But only little more than 10% of it was devoted to computational methods. This imbalance reflects the past, but not the situation in the 1990s, and certainly not the future.

The latest *Landolt-Börnstein* compendium of accurate experimental structures (1995) includes approximately 2000 gas-phase electron diffraction, microwave, etc., determinations. This data required a decade of effort by dozens of research groups and many more coworkers. But a single computational chemist, working diligently with relatively inexpensive work stations (or even the latest generation of Pentium Pro PCs), could reproduce "everything" in this compendium in a year or less.

There are many challenging areas, for example, the energies and accurate structures of polyatomic molecules of the heavy elements, where the reliability both of applicable theory (relativistic treatments are needed) and of experiment (few data are available) have yet to be established. Such cutting-edge problems are best addressed by employing both computations and experimental methods in conjunction. While this is happening increasingly, and also with large molecules whose geometries are difficult to determine reliably by experimental methods alone, a remarkable number of experimental structural research groups still don't have their own computational capabilities.

Why is it that many computational people, when they use experimental results for comparison, don't seem to care about the origin, physical meaning, and experimental error of the data they use?

This certainly is a valid complaint, but it also applies in the other direction. Many experimentalists refer vaguely to "theoretical calculations," without giving the most elementary information about the kind of computation involved. An approximate semiempirical number is not differentiated from a really serious high-level *ab initio* result. In both cases, efforts should be made to evaluate the quality of the data and to appreciate the difficulties of the research. Jim Boggs's proposed new IUPAC guidelines for computational papers include strong recommendations along the lines of your criticisms. It is not fair to compare a good computational result with rough and admittedly imprecise experimental estimates, but it is just as inappropriate to compare the best experimental data with unspecified "theoretical calculations" of unstated merit. There is a large difference between "spectroscopic accuracy" and more approximate numbers.

The physical meaning of the parameters is different too.

That's right. As you well know, equilibrium geometries (i.e., at absolute rest) are computed, but these are not what experiments measure. The data are not strictly comparable, but this usually is not the main reason for serious discrepancies.

There seems to be a proliferation of computational studies in the literature whereby people plug in some questions into their computer equipped with ready-made software, get out some numbers, and communicate them.

There are both good and mediocre scientists, whether computational or experimental. There are critical computational chemists as well as those who turn cranks on black boxes.

You are right, many published computational papers neither address significant current problems nor contribute much to the understanding of chemistry. But my complaint is just the opposite — that today's powerful computational methods are not being used creatively and imaginatively enough. Most of the possible combinations of all the chemical elements are unknown. Instead of exploring *terra incognita*, most computational chemists reexamine trodden ground and follow, rather than lead, experiment. The periodic table is like a haystack. The whole haystack can now be searched effectively, but the "needle" is an unknown, one of many unknowns. What is best to look for? What are the objectives?

What do you consider to be your most important scientific achievement?

That's a hard question. I've worked in many fields. While thinking about the answer, I'll tell you about my first contribution. In 1956, at the beginning of my Princeton career, I discovered a simple way to make adamantane in one step by rearrangement. This beautiful molecule, the C_{60} of its day, was the first cage hydrocarbon. Adamantane, discovered in Czech petroleum, had then been synthesized by Prelog in milligram amounts. My synthesis made adamantane readily available, triggered interest in other cage hydrocarbons, and propelled my career. Unlike C_{60}, which still does not have practical uses, adamantane derivatives found medicinal applications almost immediately.

Two adamantane units intertwined were dubbed "congressane" by Prelog and Barton. We quickly answered the synthetic challenge issued by the IUPAC Congress in London (1963) and made this molecule — also by rearrangement — and renamed it "diamantane." The synthesis of the next member of the series, triamantane, was achieved a year later. The synthetic side of my career led to the preparation of many more cage ring systems, but basically I am a physical organic chemist. The special structural features of adamantane were exploited for mechanistic studies. Ingold and later even Winstein didn't realize that many so-called S_N1 reactions involved the solvent, attacking from the rear. This was quite impossible in adamantane derivatives, where the cage structure precluded such solvent attack. We proposed adamantane derivatives to define real S_N1 reactivities, and reinterpreted much of the solvolysis literature.

The cage rearrangements required knowledge of the thermodynamic stability of hydrocarbons. Experimental data were not available. This is why I became interested in computations in the mid-1960s. The computers were still very primitive, and the optimizations were done point by point. The history of "molecular mechanics" calculations goes back to Westheimer, who computed the hindered rotation of biphenyls. Later, Jim Hendrickson's calculations of the conformations of seven-membered and larger rings made a big splash for me. Ken Wiberg's group had written a program to do automatic optimizations, and Jerry Gleicher, a postdoc from Dewar's group, implemented it at Princeton. Allinger already was developing his molecular mechanics programs independently. Our interest was in strained molecules and carbocation activities. My Ph.D. mentor, Paul Bartlett, had shown that strained carbocations at bridgeheads could not become planar and thus were less reactive. It was an obvious application of such calculations, and they worked beautifully, over a remarkably large range of energies.

Then I met John Pople and became his "consultant" on carbocations. I knew little about theory but was impressed that quantum mechanics didn't require any parameterization. We could compute not only carbocations but also many other inaccessible species. Pople and his group had just developed the Gaussian 70 program. I fell under his influence and became absolutely hooked on the subject. Pople had been studying, in his characteristic systematic fashion, the effect of substituents involving elements to the right of carbon — nitrogen, oxygen, and fluorine. I suggested that boron, beryllium, and lithium be included, even though there was hardly any experimental data. We soon discovered that lithium compounds were remarkable rule breakers, exhibiting planar tetracoordinate carbons, perpendicular ethylenes, and bridged acetylenes. I took my entire group with me to Munich in 1974 for a sabbatical, and for the first time in my career I had sufficient computer time for my work. By then, my group had shrunk considerably. At its height, in 1970, I had 18 postdocs at Princeton, mostly supported by NIH.

Coming back to your question, I consider that my most important contribution is to have increased chemists' awareness, by the examples of my research, of the power of computational methods and of their potential for widespread applications in chemistry.

Who else can be mentioned in this respect?

John Pople developed programs and methods intended for use by chemists generally. In contrast, most other quantum mechanicians were carrying out the best possible calculations on necessarily small systems of greater interest to molecular physicists than to the vast majority of chemists. Pople addressed real organic and inorganic chemistry problems. He used relatively primitive methods, for example, minimal STO-3G basis sets, but devised "isodesmic" equations which tend to cancel errors. His computer programs were easy to use, rather than requiring much of a day to set up a job. I consider him to be the founder of computational quantum chemistry. My contribution was mostly to demonstrate the power of these methods and to show how effectively they could be employed, in particular to discover new chemistry. Lithium compounds are good examples. CLi_6 was a totally inconceivable system: an octahedral carbon surrounded by six lithium atoms! Like many of our predictions, CLi_6 has now been verified experimentally. The power of the calculations tempted me to find ways to break all the commonly accepted rules in chemistry. It's in my nature to attack such challenges deliberately. When somebody states a rule, I set out to find an exception that breaks it.

Have you experienced any difficulty in getting your papers published?

My early computational papers were accepted readily enough, but it was much more difficult to gain the approval of my peers for this new line of research. My aggressive presentations did not help. I chose to talk about our newly discovered (i.e., computed) rule-breaking structures of lithium compounds — planar tetracoordinate carbon, perpendicular ethylenes, bridged acetylenes — at a stereochemistry conference in Kingston, Ontario, in 1978. What I thought would be a bombshell proved to be a lead balloon. There was one trivial question; afterwards, I was shunned: the worst reaction to any lecture I've ever given.

My friends thought I'd gone crazy — leaving the respected field of experimental physical organic chemistry for such computational fantasies. Who could believe such weird results? While easy to understand now, the ionic bonding of lithium compounds follows structural principles different from covalent bonding. But it was not apparent in 1975 that lithium structures shouldn't conform to van 't Hoff's rules.

Originally you were an experimentalist. How about your students? What's their background?

I protest! I am still an experimentalist. My background as a physical organic chemist was wholly experimental, and even my approach to research using computational methods is that of an experimentalist. I explore chemistry with all tools at my disposal. My Erlangen coworkers have the standard German university education, which emphasizes breadth and extensive experimental training in physical, inorganic, and organic chemistry. In my early Erlangen years, my beginning research students only wanted to undertake experimental projects, since they felt their job prospects in Germany would be better. Half of my present coworkers do experimental work, but all of them have computational projects as well. Computational chemists are now very employable, also outside of chemistry, because of their expertise with computers.

How many coworkers do you have?

About 25 in 1995, including seven or eight postdoctorals and visitors. But I am required by German law to retire in 1998, and my group is already winding down.

You are known to have had many coworkers from Russia.

I have had 13 or 14, but there are none at present.

What happened?

Before the political change, I encouraged as many visits from the Eastern Bloc countries as possible. The many fine theoretical chemists in the former Soviet Union and its satellites lacked the computational facilities we could offer. Arrangements were very difficult to make. It was impossible for East Germans to cross the Bavarian boarder, but relatively easy for Czech scientists, for whom our modest honoraria, converted to their currency, were princely sums. I also attracted some top-level Russians; being politically acceptable in the Soviet Union, they were able to travel to the West. When the change came about, such well-placed Soviets were able to get the first joint research grants with Germany to support "exchange" visits. At one point, there were a half-dozen Russian scientists in my group at the same time.

The adjustment to the West was difficult for these Russians, none of whom had been outside their country before. They did not know how to behave in the West. Their often overbearing and aggressive personalities made them highly unpopular. My experience was extensive enough, I think, to allow me to make such generalizations. These are based on the large majority, but certainly not all, of my visitors. Our facilities (beyond any they had seen before) and help from my associates (offered as personal favors) were taken for granted, and they monopolized common equipment. Instead of being grateful, they asked for even more. We had devised a sophisticated system to distribute computer time equally, but "since their work was more important," they complained that they weren't getting enough. Many were secretive, and most were excessively argumentative.

Were they well prepared?

All these visitors were first-class scientists, dedicated, and extremely hardworking. But many were often abrasive in their interactions, not only with the rest of my research group, but also in their personal dealings with the Institute staff. Finally, a delegation of my coworkers, believing that they couldn't speak to me directly, appealed to my wife to implore me never to take a Russian again. My staff underscored this request.

Just to give you one example. A just-arrived Russian associate was ushered into my office. Breaking into my conversation with a colleague, he began to discuss the project to be pursued. We had agreed on the general area in our prior correspondence, and I had given some thought to the matter. But he quickly interrupted me and exclaimed, "Oh, no, that's not what we are interested in at all." This reaction was not quite what I expected from someone I had never met before. His counterproposal had no obvious relationship to the topic. My further suggestions and scientific reasoning met with similar outbursts and resistance. My colleague asked me afterwards, "Why didn't you throw him out then and there?"

But this is not the reason that there are no Russians in my group now. I have had Russians since that time. Anticipating possible problems, I wrote detailed instructions bluntly stating how the new arrivals should behave in Erlangen. This helped.

My impression in my scientific dealings with my Russian associates was that they had poor working facilities in the Soviet Union and, as a consequence, argued a lot instead. But managing a large research group didn't allow me time for pointless debates, which could last an hour without reaching a conclusion. I normally try to convince coworkers of the merits of my research proposals. I present the logic behind the project and expect it to be carried out. But this approach was hopeless with the Russians. So I simply ordered them. This proved to be the solution.

While expecting my coworkers to address themselves to the assigned problems, I allow them much latitude in their research. Some of my Russian associates were secretive to an extreme about their work. They were not open to exchanges of data and collaborations with my other co-workers. The worst stole computer time and even data from my research group and attempted to publish a number of papers independently.

Despite all this, I have maintained contact, often quite friendly, with half of my former Russian associates, particularly those who have emigrated to the West. My attitude has mellowed.

When did you first become interested in chemistry?

My mother gave me a Gilbert chemistry set on my 5th birthday in 1935. I delighted in reacting sulfur with iron filings in a spoon over the kitchen stove. The smell of the burning sulfur was awful, but this nuisance was permitted, as the experiment was described in the manual. When I was 12, my earnings as a newspaper delivery boy enabled me to buy a larger Chemcraft set. But I was especially intrigued by the catalog, which showed beakers, Erlenmeyer flasks, and similar "real" equipment not supplied with the set. But what could be done with such apparatus? I started reading chemistry books from the public library, one after the other, voraciously. My basement laboratory grew and grew. When the time arrived for high school, there was no point in taking the junior-year chemistry course, so my mother arranged for me to try the final examination. My grade of 99.5% annoyed me because I had missed a question. So I started with the senior advanced chemistry course as a sophomore.

Do you think that today a 5-year old or a 12-year old child would find a chemistry set as attractive as you did?

No. My grandchildren were given computers. Of course, when I was young, chemistry didn't have the bad image it has today, and there was almost no safety consciousness. Dangerous experiments were part of the "fun." I escaped several serious accidents in my basement laboratory, but my right hand was badly burned by the white phosphorus "bomb" I had concocted.

Where did you spend your childhood?

In my home town, Cleveland, Ohio. During World War II, I needed an adult's approval when I wanted to buy chemicals. My grandmother was willing to accompany me. We took streetcars all across town to the Chemical Rubber Company, where all my purchase requests were honored — a quarter pound of potassium cyanide, for example. This would have been enough to poison Cleveland's water supply, but it was sold to me anyway. My grandmother hadn't the faintest knowledge of chemistry.

You were an American then?

I still am. My father's ancestors were half German and half British. The latter had come to Virginia in Colonial times. Obidiah Hardesty served first as a "whiskey boy" in the American Revolution and finally as a junior officer. After the war, he was rewarded with a land grant in Southern Ohio. Georg Schleyer reached the same area from Baden-Württemberg with his wife and young son in 1832 via the newly opened Erie Canal system. Great-grandfathers on both sides were wounded in the battle of Gettysburg; one survived in the rain for three days before receiving medical attention. He limped the rest of his life.

Mother's ancestors were born and educated in Germany and were sent to the U.S. mid-west as Evangelical Protestant missionaries. Louis von Ragué, my great-grandfather, arrived in Wisconsin in 1864 and founded many congregations. My grandfather, Hugo (Quack) Kamphausen, came in the 1880s. He was my last ancestor not born in the USA.

What about your name?

I was given my middle name, "von Ragué," in honor of my grandmother. I have visited the family house where the Ragués originated around 1550, in the small village of Grandfontaine near Basel, Switzerland, at the French border. My great-great-great-grandfather became a commissioned Prussian officer, and the "von" was added to his name ("Militaradel"). I was simply called Paul Schleyer until I moved to Germany. Then I somehow became "von Schleyer," not only in Germany but also back in the United States. The "von" has nothing to do with Schleyer. "Give the people what they want," is my wife's advice.

Schooling?

Harvard accepted me, but Princeton also offered a full-tuition scholarship ($500 in 1947) and the deserved reputation of being more undergraduate-oriented. But I did choose Harvard for graduate school and Paul D. Bartlett, a leading physical organic chemist, as my Ph.D. supervisor. Fieser, Woodward, E. Bright Wilson, Stork, Moffitt, Kistiakowsky, Rochow, and Wilkinson were among the faculty. Our class included Herschbach, Cotton, Seyferth, Saunders, Patchett, and many other now quite famous chemists.

At that time it was common to leave Harvard before completing the final Ph.D. requirements. I departed in 1954 after three years to return to Princeton as an Instructor. It took me three years to write up my thesis, which ended up being 600 pages long, with a review of the entire norbornane and related terpene literature. I finally got my Ph.D. in 1957 when I was 27.

Upon my return, the Princeton chemistry department seemed very different from the image I had retained; antiquated, no comparison with Harvard. Eventually I rose to a chaired professorship, but, after 21 years, I left for Erlangen, Germany, in 1976. Most were amazed by my decision. My motivation was twofold. One was cultural: the tradition, the beauty, the art, the history, the quality of life, and the diversity of Europe, where I had been invited as visiting professor many times, drew me. But the main reason was scientific: the availability of computer time. My interest in what today is called computational chemistry had been increasing since the late 1960s. It was simply not possible to pursue a computer-time-intensive field in Princeton in 1976. An hour of CPU time at the Computer Center cost $800 of grant money, and the speed of the machine was about that of an early PC. To induce me to stay in Princeton, I was offered two hours of free computer time a week (for two years)!

How was your German?

Mostly German was spoken at home when I was a child. Since my English vocabulary was poor, I was put into the "retarded" class upon entering primary school. Shocked, my mother and grandmother discontinued German completely. However, some family background was retained, and from the beginning in Erlangen I lectured in (grammarless) German.

When I started, the Telefunken in the Erlangen computer center was only turned on 40 hours a week, as an operator was required by the maintenance contract. The University didn't need more computer time. My arrival changed this: we got an extra shift, as promised, and then were able to use the computer over the week-end. For many years, about 70% of the total CPU time was used by my group. Computer time was both free and available!

And I hadn't fully appreciated the other benefits. As an "Ordinary" (full) Professor in Erlangen, I had many positions at my disposal, 11 assistantships, support personnel, and fine research facilities. A large research group could be supported without any outside funding. But I always had grants (easy to get relative to the U.S.) and also A. von Humboldt Fellowships for postdoctoral associates. But the best asset Erlangen offered proved to be my many fine German predoctoral students; these won stipends as well.

Please tell us about your present family.

My first wife and I divorced in 1969; she died of cancer a few years ago. Our three daughters have always lived in the U.S. I see them twice a year. My second wife, Inge, had been an administrator in Princeton and resented that I didn't arrange a similar position for her in Erlangen. As she is German, many people assumed that I moved to Germany because of her. The truth is the opposite; women are still not yet fully emancipated in the German attitude.

You are nearing the official retirement age in Germany. What are your plans for the future?

If retirement is a state of mind, I am far away. While maintaining my intellectual ties with Erlangen, I plan to move to the University of Georgia in Athens, in 1998, as full professor to begin my next career. I have been a Visiting Professor there since 1980. The new Center of Computational Chemistry will house my group along with those of Fritz Schaefer, Lou Allinger, and Phil Bowen, a molecular-modeling colleague. Allinger and I have co-edited the *Journal of Computational Chemistry* for 15 years. My collaboration with Fritz Schaefer is quite active; we have shared several co-workers. Two have actually obtained Ph.D. degrees both from Erlangen and Georgia, for two different dissertations each!

Through the years, who have been the most important influences on you?

I had two inspiring teachers. The first was Clark Bricker, a superb undergraduate lecturer in analytical chemistry at Princeton. He became a distinguished university administrator elsewhere later. My senior thesis under his direction drew me toward physical organic chemistry. We followed the kinetics of the Maillard reaction, the acid-catalyzed degradation of various sugars, by UV and polarography.

R. B. Woodward was the dominating influence at Harvard. Although I was not attracted by natural products research, his overpowering intellect was inspiring. I attended every lecture he gave during my graduate student days and also participated actively in his famous weekly group meetings, which sometimes lasted through the night. He employed mechanistic reasoning brilliantly in designing his syntheses and demonstrated, by example, how effectively organic chemists could use instrumental methods, the newly developed infrared spectroscopy in particular, to help elucidate structures.

Later in life, two colleagues changed my career decisively. I became John Pople's disciple; John's conceptions converted me to applied theoretical chemistry. Rolf Huisgen was both an inspiring chemist and a genial, cultured host. The positive experiences of my two sabbaticals and several stays with him in Munich led to my move to Erlangen.

Do you have any interest outside computational chemistry?

Of course, but none that I pursue enough to mention. I feel very strongly that computational chemistry is still not receiving adequate recognition in terms of its scientific and educational capabilities, to say nothing of its incredible future. Chemistry is a very conservative science, with a long and almost exclusively experimental tradition. The ability to gain valid chemical information computationally is recent and is not still accepted by most chemists. The strong prejudice against computational chemistry remains. Such skepticism no longer is healthy. Older chemists in influential positions have not kept up with rapid developments in computers and programs and are not well informed about the status and potential of computational chemistry, but they make prejudicial pronouncements and influence negative decisions. Most university faculties have no professors of computational chemistry. When new faculty members are hired, this area is overlooked. Consequently, the conservatism is "programmed" to continue. The importance of computational/theoretical chemistry — particularly its didactic advantages — is not reflected in the curriculum. Several undergraduate as well as graduate courses in this area should be among the requirements in every chemists' education.

The user of *ab initio* programs quickly realizes the enormous sophistication of molecules. Chemistry is revealed in more detail, and problems are conceived more basically. One appreciates how atoms are joined in a very intimate way; structures can be visualized directly, as can the complex dynamics of transformations from one conformation to the next, or from one molecule to another. How do bonds break and make? What is the nature of chemical bonding? A new chemical education awaits the user of electronic structure programs, which are far more than a black box to be manipulated.

Few colleagues appreciate this. I attempt to point out that the future of chemistry will be more and more computational, but, like your first question, this is interpreted as "Schleyer says that nobody should do experiments any more, but should only compute." The transformation is obvious. Chemistry is becoming a computational science but not rapidly enough.

No other hobby then?

I've been a passionate lover of classical music since I was 16, and I listen to good FM stations or play CDs in my office all day long. Isn't classical music better as background than silence? It's true, intense composers — Mahler is a good example — demand concentration, but most music enhances rather than interferes with the day's normal activities.

Culture, certainly one of the advantages of Europe, has always attracted me. At Harvard, I even toyed with the idea of switching to a career in art history. But even my interest in the Northern Renaissance artists has been neglected since I became fanatical about carrying out computational research personally. The creative opportunities fascinate me. The enigmatic paintings of Hieronymus Bosch require much knowledge and imagination to interpret, but they were painted by Bosch and not by Schleyer.

I like to create what I study. An idea flashes in to my mind. I imagine a new molecule with a weird structure and unusual bonding. Selected combinations of component elements give the proper size relationships, electron occupancy, and symmetry, but will the idea survive the test of a good-level quantum-chemical computation? The jobs take only a few minutes to set up and submit, and the first part of the run (the "initial guess") quickly reveals if the prediction is on the right track. If the first geometry optimization cycles still look favorable, I admire details of the structure — my own new invention — plotted on the monitor. How exciting — to study a completely new composition of matter no one else has thought of before. Refinement must follow and can lead to disappointment. But what a magnificent opportunity, to be able to test one's own ideas personally in great detail, so quickly, easily, and reliably!

You can understand that my main hobby now is to play the computer game "*ab initio*" with Bach partitas resonating in the background. I would rather do that than read what novelists have written, enjoy what artists have painted, or admire what architects have achieved.

Albert Eschenmoser, 1999 (photograph by I. Hargittai).

7

Albert Eschenmoser

Albert Eschenmoser (b. 1925 in Erstfeld, Uri, Switzerland) is Professor Emeritus (since 1992) of the Swiss Federal Institute of Technology (ETH) in Zurich and a member of the Skaggs Institute for Chemical Biology at the Scripps Research Institute in La Jolla, California (since 1996). He received his Diploma and doctorate at the ETH in 1949 and 1951, respectively, and spent his career at the ETH. I mention only a few of his exceptionally large number of honors. He is a Foreign Associate of the National Academy of Sciences of the U.S.A., a Foreign Member of the Royal Society (London), and a member of the Academia Europaea. He received the Robert A. Welch Award in Chemistry in 1974, the Davy Medal of the Royal Society in 1978, the Arthur C. Cope Award of the American Chemical Society in 1984, the Wolf Prize in Chemistry (Israel) in 1986, and the Paracelsus Prize of the Swiss Chemical Society in 1999. One of Professor Eschenmoser's latest publications reviews his recent interests [Eschenmoser, A. "Chemical Etiology of Nucleic Acid Structure." *Science* **1999**, *284*, 2118–2124]. We recorded our conversation on September 6, 1999 at the ETH.*

Looking back, which of your research results are you most pleased about.

It was the corrin and Vitamin B12 project, which lasted for 12 years and which, so I suppose, has been at the center in my life's professional activities. But I find it also pleasing to remember the beginning of my research career

*This interview was originally published in *The Chemical Intelligencer* **2000**, *6*(3), 4–11

in the late 1940s and early 1950s, when I was spellbound by the chemistry of Leopold Ruzicka. It was the critical period when natural product's chemistry — and this, at ETH, was terpene and steroid chemistry — was changing from the classical period to the mechanistic and biogenetic era. For a young organic chemist, it was a very exciting time. It was the time when you could screen the literature in terpene chemistry and — armed with fresh mechanistic views and a general biogenetic hypothesis — easily spot constitutional formulas of sesquiterpenes that had to be wrong. Then, from what was already known about their chemistry, you could deduce what their correct constitution might be. For some cases, the correctness of such proposals could be proved with relatively little experimental work. That is what I did in my doctoral thesis.

What turned you originally to chemistry?

I chose chemistry by exclusion. I studied at the Abteilung für Naturwissenschaften (natural sciences) at ETH, the division that let me postpone the final choice between the various branches of natural science. As for going into science, that choice had already been made by entering ETH and, at least in part, even before by the specific type of gymnasium I had gone to. I grew up in a small village in the center of Switzerland. My father was a butcher and there was no academic tradition whatsoever in my family. My original intention was to become a primary school teacher and my parents sent me to a secondary school in my native canton's capital. This was a vocational secondary school (Realschule) which did not point to any academic career. However, one of my teachers advised me to further my education and become a teacher for a higher grade school rather than a primary school. He sent me to the Kantonsschule St. Gallen (the canton of my citizenship), where with my preparatory training I had no other choice than to enter the so-called Oberrealschule, which, alas, was not quite a gymnasium because it did not have Latin but concentrated on natural sciences and mathematics. This, in a way, had predetermined my future in natural sciences.

Any regrets?

No, no. I have no talents whatsoever in the arts of talking or writing.

Was Professor Ruzicka the supervisor of your thesis work?

Officially, he was my academic Doktorvater, yes, but I worked in the laboratory of his former collaborator Dr. Hans Schinz and, for the whole one and a half year period of my thesis work, I do not remember ever having seen Ruzicka in my laboratory. On the other hand, toward the end of this period he became very supportive of me. Many years later, when Vlado Prelog had to introduce me as a lecturer at the 75th anniversary symposium of the Swiss Chemical Society, he smilingly referred to my independence during my thesis research by saying that the lecturer did his Ph.D. thesis under the supervision of a completely unknown young man, whose name was Albert Eschenmoser.

Toward the end of 1950, Ruzicka determined that I should finish my thesis, stay at ETH, continue doing independent research, and take Ph.D. students of my own. The first of them, Jakob Schreiber, started in the spring of 1951; after his thesis, he stayed on as a member of my research group for his entire life. When he died in 1992 at the age of 70, I lost one of my best friends.

In the 1950s, the organic chemistry laboratory of ETH had become again quite an international place, thanks to Leopold Ruzicka and Vlado Prelog. At that time, a frequent visitor, who eventually became a good friend of both of them, was R. B. Woodward from Harvard. His brilliant lectures on natural products chemistry, especially synthesis, had a great impact on us, the young generation of ETH chemists. These visits were, incidentally, the personal and scientific seeds of the Harvard-ETH collaboration on Vitamin B12 about a decade later.

Was there also competition?

I have referred to it as "collaborative competition;" I could have also called it "competitive collaboration." The agreement was that neither of us should try to rush to the final goal to arrive there first. Both partners should fully exchange all information that became available in the two laboratories along the way. This should give us the time and the opportunity on each side to explore the scientific problems of the project as extensively as possible. Within the decade that the collaboration lasted, there was indeed an enormous amount of information exchange going back and forth across the Atlantic.

The total synthesis of Vitamin B12 provided a sort of proof that nowadays chemists can, in principle, synthesize any (low-molecular-weight) natural product structure, provided they are willing to invest the necessary time

R. B. Woodward and Albert Eschenmoser at the B12 party in Zurich, 1972 (courtesy of Albert Eschenmoser).

and effort. In a sense, the accomplishment closed a chapter in chemistry that had been opened 150 years earlier with Wöhler's synthesis of urea.

The recent history of chemistry shows that problems which traditionally were part of chemistry are increasingly dealt with, even taken over, by other branches of science; an important example being elucidation of molecular structure. The perhaps "most chemical" activity of chemists is synthesis of molecules, and quite probably it's going to be synthesis that experimental chemistry will have to offer as its specific contribution to research in the natural sciences of the future. However, not even synthesis is a secure preserve of experimental chemistry; large biopolymers are made by biotechnologists.

To what extent do you expect combinatorial chemistry to replace synthetic work?

This is an important point. Chemical synthesis of complex natural products and combinatorial synthesis have very little in common. When you synthesize a natural product, you know your target in advance; in combinatorial synthesis you don't, you first synthesize, and then you find the target. The essential step in combinatorial chemistry is not the synthesis, rather, it is finding a target, based on specific properties. Combinatorial synthetic research is an adaptation to what nature has been doing all the time: synthesis followed by selection, as opposed to synthesis by design. The aim of combinatorial chemistry is primarily *discovering* a molecule, whereas that of chemical synthesis is *constructing* a molecule.

Natural product synthesis is prototypically based on design. Yet, if synthesizing a given molecular structure were to require nothing other than carrying out a pre-conceived design, this activity would not really belong to science anymore; though such a synthesis could represent a piece of art, it would belong to the realm of molecular technology. However, still today, complex natural product synthesis needs much more than mere design. Any first attempt to convert the design of a complex synthesis into reality runs into situations where science sets in: the chemist finds parts of the design incomplete, or misleading, or wrong. Then, the design is to be replaced by a search, which always will involve elements of variation and selection, be it with regard to reagents, reaction conditions, or the type of chemical transformation to be used. What a synthetic chemist does in such a situation is, in fact, a sort of confined combinatorial chemistry in order to overcome the difficulty that a less than perfect synthesis design had led to. It is through such situations that the realization of a designed synthesis creates new knowledge and is, therefore, part of science.

Would you care to comment on C_{60}?

The existence of C_{60} is an outstanding example of the power of constitutional self-assembly in the world of molecules. Self-assembly is what synthetic chemists will increasingly have to aim at in the future in order to be able to make rational use of its potential, both on the molecular and supramolecular level. C_{60} is of course also a great discovery, for chemists one of the "Waterloo type," as I sometimes refer to it.

Waterloo from whose point of view?

In the sense that, if in the concert of the natural sciences it is the chemists' task to fight the battles for new knowledge on their territory, then C_{60}, for them, was a "lost battle." A sort of molecular jewel was lying there below the surface of the chemical territory waiting to be thought of and made, or discovered, by chemists. Yet the discovery came from outside.

The other example is, of course, DNA. In retrospect, it seems clear that there was hardly any other comparably fundamental reactivity principle on the territory of natural products chemistry waiting to be discovered than that we know today as Watson–Crick basepairing.

You are saying that the discovery of C_{60} came from the outside, although the three Nobel laureates for this discovery were all chemists.

When I use the term chemist in this case, I mean synthetic chemist, and when I use the term chemist with regard to DNA, I mean natural products chemist.

Much chemical progress and tremendous discoveries have been made in biochemistry, molecular biology, and biotechnology, often by people who did not have a classical training in chemistry.

Watson and Crick is a prototypical case.

It has even been suggested that this may have been to the advantage of these discoverers. Because they were not bogged down by the burden of a classical training, they were able to arrive faster at a new frontier of research.

I have no doubt that what you are saying is correct. During the last 10 years or so, Vlado Prelog and I very often had lunch together. In discussing with him "God and the world" and, of course, matters of chemistry such as the past and the present of natural products chemistry, I more than once gently provoked him by saying: "Vlado, every year during which we did not work on DNA was a wasted year." This, of course, is an overly drastic statement and, obviously, an exaggeration. However, there is a kernel of truth in it. This becomes evident when you look at DNA today and consider the impact of this single type of natural product structure on essentially everything that concerns human existence, an impact incomparable to that of any of the organic chemist's traditional natural product families, Vitamin B12 obviously included.

Albert Eschenmoser, Vladimir Prelog, and Jean-Marie Lehn in 1989 (courtesy of Albert Eschenmoser).

At its origin, organic chemistry was meant to be the branch of chemistry that had to deal with carbon-containing compounds produced by living matter. After the explosive proliferation of the field towards the end of the last century, studying the chemistry of naturally occurring carbon compounds became the task of organic natural products chemistry. To the natural products chemists of today, it is, in a way, sobering to realize that their science in its past went along studying all those wonderful alkaloids, carbohydrates, natural dyes, terpenes, vitamins, and hormones, yet did so by virtually ignoring proteins and nucleic acids. Sobering it is, in retrospect, to be aware of that immense disparity in relevance between this single type of natural product structure, DNA, and all those natural products that have been at the center of the organic chemist's attention for more than 100 years. Furthermore, it is sobering to realize how late in the game organic natural products chemists started worrying about the chemistry of DNA.

How do you explain this paradox?

There must be a variety of reasons. One of them is certainly the overwhelming success of classical organic chemistry in studying low-molecular-weight natural products. Success tends to make people conservative. Alkaloids, natural dyes, carbohydrates, terpenes, steroids, and vitamins were all chemically exceedingly interesting, and many of them were biologically important as well as medically, or otherwise industrially, very useful. These materials could be studied by experimental and theoretical methods that chemists themselves had developed: isolation, chemical degradation, and synthesis. The size of the molecules was relatively small so that their constitution could be deduced and all those exceedingly pleasing and beautiful chemical formulas could be written down for them.

In those lunches with Prelog, we often spoke about the early 1950s. While discussing that relevance gap between the organic chemist's natural products and DNA, I also asked him: "You and Ruzicka, leading and powerful natural products chemists as you were at that time, why did you ignore DNA?" This was not, of course, meant to be a criticizm of him, no, but the question seemed to me intriguing: Why did organic natural products chemists ignore the DNA problem? After all, Avery had published the paper about DNA being the transforming principle, the genetic material, back in 1944. Prelog said that it was out of the question, one couldn't possibly do proper chemistry with it. One didn't know whether it's a large molecule, one couldn't isolate it, one couldn't purify it to the standards of organic chemistry; it was not a proper object for chemical research. I repeatedly asked him to write this down, because of his being a witness to that time, a witness to that attitude. I was gently prodding him, but he told me repeatedly, "I don't like to do it because it could be misinterpreted." However, I was increasingly intrigued by the question and thought that it was important to have his words in written form as the historical record of a witness. For about a year, I didn't get anything from him but then, one day, he came into my office and gave me a piece of paper which contained a (shortened) reply, written with a distinct touch of irony [see below, separately].

Are you working on DNA-related problems?

Yes, for the last decade or so we have been working on a special question that is related to DNA. But, in this context, let me first come back to combinatorial chemistry again. Combinatorial chemistry is far more than just a new methodology. It is changing the chemist's thinking in a more fundamental way, it widens his outlook on nature, it brings him closer to one of the central concepts of natural science as a whole, the principle of Darwinian evolution. The emergence of combinatorial chemistry was the transplantation of the evolutionary mechanism into chemistry. Life, as we know it, is a chemical life, and one of the prerequisites of both its emergence and its evolution was the chemical world's immense diversity with regard to its molecular structures as well as their chemical and physical properties. Biological evolution proceeds by "happening and selecting itself by surviving," combinatorially and not by design, Darwinian and not Lamarckian.

Interestingly enough, deep down in the minds of synthetic chemists — and I conjecture that this may apply to scientists in general — is the belief that the hidden final goal of everything we pursue in science is to eventually be capable of doing things by design. Design is the ideal at which we seem to be driven to aim. If chemists today accept combinatorial methods because they have to admit that such methods offer a faster way to progress in solving certain problems, in the back of their minds they may still feel that, in the long run, the purpose of combinatorial chemistry is to speed up the progress toward achieving that ultimate goal of being able to solve problems by design.

Could we have a glimpse into what you are doing now?

What we are doing is related, in a way, to both biological evolution and the principles of combinatorial chemistry. Our work aims at narrowing down the diversity of possible answers to the question of why Nature chose the specific structure of DNA (actually RNA) and not some alternative type of molecular structure as a genetic system.

When, as a synthetic chemist, you look at the structure of RNA, you may ask yourself how did this particular type of molecular structure enter the scene for the first time? If you adhere to the principles of evolution, you are led to conclude that the structure is the result of a selection process. Selection from what? An important possibility is that RNA had been selected (or had selected itself) from a combinatorially formed library of alternative structures. Again, if you, as a synthetic chemist, hypothesize about the type of chemistry that could have produced RNA for the first time, chemical reasoning leads you to consider that there are many chemically closely related alternative structures that could have had a comparable chance to be formed by the same kind of chemistry. In such a scenario, RNA was selected, or had selected itself, by functional criteria from a library of structurally related nucleic acid alternatives. Such a view defines a strategy for approaching the question experimentally: you systematically make such potentially natural alternatives in the laboratory by chemical synthesis and compare them with RNA with respect to those chemical properties that are known today to be relevant for RNA's biological functions. Such comparisons may tell you, for instance, that a given alternative had no evolutionary chance to compete with RNA because it lacks the most fundamental capability of an informational nucleic acid system, namely, informational base pairing. Another possibility is to find that an alternative does possess this capability and, therefore, might have been able to fulfill the role of a genetic system. In such a case, you will have to escalate the comparison to functionally more demanding properties, e.g., the capability of nonenzymic replication. Such systematic comparisons of potentially natural nucleic acid alternatives with RNA can be expected to teach you something about why we have today RNA and DNA, and not one of those alternative structures.

The project started in 1986 at ETH, and it is continuing today with a research group at the Skaggs Institute for Chemical Biology in La Jolla, which is part of TSRI (The Scripps Research Institute). The project fits the term "chemical biology" rather well: the question that is raised refers to biology, and the experiments to be carried out require a chemist. In this work, we have learned that in the structural neighborhood of RNA there exist a number of potentially natural alternative systems which Nature could also have chosen as genetic systems, at least according to the criterion of informational base-pairing. The study of some of those alternative pairing systems ought to be pushed further. On the other hand, some of the potentially natural alternatives were clearly not evolutionary competitors of RNA, because they were found to lack the capability of informational base pairing. Take, for instance, a hexose instead of a pentose: hexoses from a chemical point of view are at least as elementary with respect to their formation as are pentoses, but those hexose analogs of RNA that have been investigated thus far do not have the capability of reliable Watson–Crick base pairing.

Is it a steric effect?

Yes, hexoses are too bulky; base pairing in the Watson–Crick mode of hexose analogs of RNA runs into steric difficulties. This is a relatively solid result in this kind of work; after all, research on molecular evolution cannot be expected to reach conclusions that are as definite as those in normal chemistry. This type of inquiry and experimental strategy will gradually allow us to approach a question that science will have to ask sooner or later, namely, whether the chemical uniqueness of life, as we know it, is intrinsic or accidental. For the time being, inquiring about the uniqueness of life is premature, but this may change some time in the future.

English Translation of the Prelog Statement

Zurich, October 3, 1995

Dear Albert,

For some time you have prodded me to tell you, why the great Leopold and I did not recognize, in a timely fashion, that the nucleic acids are the most important natural products, and why did we waste our time on such worthless substances as the polyterpenes, steroids, alkaloids, etc.

My light-headed answer was that we considered the nucleic acids as dirty mixtures that we could not and should not investigate with our techniques. Further developments were, at least in part, to justify us.

As a matter of fact, for personal and pragmatic reasons, we never considered working on nucleic acids.

Yours,

Vlado

Gilbert Stork, 1999 (photograph by I. Hargittai).

8

Gilbert Stork

Gilbert Stork (b. 1923 in Brussels, Belgium) is Eugene Higgins Professor of Chemistry, Emeritus, at Columbia University in New York. Following his secondary education in France, he got his B.S. degree from the University of Florida in 1942 and his Ph.D. from the University of Wisconsin in 1945. He spent some six years at Harvard University and has been at Columbia University since 1953. He is a Member of the National Academy of Sciences of the U.S.A. and a Foreign Member of the French Academy of Sciences and of the Royal Society (London). He was the recipient of the Arthur C. Cope Award of the American Chemical Society in 1980, the National Academy of Sciences Award in Chemical Sciences in 1982, the National Medal of Science in 1983, the Robert A. Welch Award in 1993, and the Wolf Prize (Israel) in 1996, among many other distinctions. Our conversation took place at Columbia University on May 10, 1999.*

Today is the day of the annual Gilbert Stork Lecture at Columbia University. How did this lectureship come about?

When I became 70 years old, there was a celebration organized by current and former members of my group. There are over 350 of them, Ph.D. and postdoctoral. They produced some money and set up an annual lecture here. Today's lecturer is Barry Trost.

---

*This interview was originally published in *The Chemical Intelligencer* **2000**, *6*(1), 12–17

You have had a lot of interaction with Carl Djerassi. Are you also involved in biotechnology?

No. Djerassi's involvement in biotechnology is from the point of view of scientific and financial interest, but he is not doing research in that field as I define it. Neither he nor I know enough biology or biochemistry to do things at the research level.

How much do you need to know? I'm asking this because there are examples of people deeply involved in biotechnology who were trained as theoretical physicists.

Some people have been remarkable in that way. Several of my former students are now operating in biotechnology, many of them on the West Coast. One of them learned biology and biochemistry in a very short time and knows as much as anybody else who has spent all his life in the field. He had the drive and the brilliance to learn all this. It's like learning another foreign language. Some people can do it better than others. Walter Gilbert and Francis Crick are outstanding examples. Of course, Crick got involved very early when you didn't have to learn this enormous encyclopedic amount of material. It is very difficult to learn it because it is still much less structured than a more ancient discipline, like organic chemistry. In chemistry, you can tell from the title of a paper whether you should be interested in it or not. In biochemistry, that's not so; it's so diffuse that you can have, at most, an intuitive feeling. To get back to Carl Djerassi, our major interaction was during our Ph.D. work at Wisconsin, and later when he arranged to involve me in Syntex, a pharmaceutical company, which he really put on the map, and for which I became a consultant in chemical synthesis.

Is synthesis as important in pharmaceutical chemistry today as it was in those days, say, 40 years ago?

Yes. A few years ago, there was a general feeling that combinatorial work would change synthesis in the direction of automation. For a while, pharmaceutical companies even stopped hiring chemists. They did find some interesting molecules by these new methods. Then there was no one who could make them. So, all of a sudden, they became frantically interested in acquiring chemists and now there is a shortage of them. Of course, these things go in cycles.

Is natural products chemistry as important as it used to be?

There is no question that people keep finding interesting structures, and to a great extent the initial lead comes from natural products. This is perhaps not that strange because we have a common evolution; there was interaction of the various living systems and there is a connection between these molecules of nature in biological receptors of plants and people. It is amazing how many molecules have biological effects. Whether they become drugs is a different question.

Would you please single out one or two of your most important research achievements?

It would be easier if you asked me about someone else. The enamine reaction is obviously our most widely recognized contribution.

The enamine alkylation is called the Stork reaction.

Well, I have a tentative hypothesis that if there is a name attached to a reaction, it was probably discovered by someone else. I like to think that what some have called the Stork reaction may be an exception. It originated with wondering what might be going on in nature at a very primitive level. The question is how does nature manage to make carbon–carbon bonds. That must be, obviously, a much milder process than we normally use in the lab. We take an enolizable carbonyl compound, treat it with a very strong base at very low temperatures, and alkylate the resulting enolate with an alkyl halide. This is a violent operation and this is clearly not what goes on in nature. So I wondered about that and I thought that maybe there was some sort of a reaction, not of an enolate of a ketone, which is what we chemists use, but an equivalent, which might be a nitrogen analogue of the enolate, such as an enamine.

Nitrogen has an unshared electron pair so an enamine could function as an enolate ion. Many people had had this idea before, but nobody had done anything synthetically useful with it. We were extremely lucky because the kind of reaction we tried, the monoalkylation of a ketone enamine with an alkyl halide, is actually poor with enamines with one exception, and, by chance, it was that exception that we tried first and it worked very well. The particular molecule was a so-called β-tetralon. These ketones give 90% yield by the enamine alkylation process. This was in the mid-1950s and we published a general review in 1963 [Stork, G.; Brizzolara, A. H.; Landesman, H. K.; Szmuszkovicz, J.; Terrell, R. "The enamine alkylation and acylation of carbonyl compounds." *J. Am. Chem. Soc.* **1963**,

85, 207]. Since then, enamine chemistry has made its way into elementary organic textbooks.

While we are on the subject of specificity and control in substitution adjacent to a carbonyl, there is another contribution, which I would like to mention because it has turned out to be very important. It is what I refer to as regiospecific enolate formation and trapping. It made possible the introduction of an electrophile, including a carbon chain, on one or the other α-methylenes of a ketone.

I would like to tell you about another example that was important to me, though less willingly recognized by the public. I have to start with William S. Johnson, Bill Johnson, who died only a few years ago. He started his academic career at Wisconsin and was eventually Professor of Organic Chemistry at Stanford. He achieved spectacular, stereocontrolled constructions of the polycyclic systems of steroids and triterpenes, based on his imaginative and daring expansion of a concept for which he gave credit to Albert Eschenmoser and me. Some people call it the Stork-Eschenmoser hypothesis. It has to do with the cyclization of certain polyenes in which the double bonds are separated by a couple of saturated carbons, as a repeating unit. These are commonly found in acyclic terpenes. The hypothesis was that the structure and stereochemistry of many polycyclic systems, like sterols, arose from a certain mode of acid-catalyzed cyclization of such polyenes. A terminal double bond would interact with a proton, thus becoming a carbocation, which would then interact with the next double bond, simultaneously cyclizing and forming the next cation. The process continues, making rings of predictable stereochemistry, as found, for instance, in cholesterol. I thought it was a great idea. I was pretty young, and I was impressed with myself. I presented these ideas in a colloquium I gave at Harvard in March 1950. This was a pure concept; no experiments had been done at all. I just presented this idea that polyene cyclization could make rings of specific stereochemistry, and this might be the way steroids are formed in nature. This was eventually shown to be true, but we didn't get that much credit for it. The beautiful work of Konrad Bloch established the origin of the various carbons of cholesterol from its known acyclic squalene precursor. The cyclization process was in accord with the ideas I had presented in my Harvard colloquium. Bloch was unaware of this, but what I did find irritating was that Woodward, who co-authored Bloch's paper, which appeared after I had left for Columbia, had been present at my colloquium but did not make reference to it. Woodward was, of course, an extraordinary individual, but he had an implicit belief that if he did not produce or suggest something, it had no particular importance. That is to say, until the Pope had given his approval to something, then it didn't exist. So he didn't have to give too much credit to previous workers because, by definition, until Woodward said it was all right, it was not.

The fact is that I had great admiration for Woodward, who, quite aside from his chemical brilliance, was a man of considerable wisdom. Perhaps this story illustrates it. George Wald had just received the Nobel Prize (in 1967) for his work on vision, and he and Woodward were in a car together. Wald was making political-philosophical statements about various things and Woodward's comment was something like this, "You know, George, now that you're a Nobel laureate, your ideas are not any more sensible than they were before."

Coming back to the cyclization reaction, Bill Johnson was the first who called it the Stork-Eschenmoser hypothesis. Eschenmoser was working in Zurich, independently, on related ideas. Without Johnson, nobody, including Eschenmoser and myself, would have seriously considered that practical constructions of structures as complex as those of the tetracyclic steroids could be achieved in the laboratory by processes based on the cationic cyclization of polyenes. Eschenmoser even said so in print, and Johnson took great delight in pointing this out whenever he gave a lecture on his own (successful) work. Obviously, even a great idea requires great experimental design to be made operational. Johnson did it.

In any case, it was a great colloquium. Derek Barton was there, being a visiting professor at Harvard at that time, Bob Woodward was there, and George Buchi and other MIT people came. My suggestion was sufficiently unusual at that time to prompt Professor Louis Fieser to make a rather critical statement, which is of historical interest. There was a group of compounds, known as the "essential fatty acids," which we know today to be the precursors of the prostaglandins. At that time, however, the prostaglandins had not been discovered. So the question was, why are these fatty acids essential in nutrition? Fieser firmly believed that these fatty acids were essential because steroids originated from them. I still remember as, at the end of my colloquium, Fieser said, "I don't believe that sterols come from polyene cyclization. They probably come from the essential fatty acids."

Weren't you afraid when you presented your ideas before doing the experiments that somebody might scoop you?

It's true that chemistry experiments don't take very long. In some cases you can find out in two days that something is not going to work the way you thought it would. On the other hand, it often isn't that fast. A synthesis may take quite some time. But it's not that easy to steal an idea when the research is expected to take a long time because the initiator has a head start.

Sometimes people do engage in intense competition. Some like it, and some believe that advances in science are not unrelated to competition. Others disagree. It may play a role. There have been famous competitions, between Woodward and Robinson, for instance, in the steroid synthesis, or in the determination of the structure of strychnine. Some people love the type of competition common in sport: "Who will make cholesterol first?" At one time, everybody was fascinated by who would be the first to achieve a four-minute mile, and I must admit that I remember that it was Bannister. Today, many people run the mile in considerably less than four minutes. Is it important who did it first? It's not but it is striking.

You had a record number of students. Would you be willing to mention some names?

I have had the good fortune to have many outstanding associates. Some have had, or are making, successful careers in industry, others in academia. To mention a few, in no particular order, who became professors in universities around the world: Paul Wender, Paul Grieco, Samuel Danishefsky, Ian Paterson, Clayton Heathcock, Clark Still, Jacqueline Ficini, Jiro Tsuji, Minoru Isobe, Takashi Takahashi, and many more.

Gilbert Stork and Robert Robinson in 1967 (courtesy of Gilbert Stork).

There was an era when many Japanese came over to the United States for two reasons. One was because the U.S. had become a preeminent power and there was the image that that was where the future was. The other reason was that we were ahead in chemistry at the time. This is debatable today. In several fields the Japanese are ahead. The popular conception before the war was that the Japanese could only make junk unless they copied something. They made all sorts of toys that you cut your fingers on. After the war, that image lasted for some time until, eventually, people in Detroit learned to their surprise that that wasn't necessarily true and that the Japanese made cars that were better, and lasted longer, than the American cars. The same thing happened in science. It was a convenient thing to believe that the Japanese couldn't do original research, although they obviously worked very hard. One didn't like to think that they both worked hard and were also creative. We held onto that belief for a long time, way beyond any rational justification. That's no longer true. Many important contributions in organic chemistry, such as in the control of asymmetry and in transition-metal chemistry, have come from Japan.

Would you care to single out any of your teachers who had an important impact on your career?

I had a very good chemistry teacher in high school in France, where I was brought up. He was influential not because of his chemistry but because of his personality. Most of the other teachers were too good for us, and we defended ourselves by making their lives as miserable as possible. In college I was lucky that I went to the University of Florida, in Gainesville. It was not very good at that time, although today it is a very good school. In those days it was not, and for me this turned out to be very fortunate. I probably wouldn't have made it if I had gone to a really good school, to Caltech or Harvard or MIT. I might have even become an economist. I was at Florida for two and a half years only, because I managed to get a lot of credits for Greek, which I had barely studied and did not know. This was poetic justice because I received almost no credits for French in which I legitimately should have received a lot. During those two and a half years, I had my own lab, and the glassblower made anything for me. I almost killed myself because I didn't know what I was doing, but I was learning a lot of chemistry. I was so happy there that I wasn't willing to go anywhere else for a Ph.D. Fortunately, the Professor of Physical Chemistry convinced me to go somewhere else. By then, I was interested in a particular area of chemistry and I decided to go to the University of Illinois because Roger Adams was there. Hard as it is to believe, I didn't realize that you had to apply for admission. The idea that they wouldn't immediately give me a lab didn't occur to me. I went to Illinois and demanded to talk to Roger Adams. A secretary told me that he was busy and could not see me. I thought this was outrageous and took the train to Wisconsin because I thought that Professor Samuel McElvain there had interests, which I thought were compatible with mine. He was not a famous chemist, but he was very good, one of the first in this country to teach the merits of what was then known as "the electronic theory of the British School" (Robinson's curved arrow pushing). His judgment was very good, he was very supportive, and, although he let me do what I wanted to do, he was not very tolerant of nonsense. I thought of him as the Rock of Gibraltar. Bill Johnson had started at Wisconsin just a couple of years before I got there. I worked for McElvain, but Bill and I had a very good relationship because he was working on steroid total synthesis, and I was very interested in synthesis at that time. Another young faculty member at Wisconsin was Alfred Wilds, with whom Djerassi was doing his Ph.D. Wilds had been involved, working with Bachman at Michigan, in the first American total synthesis of a natural product of some complexity, a steroid called equilenin. After Wisconsin, I worked for a year in a small pharmaceutical place, Lakeside Laboratories, in Milwaukee. The main reason was that I was still on a visitor's visa and the Research Director of the company, who was one of McElvain's former students, was the only one willing to hire me (at low pay). During that year I made progress regularizing my immigration status and would visit Johnson, who was still at Wisconsin, from time to time (he eventually left to go to Stanford as chairman of its chemistry department, which he proceeded to make into one of the top chemistry departments of the world). Johnson convinced me to apply for an independent postdoctoral fellowship, which had become available at Harvard. I applied, suggesting a possible stereo-controlled total synthesis of estrone. It has remained untried to this date, but an unexpected and extremely fortunate event took place, which changed the course of my life. I still remember the phone call, which reached me in the lab. It was from Paul Bartlett, who was the Chairman of the Chemistry Department at Harvard, asking whether I would accept an Instructorship (what the lowest rung on the academic ladder was then called) rather than the independent fellowship I had applied for. That same afternoon I resigned from Lakeside. I stayed at Harvard from September 1946 to January 1953, and I've been at Columbia from then on. I owe much to Jack Roberts (Caltech) for that move. Louis Hammett, a major figure in chemistry, was chairman of the department and Roberts convinced him to bring me to Columbia.

Can we go back to your beginnings?

My mother was French, from Lorraine in Northeastern France. My father was Belgian, and I was born in Brussels. We lived there only the first nine months of my life. We then moved to Paris where I went to school. My father's father had a jewelry store in Brussels. He had three sons and one daughter, and the three sons continued the jewelry business. My father's involvement was in negotiating the purchase of precious stones from an office in Paris. Nobody in the immediate family was in science, although a relative of my mother whom I once met was Professor of Chemistry in Nancy.

What turned you to science?

I'm still not very much in science because I don't think the kind of chemistry I do has all that much to do with science as one conceives it. It was essentially a process of elimination. I was good at French literature, and I was even selected to represent my lycée in a nationwide high school competition in French writing. I was not terribly self-confident, however, and did not think that I could get a job in what I liked to do. So I was actually considering getting some safe government position. Something in French Indochina seemed especially attractive to me. Things took a different turn. In 1939, my father became very concerned about what was going to happen in Europe and decided to emigrate.

Jewish?

Yes and no. I don't remember having ever been in a religious establishment except for somebody's wedding.

I had just finished my secondary education (lycée). I was not yet 18 when my parents moved to the U.S., but the next thing was obviously to pursue my education. I first had to learn to speak English. I had studied English literature in France, and I could dissect a paragraph from Shakespeare in 15 minutes, but I couldn't speak the language at all. I spent a lot of time in the public library in New York to find out what I should do, where I should go. And this is how I decided to go to the University of Florida. My parents also moved to Florida with me and my sister, Monique, who is five years my junior. She is French and now lives in France. Fortunately, the chemistry exams were "multiple choice" so I didn't have to speak English to pass them.

Your present family?

My wife, Winifred, died seven years ago, after 48 years of marriage. We have four children, three daughters and one son. He is the youngest, got an MD, and is now a molecular biologist at the Vollum Institute in Portland, Oregon. My oldest daughter, Diana, got a Ph.D. in business and teaches business ethics in business school. Linda, an MD, is a pediatric oncologist at Denver Children's Hospital. The youngest daughter, Janet, works on a variety of projects in this country and abroad, concerned with elementary school teaching methods, and on the design of elementary schools. I have four grandchildren.

You seem to be a happy man.

On the outside, this is probably true. I tend to worry too much and not in an operationally useful way. People have been nice to me here at Columbia and, although I am now Emeritus, I am able to continue some research. I now have four postdoctoral associates and hope to continue my work for a few more years. I'm now 77 years old, which is not trivial.

Would you like to add anything?

One of my main concerns is that the major support of organic chemistry comes not from the National Science Foundation (NSF) but from the National Institutes of Health (NIH). This is, in a way, very fortunate because NIH has a lot of money, but it is a concern because although, over the years, NIH support for organic chemistry has been generous, that support is obtained on the basis of perceived relevance — broadly interpreted, to be sure — to public health. Important research not so perceived presents a problem. NSF is much too underfunded, given its multiple constituencies, to make more than a marginal contribution to the support of chemical research. The difficulty in obtaining adequate research support has had the result that a number of investigators have found the time and effort required to be prohibitive and this has led to their withdrawing from the contest. The resulting drop in applications has led NIH to conclude that it reflects a decrease of interest in organic chemistry, and that less money need, therefore, go into that field. These problems are particularly serious because NIH grants are essential to academic research, and because NIH has acquired significant *de facto* control of academic tenure decisions in organic chemistry. The potential long-term effects of these problems, and what can be done to alleviate them, seem to require serious thought.

Endre A. Balazs, 1999 (photograph by I. Hargittai).

9

Endre A. Balazs

Endre A. Balazs (b. 1920, Budapest, Hungary) is Malcolm P. Aldrich Research Professor Emeritus of Ophthalmology, College of Physicians & Surgeons of Columbia University, New York; Chief Executive Officer, Chief Scientific Officer and co-founder of Biomatrix, Inc., Ridgefield, New Jersey; and President of the Matrix Biology Institute. He received his MD degree from the University of Budapest in 1942. He did further studies for two years at the First Surgical Clinic of the medical school and in the Department of Clinical Chemistry and Pathology at the University Hospital. He also studied physical chemistry at the Faculty of Science between 1944–1946, in Budapest; biochemistry at the Wenner-Gren Institute of the University of Stockholm between 1947–1948, and electron microscopy at the Cavendish Laboratory of Cambridge University, England, in 1949.

He emigrated to Sweden in 1947 and moved to the United States in 1951. Between 1951–1975 he held teaching positions at the Department of Ophthalmology of the Harvard Medical School in Boston. He was Associate Director of the Retina Foundation in Boston between 1951–1961 and was President of the Institute of Biological & Medical Sciences of the Retina Foundation between 1962–1963 and 1965–1968, during which time (1962–1975), he was the Research Director of its Department of Connective Tissue Research. He was the first director of the Boston Biomedical Research Institute. During three sabbatical years, he was Visiting Scientist in Sweden, France, and England.

At the College of Physicians & Surgeons of Columbia University between 1975–1985, Dr. Balazs was the Malcolm P. Aldrich Research

Professor of Ophthalmology and between 1975–1982, he was the Director of The Eye Research Institute.

We recorded our first conversation in his home in Fort Lee, New Jersey, on March 17, 1999. It was a full-day recording, which was then augmented by a shorter session in Fort Lee in late 2000, and the text was finalized in the summer of 2001.

I was born in Budapest in 1920, so I'm a member of the post-World War I generation. My parents married in 1919. My father came back from the war without injury where he had been at the Bosnia-Hercegovina front as a first lieutenant of the mounted artillery in the Austro-Hungarian Army. My mother's oldest brother, a brilliant man and a professional officer, exactly my father's age and his best friend, died during the first week of the war. I grew up in a family environment where having a professional background (mostly legal and military) was important, but no one was a physician. My father was a civil engineer with a degree from the Budapest Technical University. He specialized in city water supplies, and eventually rose to the position of Technical Director of the Budapest Water Works. He served in this position through World War II and after the war. Then in 1949, he was retired because he refused to join the Communist Party.

My first years of elementary education were taught at home by my mother, who was a school teacher. My interest in medicine and science was stimulated by a very good science teacher in the gymnasium (high school) when I was 12. He made a strong point that those who were interested in science should do more than just attend classes. In the gymnasium, we had a "science club" and a "literary club." I enthusiastically participated in the activities of the science club, writing essays on various subjects such as how animals fly. Eventually I became president of our science club, and later of the literary club also. I was very interested in biology. My science teacher's influence was much more important for me than I realized at that time, and I regret that I never had the opportunity to tell him about it.

There was another influence, a family friend who was a gynecologist. He asked me what I would like to do and I learned from him about the medical profession. By the time I was 15, I decided that I would become a doctor. I had also read Paul de Kruif's *Microbe Hunters*, which was a very popular book, the first of its kind. I wanted to discover things in medicine. From the very beginning I had a goal: to cure people by scientific discoveries.

I graduated from high school at 17, and started medical school at the University of Budapest. My research career actually started when I was a first-year medical student. Tivadar Huzella was the Professor at the Institute of Histology and Embryology and he had an international reputation in the field of research on the substance between the cells. His laboratory was also famous, because it was one of the few places in the world where the new research technique of growing cells in a test tube was used. Huzella was also a linguist who spoke several foreign languages. He was tall, imposing and friendly, and he came from a wealthy family. He had a family estate on the shores of the Danube not far from Budapest where he had established a small, private research laboratory. He invited his research associates and his students there during the summer months, and created an exciting intellectual atmosphere. Huzella was, first of all, a biologist whose aim was to understand and to integrate the structure and function of the substance between the cells with cellular function. I learned from him how to conceptualize ideas and how to ask questions in research in such a way that the answers could fit into a biologically meaningful picture. I also learned tissue culture techniques and microcinematography from his associates. Microcinematography is a method by which one can film, by a time-lapse technique, the movement of cells *in vitro*. I also learned histology and embryology and later became a teaching assistant in his department.

The very first research project Huzella gave me was to study the intercellular substance between plant cells with the latest histological techniques used in his laboratory. He and his associates were concentrating on the study of the substance between the cells in animal tissues. Their primary interest was collagen, the fibrous, water-insoluble protein that is the main building block of the intercellular substance or intercellular matrix, as we call it today.

Huzella and his coworkers believed that a fine, reticular structure between the collagen fibers, which they called argyrophil fibers, was the precursor of collagen and made of a collagen-like protein. The name "argyrophil" comes from the word "argentum" (silver), because the only histological technique that could demonstrate these fibers was a process by which silver particles were deposited on the surface of the fibers, producing what looked under the microscope like a fine, filamentous, reticular network.

In my first research project, I demonstrated that indeed argyrophil fibers are present between the cell membranes of plant cells and the extracellular cellulose fibers. I also demonstrated that this argyrophil fiber system could be artificially produced by applying the same histological techniques to pectin or plant gums. Pectin, a pure polysaccharide like cellulose, is present between the cell membrane and the intercellular cellulose fibers, but not in the form of the argyrophil reticular filamentous network. I thus concluded that these fibers, at least in plants, are artifacts, caused by the histological technique itself. Huzella liked my discovery very much and instructed me to report it at the next International Anatomy Congress, which was to be held in the summer of 1938 in Budapest. This finding on the plant intercellular matrix was the subject of my first lecture (in German) at this congress and eventually the subject of my first published scientific paper.

Huzella included my finding, with illustrations, in his book published in 1942 and titled "The Organization of Cell Assembly: The Intercellular Theory." He concluded in his book that the existence of argyrophil fibers, shown both in plants and animals, proved a broad biological principle. This principle was that there is a structurally rigid intercellular system in animals made of collagen, which in plants corresponds to a similar support system made of cellulose. But second, and most importantly, a softer, finer reticular system of argyrophil fibers fills the space between the collagen in animals and cellulose in plants. The function of this argyrophil system was to provide an "elastomechanical" property for the intercellular substance. I fully agreed with him about this conclusion, but wondered about the composition of these artificially produced argyrophil fibers in the intercellular substance of animals. Since in plants these fibers were made of pectin, a polysaccharide, could they also be made from a polysaccharide in animals? Because none of the textbooks in histology and biochemistry mentioned extracellular polysaccharides, I started to read the recent German, French and English scientific literature. The library associated with the British Embassy was a great help in finding the latest scientific and medical journals published worldwide. Finally I found Karl Meyer's papers published in the late 1930s on hyaluronic acid. He described a pure intercellular polysaccharide present in large quantities in the joint fluid and in between the collagen fibers in the vitreus of the eye and the umbilical cord ["vitreus" is used as a noun and "vitreous" as an adjective].

Karl Meyer was a physician and biochemist who emigrated from Hitler's Germany to America and worked at that time in the Eye Institute of the College of Physicians and Surgeons, Columbia University, in New York. He did not discover hyaluronic acid, which was known under the name of mucoitin sulfate, but he verified its chemical structure and discovered that it contained no sulfur. But, most importantly, he gave a new, more appropriate name to it: hyaluronic acid. After reading his papers, I concluded that what pectin is in the intercellular matrix of plants must be what hyaluronic acid is in the intercellular matrix of animals. I decided to investigate the fluid in joints, which I collected from dead cattle at a Budapest slaughterhouse. The joint fluid contained relatively large amounts of hyaluronic acid, which makes it viscous and stringy, with an elastic appearance and a consistency similar to that of egg white. Actually, the scientific name of joint fluid is "synovial fluid" which comes from the Latin term meaning "egg white-like." Paracelsus, the 15th century physician who gave the name "synovium" to the joint fluid, discovered the similarity of joint fluid and egg white. I made good use of my collected joint fluids and showed that one could produce argyrophil fibers from hyaluronic acid, just as one could from pectin. I also obtained my first patent at the age of twenty-one on the practical discovery that cattle synovial fluid could be used as an egg white substitute in cooking. Needless to say, this invention was never commercially exploited.

By then Huzella had completely lost interest in my project which proposed that argyrophil fibers were artifacts made of this obscure polysaccharide called hyaluronic acid between the collagen fibers. Huzella never accepted this conclusion because he had absolutely no interest in polysaccharides. But, being a gentleman, he let me continue my work. It was important for my future work that I started in histology and learned tissue culture techniques, and through my research on hyaluronic acid, I also learned biochemistry.

After I received my medical degree in 1943, there was more time for research and I began to use hyaluronic acid prepared from the synovial fluid to investigate its effects on cells and on tissue regeneration. For this I needed sterile preparations. While I was still in medical school, I had contacted a German drug company through their Budapest representatives, and they were very interested in supporting my work. They gave me a large sterile filter, a truly state-of-the-art device. This made it possible for me to prepare sterile hyaluronic acid solutions for the first time in 1941. I injected this sterile hyaluronic acid preparation into dog joints, and placed it between healing bones in rats to study how it affects healing and bone regeneration. Thus my interest in using hyaluronic acid in the joint for therapeutic purposes dates back to that time.

Then the war reached me. I had to report for military training in the Spring of 1944. After the training, I was assigned to the First Military Hospital in Buda. The Hospital Commandant selected me to be his administrative assistant, which meant a lot of paper work, and he also put me in charge of a small infectious disease department. This clearly shows the shortage of experienced doctors in the military at that time. In March 1944, Germany occupied Hungary and in October, the Hungarian Nazis, the Arrow Cross Party, took over the government. In the meantime, the Russian army was advancing rapidly toward Budapest. Budapest was a mess; Hungarians of Jewish faith from the country had already been deported to Auschwitz and the killing of Jews intensified in Budapest. The Nazis took the governor of Hungary and the members of his government to Germany.

Just before the Russian army reached the outskirts of Budapest, the Hospital Commandant gave us young doctors the order to leave Budapest and re-assemble in Vienna in two weeks' time to be available to support the war effort. In the meantime, we could consider ourselves on a vacation. We took the vacation, but none of us had any intention of reporting in Vienna. I extended the vacation and eventually became a permanent deserter from the Hungarian army which by then did not exist. I went north to a little town where the parents of my future wife, Eva Tomas, lived. I remained in hiding until the Russian army took over the town from the Germans. After the liberation by the Russian army, we married, and I stayed for a few months to practice medicine with my father-in-law, a dermatologist, before Eva and I returned to Budapest. Of course, one could not earn a living doing research at that time. Back in Budapest, I was offered and accepted a paid training position in the Department of Pathology and also started to study physical chemistry at the University. I still had my unpaid assistant professorship at the medical school with my own laboratory there, and taught histology and embryology to medical students.

My major goal at this point, however, was to get out of Hungary as soon as possible, as Hungary was now under Russian occupation and it was clear that a communist dictatorship would be established soon. The opportunity to go to the West came in the spring of 1947 when I was accepted as a member of a Hungarian scientific delegation that was sent abroad to participate in two international medical congresses held for the first time after the war. This was a large delegation, and it included young scientists like me and famous professors like Albert Szent-Györgyi. I left Hungary with one suitcase and one hundred dollars in my pocket, and decided never to return to Hungary. The plan was that my wife would join me after I found a job. The first congress, held in Stockholm, was the IV International Congress of Cytology. At that time biochemists did not have their own international organization; therefore, all biochemists mingled with the histologists and cytologists at this meeting. The second congress, just a week later, was held in Oxford and was the XIV International Congress of Physiology, where biochemists mingled with physiologists.

At the meeting in Stockholm, I gave a lecture about my work during the past years on how hyaluronan affects the growth of embryonic cells in tissue culture. By the way, I use the name hyaluronan because this became the official name for hyaluronic acid, at my initiative, in 1986. There were four Swedish professors in the audience to whom my lecture meant something. Erik Jorpes, Professor of Biochemistry; Hjalmar Holmgren, the Professor of Experimental Histology; and Bengt Sylvén, Associate Professor of Pathology, all from the Karolinska Institute, the medical school of Stockholm. The fourth was Gunnar Blix, Professor of Biochemistry at Uppsala University. They all appreciated my lecture because they worked on heparin, mast cells, and hyaluronan. During the coming weeks, while supported by a fellowship from the Swedish Institute, I was able to remain in Sweden to have several discussions with them, not only on scientific subjects, but also concerning a research job in Sweden. I received three offers, including one at Gunnar Blix' Department in Uppsala, but I decided to join Hjalmar Holmgren in his newly established Department of Experimental Histology at the Karolinska Institute. I also received a Wenner-Gren Fellowship which paid my salary for the rest of my stay in Sweden. Since I was to work in a new building of the Karolinska Institute that was under construction at the time, I had the opportunity to design my own laboratories, which included tissue culture time-lapse microcinematography and biochemistry laboratories.

During my first year, while my future laboratories were still under construction, I worked at the Wenner-Gren Institute in the Department of Biochemistry and Metabolic Diseases. When my laboratory was ready, I left the Wenner-Gren Institute and moved over to the Karolinska in the Department of Experimental Histology. We were already well established in my laboratory when, in September 1948, Holmgren introduced me to a young first-year medical student, Torvard Laurent, who wanted to work with us. He worked in my laboratory for two years and we co-authored three pioneering papers on hyaluronan. He was to become a pre-eminent hyaluronan researcher throughout the rest of his career. He and his wife, Ulla, who is also a research scientist on hyaluronan as well as an ophthalmologist, remain my good friends to this day.

Holmgren had never been to America and encouraged me to spend a few years doing research there. The plan was that I would go to America

Endre Balazs using an in-house constructed microcinematography apparatus to take time-lapse movies of the movements of cells growing in tissue culture, in the 1960s (courtesy of Janet Denlinger, Fort Lee, New Jersey).

and, after a few years, return to his department. Holmgren was in his forties and full of energy, and with great plans to build his department. However, things were not to be as we planned. Holmgren was diagnosed with colon cancer in September 1950; I left for America in December, he was dead by February the next year and I decided to stay in the United States.

I accepted a job in Boston offered by Edwin Dunphy, Professor and Head of the Department of Ophthalmology at Harvard Medical School, to establish a new research laboratory associated with his department. His idea was to start a new research unit that focused on fundamental eye research using biochemistry, biophysics and "fine structure" methods, which was the name at that time for research using the newly available high resolution microscope, called the electron microscope. A pathology laboratory that was devoted to the physiology and pathology of the eye existed, but there was no one working in Dunphy's department on the biochemistry and fine structure of the eye tissues. Dunphy cleverly combined his plan with the ambitions of two young ophthalmologists who had recently moved to the United States: a British-trained Belgian, Charles Schepens, and an

Italian, Antonio Grignolo. Both were interested in retinal detachment, a blinding eye disease, and wanted to contribute to the understanding of the cause of the disease and to improve the diagnostic and surgical tools available for therapy. Both of them were in Dunphy's department at the Massachusetts Eye and Ear Infirmary and were actively involved in research.

Grignolo organized a research team which included the electron microscopists at the Massachusetts Institute of Technology (MIT). The American electron microscope (made by RCA) was invented during the war at MIT, and this was the only place where this instrument was available in the Boston area for biomedical research. Grignolo wanted to know what the collagen fibers of the vitreus looked like (the vitreus being the transparent gel filling the center of the eye), because he and others at that time believed that the changes in the structure of these collagen fibers caused the detachment of the retina. Schepens, on the other hand, was already in the process of developing a new ophthalmoscope using binocular observation. One year after my arrival, Grignolo returned to Italy where he became a famous ophthalmologist, ending up in the most prestigious position in ophthalmology in Italy, Professor of Ophthalmology at the University of Rome. Schepens' ophthalmoscope, diagnostic and surgical tools, coupled with his surgical skill, earned him international fame and fortune in the decades to come.

Accepting a research job in the Department of Ophthalmology at Harvard University did not create a problem for me, though I was never interested in ophthalmology *per se*. For years, however, I wanted to know more about the structure of the vitreus. For a half century it had been known that, in addition to collagen fibers, the vitreus contained a unique polysaccharide, the same found in the joint fluid and the umbilical cord. This polysaccharide was hyaluronan, the same molecule I had studied for more than ten years in the joint fluid. I had concluded, during my years in Sweden, that in order to understand the structure of the intercellular matrix one must analyze the simplest matrix of all: the vitreus. It is the simplest matrix, because it was already known that it contained thin, very regular in diameter but randomly distributed, collagen fibers, and that the space between these fibers was filled with hyaluronan. The entire vitreus, it was believed, in the healthy adult body was free from cells. The opportunity to study this natural model of the simplest of solid intercellular matrices fit perfectly into my long-range research plans.

In December of 1950, I arrived, with my small family (my wife and daughter of less than one year old) in Boston, knowing that I had to set up a laboratory and find coworkers and funds for my research. What I did not know was that the building the new research organization, called the Retina Foundation, had purchased was a 100-year-old, five-story apartment house without elevators that still had tenants on two floors. In the basement was a machine shop where a machinist worked on developing Schepens' ophthalmoscope. The first and second floors were empty. I also did not expect that there were no funds available to rebuild this house to make it a laboratory building. Coming from the most advanced and most modern medical research laboratories of the Karolinska Institute, I was bitterly disappointed. There is no question in my mind that if Hjalmar Homlgren had been alive, I would have turned back immediately to Stockholm.

I stayed, however, and started to raise money, looked for coworkers and did the best I could to continue my research. Schepens turned out to be a superb fundraiser. Dunphy and the President of the Retina Foundation, Ralph Lowell (the senior), a prominent Boston banker, philanthropist and member of most hospital boards affiliated with Harvard University, and the treasurer of the foundation, Holland Warren, a young lawyer with great enthusiasm for our cause, all provided essential help. Before the end of 1951, I received an NIH grant from the National Cancer Institute. The American Optical Company gave us an electron microscope, thanks to the help of Paul Boeder, who was the Director of the Visual Science Office of this company. By 1952, the Retina Foundation was in full swing. We had biochemistry, physical biochemistry, electron microscopy, tissue culture and histology laboratories. I managed to convince Marie Jakus, from MIT, to join us as the electron microscopist and also persuaded some of my friends from Stockholm to spend a few years with us. Torvard Laurent, after completing his doctoral thesis, joined me for three years and his new wife, Ulla Laurent, started her doctoral thesis in my laboratory.

The Laurents became very important in my life and in the life of the hyaluronan molecule. Torvard's doctoral thesis for which I was in part his advisor, was on the physical chemistry of hyaluronan, and contained some of our early work together. He and Ulla, who also made her doctoral thesis on hyaluronan, continued to do research on this molecule for decades to come. Torvard became one of the youngest professors and head of the Department of Medical Chemistry in Uppsala and concentrated his own work and that of most of his doctoral students on research on hyaluronan and other molecules of the intercellular matrix. Several young Swedish physicians worked for years in my laboratory on vitreus and cornea of the eye in preparing their doctoral thesis.

The Laurents, their coworkers and we (Janet Denlinger and I) with our coworkers, worked in parallel on many aspects of hyaluronan research, and we all devoted a good part of our lives to it. I am probably not exaggerating when I say that maybe as much as half of the new knowledge generated during the past half century on hyaluronan originated from the four of us, and from our students and coworkers. I am very grateful for this to the Laurents and for the close relationship on the private and professional level that greatly enriched my life.

The Retina Foundation was a unique place in the late 1950s. The five-story, old tenement house was rebuilt and filled with the most modern biomedical research equipment available. We had analytical ultra-centrifuge, free electrophoresis machine, nuclear magnetic resonance machine, equipment to determine C-14 in the gas phase (before scintillation counters existed), electron microscope, tissue culture equipment with microcinematography, and a superb instrument shop to make new research tools.

My work evolved in two directions. In cancer research, I continued to explore the antigrowth effect of sulfated hyaluronan and other sulfated polysaccharides. I also started my work on the structure and function of the vitreus. Although it was known that the vitreus has liquid and gel components in certain animals, including humans, not much was known about why there should be gel and liquid components or the determining factors that produced them. The gel component has a specific microstructure which made this extracellular component very different from the liquid component. We discovered cells in a certain part of the vitreus that produced hyaluronan. As I became more and more interested in the biosynthesis of hyaluronan, I chose the rooster comb as the tissue to study it. In the rooster comb, the synthesis of hyaluronan is regulated by testosterone. With my colleagues, especially with Janos Szirmai, a Hungarian histochemist whom I knew from my medical school days and who subsequently emigrated to Holland, we spent years in my laboratory investigating how hyaluronan synthesis is regulated in the combs of growing roosters. With John Gergely, a physical chemist at Harvard Medical School and an old friend from Hungary, we studied the molecular structure of hyaluronan and the structure of water around DNA and polysaccharide molecules. John later joined the Institute of Biological and Medical Sciences of the Retina Foundation and in 1968, he and I founded the Boston Biomedical Research Institute.

In 1959, the city quarter in which we were located became part of the first urban renewal project in the U.S.A. We received land from the city of Boston to build a new research building. I was happy and excited, because now I could build an even better research laboratory than the one I admired so much ten years earlier in Stockholm. I was in charge of the planning, design, and construction of the new building, which became known as the Institute of Biological and Medical Sciences of the Retina Foundation.

In 1962, the new building was opened and I became the President of the new Institute which had now grown to 4 departments (including my own Connective Tissue Research Department) and more than 50 researchers. My research during the next years focused on the structure of the intercellular matrix and the biological role of hyaluronan in this matrix. I used the vitreus as a model of a simple intercellular matrix. To understand the structure of this simple tissue, I studied the vitreus of many animal species including fishes, aquatic mammals, (whales), large land mammals (giraffes and elephants) and all kinds of birds, including the one with the largest eye, the great horned owl. Naturally, I was also interested in primate eyes (monkeys and humans). This comparative approach, using both morphological and biochemical techniques, proved to be very fruitful. I discovered that the vitreus of various species is very differently constructed and contains different amounts of hyaluronan and collagen. We also studied the development and aging of the vitreus in various species. To carry out these broad investigations I traveled to the Bahamas to collect fish eyes, to West Africa to collect land mammal and monkey eyes, and to Norway to arrange for whale eyes. Comparing the molecular structure and chemical composition of the vitreus of animals through evolutionary lines resulted in an insight into the role of the extracellular matrix and extraordinary ways to elucidate the function of hyaluronan in the matrix. As a result of this ten-year study on the vitreus, I accomplished my goal: the simplest of all tissues became one of the most thoroughly investigated and known in the body.

I did not leave the U.S.A. until I became a citizen in 1958. Then I decided to return to Sweden and spend a sabbatical year in 1959–1960 at the Nobel Institute for Cytology in Stockholm. The Institute's Director was Tornbjorn Casperson, one of the pioneers of nucleic acid cytochemistry. I worked there closely with Gunnar Bloom, a well-known fine structure researcher. I needed to learn more electron microscopy because, upon my return, I planned to start using the electron microscope to explore the structure of hyaluronan and the cartilage and the soft tissues of the joint, as well as the cells of the vitreus which I named *hyalocytes*, because they produce hyaluronan.

After my return to Boston, I extended my work on hyaluronan from the vitreus and the rooster comb to the synovial fluid. The physical properties, such as viscosity and elasticity, of synovial fluid were poorly understood, especially in human joints. I started to investigate the physical properties of cattle synovial fluid. As I had done 20 years earlier in Budapest, I went to the slaughterhouses of the Boston area and collected joint fluids. This time I was interested in how aging affects the physical properties and hyaluronan content of the joint fluid. The physical properties or, in more scientific terms, the rheology of the fluid is important because its role is to protect and lubricate the tissues of the joint. I traveled to Iowa to collect synovial fluids from cattle herds fed various diets that contained hormones. I hoped to discover a hormonal control of hyaluronan synthesis, similar to the one Szirmai and I had found that controlled hyaluronan synthesis in the rooster comb.

More importantly, I wanted to study human synovial fluid in healthy and in arthritic joints to determine the size of the hyaluronan molecules and their rheological properties. To measure the rheological properties of a fluid like the synovial fluid or hyaluronan in the early 1960s was not an easy task, and collecting synovial fluid from healthy humans was even more difficult. Two people played a critical role in the realization of my objectives: David Gibbs, who had just started to work on his doctoral thesis at the Massachusetts Institute of Technology under Professor Edward W. Merrill's guidance and was in the process of building a rheometer to measure the elastoviscous properties of fluids. Merrill and I agreed that Gibbs' thesis should be on the rheological properties of hyaluronan and synovial fluid. The other person who played a critical role in this endeavor was Ivan Duff, Professor of Rheumatology at the University of Michigan in Ann Arbor. He collected human synovial fluid for us, not only from arthritic patients, but also from healthy human volunteers of various ages. During the next decade, I repeated on the joint fluid what I had done the decade before on the liquid and gel vitreus — a complete analysis of a tissue and its hyaluronan content during development and aging in various species and in health and disease.

In 1969, I became Visiting Professor at the Chemistry Department of Salford University. This was at the invitation of Glyn Phillips, a great Welsh patriot who had recently moved from Cardiff to Salford to head the Chemistry Department. We collaborated for many years on projects related to radiation damage of polysaccharides, like heparin, hyaluronan and other polysaccharides. I had been interested in this subject since 1948 when I discovered the extreme sensitivity of these polysaccharides to ultraviolet light. Phillips, among others, was an expert in the radiation chemistry of polysaccharides and in radiation sterilization techniques. We and our students, working together since the late 1960s, contributed significantly to understanding the mechanism of the radiation sensitivity of polysaccharides. I am very grateful for his continued interest in hyaluronan and, over the past two decades, in hylans. Phillips also recently organized in Wales the first week-long international congress, *Hyaluronan 2000*, entirely devoted to the structure and function of hyaluronan and its therapeutic uses.

In the late 1960s, I also renewed my interest in the effect of hyaluronan on cell activities. Nearly 20 years had passed since I discovered that enzymatically-degraded hyaluronan has a stimulatory effect on the division of embryonic cells cultured *in vitro*. Now I had sterile, non-inflammatory hyaluronans of various molecular weights to be tested on these cells. To learn modern methods in cytology, I decided to spend a sabbatical year (1968–1969) in the Tumor Biology Department of the Karolinska Institute in Stockholm. George Klein was the professor who with his wife, Eva Klein, were old friends from my Budapest days and both outstanding researchers. They were involved in cancer research, mostly using a tumor model called "ascites" tumors. This is a mouse model of tumor growth in which malignant cells multiply in the abdominal cavity of the animal. Naturally, many other cytological tools and methods were available in their Institute and I was eager to learn how to use them in my research on hyaluronan and other polysaccharides (known as glycosaminoglycans) present in the intercellular matrix. There, I met Dr. Zbigniew Darzynkiewicz, a Polish immigrant physician and researcher in immunology. His interest was in nucleic acids, but he was willing to digress and, after my return I sponsored his and his family's immigration to the U.S., and he spent four years in my laboratory.

After my return from Sweden my laboratory was buzzing with activity. I had a fantastic team of young, talented and ambitious associates, some of whom had already worked with me for several years like Bernard Jacobson and Gerard Armand, both biochemists, and David Swann, also a biochemist, who emigrated to the U.S.A. from England at my invitation after completing his doctoral studies. Young ophthalmologists came who wanted to work on the retina, cornea, vitreus or lens of the eye and spent a year or two in my laboratory: Michele Testa from Naples, François Regnault from Paris, Rudolph Klöti from Zurich, Bo Phillipson and Peep Algvere from Stockholm, Mel Freeman from New York and, later on, John Forrester from Glasgow and Ali El Mofty from Cairo, Egypt. We learned not only more about the structure and function of these eye tissues in health and disease, but also, most importantly, we learned about hyaluronan and its biological activities.

During these years, our work laid the foundation for future studies on the effect of elastoviscous solutions of hyaluronan on the migration, proliferation and function of white blood cells, and how such solutions affect some functions of the immune system. We discovered that hyaluronan regulates the synthesis of glycosaminoglycans in *in vitro* systems, and regulates inflammation and tissue regeneration in adult animal models and tissues. All these studies laid the theoretical foundation for the therapeutic applications of hyaluronan.

Since my medical school days I had been fascinated with the possibility of using hyaluronan as a supplement for the joint fluid in arthritis, but now I also saw a new role for hyaluronan: to be used during retinal detachment surgery to push back the detached retina and after surgery to replace the lost vitreus. Using hyaluronan in eye surgery and as a supplement for joint fluid and the vitreus required a very pure preparation of hyaluronan. This high level of purity meant that the hyaluronan to be used therapeutically could not cause any reaction in the sensitive tissues of the eye or the joint. The test of purity could not be limited to test tube methods, but had to be expanded to a test using the eyes and joints of living animals. I selected the horse joint and the eye of a South American monkey, the owl monkey (*Aotus* species), as models for this new test method. The Institute had purchased a farm in Northern Massachusetts where I kept retired race horses with arthritis and owl monkeys. To my great surprise, as I developed the new test, our purest high molecular weight hyaluronan was not usable for therapeutic purposes because it caused acute inflammation in the horse joint and in the monkey eye. A new purification process had to be discovered and developed to make hyaluronan, both from rooster combs and from human umbilical cords, usable as a therapeutic agent.

To begin this project I set up a company, Biotrics Inc., using my life savings. I did this in order to keep my academic research at Boston Biomedical Research Institute separate and independent from my industrial development work. I purchased manufacturing and testing equipment and hired people, and rented a space in Arlington, Massachusetts, not far from my home. My goal was to produce a high molecular weight hyaluronan which did not cause acute inflammation in horse joints and monkey eyes. By 1971, I had a preparation of hyaluronan that was free of inflammation-causing agents and I called it the non-inflammatory fraction of Na-hyaluronan

(NIF-NaHA). My childhood dream was fulfilled; I had a new therapeutic agent to be added to the armamentarium of medicine. With the approval of the regulatory agencies in various countries, including the U.S.A., we started clinical trials with this product which I registered under the name Healon®. These clinical trials covered a broad spectrum of medical indications, such as retinal surgery, corneal transplantation, to decrease pain in knee osteoarthritis in humans and in the traumatized arthritic knees of race horses, as well as to reduce adhesion formation and to decrease excessive scarring in animal models and humans prone to keloid (excessive scar) formation.

These early clinical studies with NIF-NaHA were carried out internationally. The clinical investigators were my friends who had worked with me in my laboratory for years and then returned to clinical practice. The use

Owl monkeys (photograph courtesy of Dr. Janet Denlinger). The eye of the South American owl monkey (*Aotus* species) was selected as a model for purity control of hyaluronan.

of Healon® in ophthalmology was tested in Boston by Robert Brockhurst; Zurich, by Rudolph Klöti; Essen, by Gerd Meyer-Schwickerath; Copenhagen, by Jens Edmund; and Paris, by François Regnault. Its use in human and horse arthritis was tested in Ann Arbor, by Ivan Duff; Boston, by Charles Weiss; Glasgow, by Richard St. Onge; Göteborg, by Nils Rydell; Paris, by Jacques Peyron; and Cape Town, by Arthur Helfet.

During the last part of the 1960s, the Retina Foundation went through a major transformation. The trustees of the Foundation decided that the future of scientific research would be best served if the Foundation were divided into two new entities. One of these was to continue as a research institute devoted entirely to research on the eye and medical problems in ophthalmology. The other was to continue as a biomedical research institute devoted to a broad program of research and not limited to any organ or medical specialty. The first institute is known today as The Charles Schepens Eye Institute, and the second as the Boston Biomedical Research Institute (BBRI). As a co-founder of BBRI I became its first Director.

By the end of 1971, it was firmly established that Healon®, developed and tested by Biotrics, Inc., was a safe and effective surgical tool in ophthalmology and a long-lasting analgesic in arthritic horse joints. I also ran out of funds for Biotrics, and started to look for a partner to take over the commercialization of Healon® for the various medical therapeutic applications I had foreseen. After negotiations with half a dozen international drug companies, I selected Pharmacia AB, a Swedish drug company to take on this task. I licensed the technology, know-how and patent application to Pharmacia and gave them the right to manufacture and sell Healon® for all tested medical applications in most of the world. I retained all rights for North and South America. Consequently, Biotrics continued manufacturing and testing Healon® in the United States for four more years, and then in 1976 I sold the rights to Pharmacia AB for the use of Healon® worldwide.

In 1975, I left the Boston Biomedical Research Institute. Although I never specialized or practiced ophthalmology, because of my many fundamental discoveries on the eye tissues, especially the vitreus and the cornea, I was invited to be Director of Research and Research Professor of Ophthalmology at Columbia University in New York. Columbia had a large Eye Research Institute affiliated with its Department of Ophthalmology at its medical school, the College of Physicians and Surgeons. The Eye Research Institute was separately funded and operated independently from the clinical ophthalmology department. As the Director of this Institute,

Endre Balazs and his wife, Janet Denlinger in their laboratory, 2001 (courtesy of Janet Denlinger, Fort Lee, New Jersey).

I also held the Malcolm P. Aldrich Research Professorship. At that time it was the largest eye research institute affiliated with a clinical ophthalmology department in the country. The "Columbia Connection" held a special place of importance for me, since Karl Meyer had started his own research on hyaluronan there about 40 years earlier. Another famous biochemist was still at the Institute, Zacharias Dische, who made his fame studying glycoproteins. A few years later in 1977, I invited Karl Meyer to move back to the Eye Institute from Yeshiva University, which just closed its graduate program. He continued his research at the same place where he had started it 45 years previously until his death.

My interest then turned to the metabolism of hyaluronan in the eye and in joints. Since we were going to use this molecule as a therapeutic agent in these tissues, we had to know more about how long this molecule could remain in these tissues. My new wife, Janet Denlinger, focused her research in our laboratory on this subject which later became the topic of her doctoral thesis.

In 1976, I received a telephone call from a former clinical fellow at the Retinal Foundation, David Miller, who was now the Head of Ophthalmology at Beth Israel Hospital in Boston. He remembered that in the late 1960s, we tested Healon® in eye surgery during corneal transplantation to protect the thin layer of endothelial cells on the inside of the cornea from mechanical damage. These cells die when mechanically injured and do not regenerate in adults. Their function is to pump water from the cornea to the inside of the eye (the aqueus). When a certain percent of these cells die, the cornea swells and becomes opaque and the patient becomes blind. We tried elastoviscous Healon® solutions many years earlier to protect this endothelial cell layer during corneal transplantation. In Boston and Copenhagen clinical trials were conducted and published on the use of Healon® for this purpose. I explained this protective role of Healon® during surgery to David Miller and his young associate, Robert Stegmann (from South Africa). He was interested in using Healon® in a new application: to protect the corneal endothelium during intraocular lens implantation. The development of the technique for implantation of plastic lenses after the cataractous lens was removed from the eye is an interesting story, because it shows how new ideas in medicine are often accepted only very slowly.

The idea of implanting plastic lenses after the removal of a cataractous lens was developed over many years and in many variations by several ophthalmic surgeons during the 1950s, and by the 1970s was widely used in Florida and Southern California where there was a large elderly patient population with cataracts. At the same time most university clinics in the Northeast did not use and had not even tested this new technology. The reason for this was the concern for potential damage to the corneal endothelium which could result in blindness. The leading ophthalmologists in most academic hospitals were not willing to accept the risk, when the removal of a cataractous lens could be easily corrected by eyeglasses without any risk of blindness due to corneal damage. But since intraocular plastic lenses were being inserted after cataract surgery in thousands of patients in Florida and California, leading ophthalmologists in other areas had to try the procedure. They did so reluctantly in Boston, as well as in the ophthalmology department at Columbia.

I arranged for Pharmacia to provide Healon®, which was at that time on the market in some countries for the treatment of arthritis in veterinary medicine under the trade name Hylartil®Vet. Healon® had already been tested extensively for the use in retinal detachment and corneal transplantation. Within two years enough experimental and clinical data was produced at various clinical research sites to present to the Food and Drug Administration to ask for permission to make Healon® available for patients undergoing intraocular lens implantation and retinal surgery. Since I already had years of experience with the Food and Drug Administration on various issues related to the approvals for the use of Healon® as a therapeutic agent, Pharmacia asked me to lead the effort to obtain FDA approval both for the use of Healon® in ophthalmic surgery and for the treatment of arthritis in veterinary medicine in the U.S.A.

The FDA approval of Healon® for eye surgery was relatively quickly obtained. Thus, for the first time an elastoviscous hyaluronan solution was available for therapeutic use. This use, which I called *viscosurgery*, to protect sensitive tissues of the eye and to create space for the insertion of the plastic lenses, revolutionized this surgical procedure, and the use of intraocular lenses exploded. Healon® was also used to manipulate the retina in retinal detachment surgery. Almost 20 years after the introduction of *viscosurgery* in ophthalmology, Healon® is still a leading product worldwide. The use of viscosurgery, which means the use of elastoviscous solutions made of hyaluronan or its derivatives as surgical tools, helped develop better and safer methods for replacing a cataractous lens with a plastic lens. Viscosurgery also helped the surgical repair of traumatized eyes.

I was very happy with this success but was disappointed that the success in ophthalmology slowed down and eventually stopped progress toward the marketing of the same material for the treatment of human arthritis. Pharmacia could not manage to bring to the patients the various applications of Healon® in medicine over a broad spectrum of disease conditions.

In the fall of 1981, my wife, Janet Denlinger and I went to Paris to spend a year in Laszlo Robert and Jacqueline Labat-Robert's Biochemistry of Connective Tissue Laboratory at the Faculty of Medicine of the University of Paris. Laszlo and his wife Jacqueline are well-known researchers in the connective tissue field. The objective was for Janet to carry out research in the collaboration with the Roberts and write her doctoral thesis on the metabolism of hyaluronan in the vitreus and joint. She had already completed her course work at Columbia University. Professor Jean Montreuil, a polysaccharide chemist and Head of the Department of Chemistry at the University of Lille, along with Professor Robert, accepted the sponsorship of Janet's doctoral thesis which she presented in May of 1982 in French, earning a summa cum laude mention along with her degree. Her thesis work laid the foundation of our understanding of how hyaluronan is metabolized in the vitreus and joints of various animal species such as horses, rabbits and monkeys.

During my sabbatical year in Paris, I had time to plan for the future. I became convinced that it was time to start a concentrated effort to modify hyaluronan chemically for a diversity of therapeutic uses. First of all, to increase the elastic properties of the molecule in solution, the molecules themselves had to have a larger molecular mass. A second goal was to create soft solids from this molecule, that is, water insoluble gels and other objects to be used as therapeutic agents, first of all for preventing adhesions and excessive scar formation. Third, I wanted to create sulfated forms of hyaluronan, that is, heparin-like molecules to control cell division. These new molecules were necessary to continue my work begun in the 1940s when I found that sulfated hyaluronan could control multiplication and growth of transplanted tumors in animals.

For several years I tried to convince Pharmacia to work with me to develop new derivatives of hyaluronan and test them on a broad spectrum of medical therapeutic applications. But Pharmacia was not even ready to market my high molecular weight, non-inflammatory fraction of hyaluronan for the treatment of human arthritis, despite the fact that they had many successful clinical trials and that they were selling the material worldwide for treatment of arthritis in animals. I believe that this was because they were not ready to invest in a substantially larger manufacturing facility for hyaluronan. They were, however, ready to support me financially in my research and development effort to develop new hyaluronan derivatives.

Once again, I was faced with the challenge of how to bring a product with great medical potential into the hands of physicians for the care of patients. In the early 1980s, Janet and I decided to start a company to develop new hyaluronan derivatives and continue our work in the field of human arthritis. We called the new company Biomatrix, Inc. and it was incorporated and started again the same way as Biotrics, namely using our savings to finance the company. My son Andre soon joined us to help take care of the business aspects of the venture. I saw an immediate opportunity for using a certain preparation of very elastoviscous hyaluronan solution in the field of dermatology as a hydrating and moisturizing substance on the surface of the skin. Now we had a new and quick revenue line, namely selling hyaluronan to the skin care industry. The first activity of the company was to produce my patented hyaluronan-protein complex from rooster combs and test it as a moisturizer for dry skin conditions. Since the most difficult dry skin condition to manage is psoriatic skin lesions, with the help of Gerald Lazarus, a well-known dermatologist, and Chief of the Dermatology Division of the Department of Medicine at Duke University, we tested the effectiveness of our product on this skin condition. The results were very positive: psoriatic patients, after applying this hyaluronan-protein complex on the lesions, felt less pain because the skin was kept well hydrated. Our product did not cure psoriasis, but it made the patients more comfortable. With this result we convinced the product developers at Estee Lauder to use our product, which we later called Biomatrix®, as a moisturizer and a delivery system for water-soluble ingredients to the skin in Lauder's next product: Night Repair®.

But Biomatrix' main goal was to discover new derivatives of the hyaluronan molecule for therapeutic applications. I experimented, during my Boston days in the late 1960s, with formaldehyde cross-linking of hyaluronan. I could produce water-insoluble crosslinked hyaluronan with an aldehyde crosslinking method, a well known procedure for crosslinking another polysaccharide-cellulose. But this time we were looking for a crosslinked hyaluronan that looked like a gel and was very hydrated, unlike the dry fibers made decades earlier with formaldehyde crosslinking. We experimented with well-known crosslinking systems, including vinyl sulfone. Finally we decided that this method produced what we wanted, that is, a soft, very hydrated gel. The most important finding was that this gel, which we later named generically hylan B, was very biocompatible. That is, tissues did not react with inflammation or with allergic reactions to this new molecule. This was extremely important because we did not want chemical modification to destroy the greatest advantage of the hyaluronan molecule in therapeutic applications, namely being completely non-foreign and noninflammatory when introduced into the body.

Aldehydes, first of all formaldehyde, have been used for centuries to preserve tissues, which means to make the proteins of the cells and the intercellular matrix insoluble in water and alkali solutions. We also knew that hyaluronan was not changed by this process, which means it could not be made water- or alkali-insoluble. Therefore, we tried to extract hyaluronan from tissues after aldehyde treatment. This worked because the extracted hyaluronan was very pure and had a very high molecular weight. We could prove that this high molecular weight hyaluronan was covalently bound to a very small amount of protein that was already loosely associated with the polysaccharide chain. This new molecule, with a molecular weight much greater than that of any other known hyaluronan, formed very elastic solutions in water. We called this new polymer generically hylan A.

In the discovery and development of hylan A and hylan B, of course, the most important criteria were their purity and biocompatibility. We could never have used these two molecules as therapeutics without the availability of the test that I developed in the late 1960s to monitor the purification process for the native hyaluronan, and producing NIF-NaHA. This test method was accepted by many regulatory agencies as a standard way of proving the biological compatibility of elastoviscous fluids used in ophthalmic viscosurgery. In my opinion, it should be used for all fluids and gels to be tested in any part of the body as therapeutic biopolymers. Our products have always been tested using this method before being applied in animal experiments and, most importantly, before the material is accepted for human clinical testing.

The discovery of hylan A and hylan B was the result of a teamwork with our young scientists — biochemists and chemists — who joined us in the early days of the company: Nancy Larsen, Phil Band, Adele Leshchiner and the late Arnold Goldman and Ed Leshchiner. Their discoveries opened the way to use these derivatives of the native molecule for a very broad spectrum of therapeutic applications. My original goal in the early 1970s now could be pursued with these new biopolymers. Because hylan A produced a much more elastoviscous solution in water than my original discovery 20 years earlier, Healon®, it could be used even more successfully for viscosurgery and for supplementing the joint fluid in arthritis. Hylan B as a gel or as a dry membrane or film could be used to prevent adhesions or to fill tissue spaces and correct depressed scars and wrinkles, fill the vitreous space of the eye after surgery, and to be formed into tubes to guide the connection of severed nerves. We also tested hylans to cover healing wounds, to protect them from water loss and infection and to promote healing. We coated and impregnated polymers used for replacing blood vessels. We found that hylans are very blood compatible, which means that hylan-coated surfaces do not denature blood proteins or affect blood cells.

We, of course, worked to develop all these therapeutic applications of the hyaluronan derivatives we created. But my business goal was to create a company that could become profitable as soon as possible. I wanted to avoid the criticism of investors that scientists who found and run biotechnology companies are interested only in technical development and the medical success, considering financial success only secondary. Therefore, I decided to focus on a few applications and leave the rest of the development for the time after profit was generated and the earnings could finance the development of more products. We were successful in achieving this goal for Biomatrix. We opened our first manufacturing plant in Montreal in 1992 and marketed our hylan product, Synvisc®, for alleviating pain in osteoarthritic knee joints in Canada in 1995 and later in Europe and the U.S. Today Synvisc® is available to arthritic patients in 58 countries. Hylaform®, our tissue augmentation product to correct facial wrinkles and depressed scars, is available in 35 countries. After 3 years of profitable financial performance, Biomatrix, in July 1998, was listed on the New York Stock Exchange. Very few biotechnology companies achieve this goal in such a short time. This success meant very rapid growth. In 1998, we doubled our size in personnel and in manufacturing capacity. In 1998, we also opened our second manufacturing plant in New Jersey. During this year and the next we nearly doubled product sales. Naturally, the value of Biomatrix increased substantially.

How do you see the future of the role of hyaluronan and hylans as therapeutics?

The therapeutic use of hyaluronan and its derivatives today is based on the function of this molecule in the healthy body as a stabilizer of the molecular structure of intercellular matrix and as a protector of cells and pain receptors. But we have known for a long time that hyaluronan has other biological functions. We know that elastoviscous solutions of hyaluronan and hylan inhibit certain cell functions by stabilizing the cell membranes. Hyaluronan also interacts with certain proteins that have important functions in the regulation of cell activities. In the future, as we learn more about the function of these specific proteins and how hyaluronan may turn on and off the signals given by these molecules to the cells, we may extend the therapeutic uses of hyaluronan significantly.

We have carried out experiments since the late 1980s to prove that the molecular matrix of hylan B was especially suitable to contain drugs and deliver them by slow release in a targeted manner. We also developed methods to bind certain drugs, enzymes or other biologically active molecules covalently to the hylan B chain so that active molecules could be delivered to the tissue sites where their action is needed. This mode of delivery of anti-tumor agents before and after surgery to the immediate vicinity of the tumor will be very useful in the near future.

Now that you have turned 80, do you have any plans to retire?

I certainly have plans to change my professional activities. As you know the Board of Directors of Biomatrix decided to sell the company to Genzyme Corporation and this deal was consummated in December of 2000. Genzyme is a Boston-based company with great interest in the therapeutic uses of hyaluronan. They developed, along different lines, cellulose-based, hyaluronan-containing anti-adhesion products, which are available to patients worldwide. I have a very high opinion about the past performance of Genzyme and hope they will continue to exploit and develop our inventions.

Since the merger, I have continued my research in a non-profit organization, the Matrix Biology Institute (MBI). This research institute that I founded in 1986 carries out fundamental research on the function of the intercellular matrix in general and specifically, on the role of hyaluronan in this matrix.

I donated some of my Biomatrix shares to the Institute and try to attract charitable donations from interested individuals, institutions, and industry to provide more funds. The Institute has some laboratories of its own, and some senior researchers will join me to form a nucleus of research activity. I foresee the Institute's role as a catalyst and facilitator for scientists worldwide interested in hyaluronan and its role in health and disease. The Institute's staff and resources will be available to those who may need it for purposes determined by its charter.

Your professional life has centered around this molecule — hyaluronan.

This is true, but with one qualification. It is not the molecule that intrigued me for more than half a century. It was the substance between the cells, what we call today the intercellular matrix that has fascinated me all the time. Since I met Huzella and his staff in 1938, I was fascinated with the regulatory effect of the intercellular matrix on cells. Hyaluronan happened to be the most fascinating and unique molecule in this matrix. When I entered science 60 years ago, the cells were considered to be the only important component of the living body. What was between the cells was regarded as a non-living, randomly distributed product of cells and was therefore an utterly unimportant cellular waste product. All my life I fought this concept and promoted and contributed to the idea that the space between the cells is even more strictly organized than the space inside the cells. In my opinion the intercellular matrix is in a near-zero entropy stage, which means that it is well-ordered and that this order is strictly kept. Hyaluronan is a unique molecule that is the key to the structure and function of the intercellular matrix. In my mind, hyaluronan has an important role in the matrix in a somewhat similar way as nucleic acids have for cells. I see the hyaluronan systems as stabilizers, preservers and re-constructors of the intercellular matrix. The hyaluronan system forms a defense mechanism against internal injuries and wounds, and against the destructive effect of dead cells and malignant tumor cells. It is biological equivalent to our immune system, which represents the defense against external noxious attacks, microorganisms, viruses, and foreign bodies. Hyaluronan represents the defense system against internal noxious effects: injuries and transformed cells that can destroy the body.

Alfred Bader, 1995 (photograph by I. Hargittai).

10

ALFRED BADER

Alfred Bader (b. 1924 in Vienna, Austria) went to Britain in 1938 in a children's refugee group. In 1940, he was interned and sent to Canada. In 1941, he entered Queen's University and earned an M.Sc. degree in engineering chemistry. He did his Ph.D. work in synthetic organic chemistry at Harvard University under the supervision of Louis F. Fieser between 1947 and 1949. He co-founded the highly successful Aldrich Chemical Company in 1951. Following an eventual merger with the Sigma Chemical Company, Sigma-Aldrich is now the world's largest supplier of research chemicals. After 40 years of an extraordinarily successful career, Dr. Bader was forced out of the company. Since then, Dr. Bader has continued to pursue his other long-time interest as an art collector and dealer. His autobiography has been published [Alfred Bader, *Adventures of a Chemist Collector*. Weidenfeld and Nicolson, London, 1995]. Our conversation was recorded in Dr. Bader's gallery and in the Baders' home in Milwaukee on November 8, 1995 and Dr. Bader added a note in February 1998.*

I would like to ask you about your work in chemistry. Not only did you create the famous company Aldrich, found Aldrichimica Acta, *and build up a remarkable art collection, but you have also done a considerable amount of research in organic synthesis.*

---

*This interview was originally published in *The Chemical Intelligencer* **1998**, *4*(3), 4–9

I first became interested in research in my senior year as an undergraduate at Queen's University. There was a very good teacher, Professor Norman Jones, a famous spectroscopist. He allowed me to do a research project, which I enjoyed. Then I got a very fine job with a paint company in Montreal, and a year and a half later, the president of the company suggested that I go on with my studies and offered me company support. I did my Ph.D. studies at Harvard in 1947–49 with Louis Fieser, who traveled so much at that time that his students saw very little of him. But there were many able chemists, students and faculty, who were very helpful.

Louis Fieser simply said to me, "Here's a quinone; in alkali it turns red, overnight it turns yellow. Find out what happens." A year and a half later, he come into my lab and asked me "How is that project going?" I said I'd solved it. Fieser said "Give a seminar." He was satisfied, said, "write it up for a paper," and there it was. So all went well, but I realized one thing: I was not a world-class chemist. I was a very good experimentalist, but there were many things I didn't understand in theoretical chemistry.

I felt obligated to go back to the paint company in Montreal, but it had been bought by the Pittsburgh Plate Glass Co. (PPG), whose paint research was concentrated in Milwaukee. So that is how I came here in January 1950. The company gave me a job but had no idea what to do with me. All the research they did was oil chemistry — linseed oil, soybean oil — and here was a Ph.D. chemist from Harvard trained in synthetic chemistry. But they left me alone, and I got interested in producing monomers from inexpensive starting materials — phenol with butadiene, cyclopentadiene and isoprene, for instance. The literature said that chemists had tried the reactions and they didn't work. But I found that if I controlled the catalyst concentration carefully, I could make them work. We had a whole series of unsaturated phenols, easily made from starting materials costing pennies.

One day a salesman from Quaker Oats stopped by and said levulinic acid would soon be available very inexpensively from Quaker Oats. The moment I heard that, I made the bisphenol by reacting levulinic acid with phenol to make what is now called diphenolic acid. A few weeks later, I sent a note off to *JACS*. Soon Johnson Wax wanted to buy the patent for which we'd applied. Our director of research asked me, "What should we charge for this?" I said "It was two days work; if we got ten thousand dollars, we would be well paid, but they must want it very badly, so ask for a million!" Well, he got it. Then, of course, lawyers descended on my lab to make sure everything I worked on was really patented, but I had already seen to that.

In 1954, PPG decided to move its research to Pittsburgh, but I didn't want to go there, left the paint company, and devoted myself full time to Aldrich. Shortly after arriving in Milwaukee, I had asked my director of research whether he would allow me to start a small fine chemicals division. At that time, Eastman Kodak had a monopoly but offered only 4000 compounds. I thought we could easily make 250 compounds a year that were not in the Kodak catalog. I was told that PPG was not interested; nobody could compete with Kodak. Yet today Kodak is no longer in the fine chemicals business. I felt that Kodak was not doing a good job and that chemists would be glad to have another source. So a lawyer friend and I set up a small company, Aldrich, very much a weekend operation with a capital of $500.

In the mid-1950s, when I visited customers, they would say, "Aldrich, you're the cheap people." Kodak was very expensive, and they didn't really care about their fine chemicals business. They even had an ad in which they admitted their service was poor. As a response, we published an ad saying "Please, bother us! We hope we never get so big that you can't talk to us."

You've built up a library of chemicals. Where is it now?

At Aldrich. It's close to 100,000 compounds. We have many famous chemists' research samples. For instance, Louis Fieser's, Bob Woodward's, Tadeus Reichstein's, you name them. A great many good things have come out of this library. Suppose that a medicinal chemist has an idea that a given bromolactone has some medicinal effect. He can contact Aldrich and get a computer printout of every bromolactone and order whichever he wants at $40 a sample.

Some years ago, a medicinal chemist in California, Dr. Summers, believed that acridines might help in the treatment of Alzheimer's patients. He asked for our list of acridines, ordered some of each, and then he came back wanting 100 grams, then a kilo, then 10 kilos of one particular acridine. Years earlier, a Viennese chemist, Dr. Pickholz, working in London, had said to me, "I've got this acridine, aminotetrahydroacridine; I think it's important in brain chemistry, I don't know how, but take a little, and we'll teach you how to make it." For years, there was little interest in it until Dr. Summers came along. The rest is history: THA was, I believe, the first drug licensed by the FDA in the treatment of Alzheimer's disease.

Should there be a universal library of all compounds?

If there were a universal library, it would be run by some government agency, and it would be much more cumbersome to get samples. With Aldrich, it's very simple. It used to be called the Alfred Bader Library or the ABC Library, but when I was thrown out of the company, they dropped the Alfred Bader. Even ABC was too close; it's now called the Sigma-Aldrich Library, but it's still the same. Sales now exceed 2 million dollars a year, with many thousands of samples going out.

What are the criteria for a compound to get into this library?

It must not be in the Aldrich catalog, and it must be clearly labeled. We don't analyze for the library, but we buy only from reputable researchers. Maybe 2 or 3 percent of the compounds are not what the researcher thought 20 or 30 years ago, and about 10% are not very pure.

You said you didn't want to be so big. Why then did you merge with Sigma?

It seemed to me in the late 1960s, early 1970s — but now I think I was mistaken — that organic chemistry had peaked. Woodward had synthesized strychnine; what more was there to do? This was before I really understood H. C. Brown's hydroboration. This was before Barry Sharpless's epoxidations. There's so much that has been done since. But I thought that most of the research funds would come to biochemistry and biomedical applications. I visited Sigma in St. Louis in the late 1960s and proposed a merger, and they almost threw me out physically. Then Sigma went public in 1972 and had some bad publicity and its stock dropped from 22 to 11. They were very thin in top management. At that point they listened to me, and the merger was very good for everybody. Only Dan Broida, the head of Sigma, was against the merger. He was very dictatorial, but immensely hard-working, and you could call him day and night, collect. He'd built Sigma sales to about 30 million dollars a year. Our combined sales then were about 45 million dollars. This year the sales of Sigma-Aldrich will be close to 1 billion dollars.

You are still a stockholder; does it give you any say in company matters?

I'm the largest individual stockholder, but it doesn't give me any say whatever.

For decades you've been visiting synthetic chemistry laboratories in the best universities. Can you share some of your experiences with us?

I don't see great differences among the countries where I used to go. Funding is very much more difficult in Britain than in either Germany or the United States. The laboratories and the equipment are very much better in Germany and in the United States than in Britain. If you compare 1995 with 1950, a good chemist nowadays could finish my Ph.D. work in a month; it took me a year and a half. There's so much more equipment around now. Then NMR was unknown, and there was no mass spectrometry. We had an infrared spectrometer at Harvard; it was a big instrument and was on the blink 20% of the time, but it worked. It was amazing to listen to Bob Woodward or Gilbert Stork interpreting the spectra.

Today, synthetic work can be done in so much smaller quantities than then. In our library of research chemicals, I can tell by looking at Woodward's students' bottles when the compound was made. In the 1940s and 1950s, you had gram quantities; in the early 1970s, you had 100-milligram quantities. Today 5 or 10 milligrams suffices.

What do you think about the image of chemistry?

There has been a tremendous change, much for the worse. If you had gone to Harvard Square in 1947 and asked people on the street, "What's your first reaction to chemistry?", they would have said, "vitamins, new plastics, new drugs." Do it today and they answer, "pollution and carcinogenic chemicals." We have not done a good job pointing out how much chemistry does. This is terrible — not for chemistry but for the world. When young, brilliant students see, for instance, the world sinking into a cesspool of toxic chemicals on the cover of *Time* magazine, why should they want to become chemists? We rely on the media, and the media have done badly. I remember a headline in the *Milwaukee Journal* saying "Benzene Found in a Well Near Chemical Company." But the details in the text were that two parts per billion of benzene had been found. This is crazy. It's next to nothing. Our analytical ability has changed so that today we can show that everything is everywhere.

It's also true that there have been chemical industries that have done very wrong things, and then tried to hide them. The word *chemistry* has become an ugly word, but everything is chemical.

To how many addresses is Aldrichimica Acta *delivered?*

It's free of charge, of course, and it goes to about 200,000 scientists worldwide. We have had very good papers in it. Over the years, we have established a number of awards in America, Canada, Britain, and the Czech Republic. We used to ask the award winners to submit papers on their award addresses.

Is there any connection between your collecting chemicals and collecting paintings of old masters?

These are completely separate matters. Take the chemicals first. No chemist wants to throw out his research samples. Unfortunately, when a professor dies or retires, these are often discarded. I could tell you story after story. For instance, Professor Haworth's compounds at Birmingham — I feel so unhappy thinking about this every time I go to Birmingham. Here were thousands of his crystalline sugars, so difficult to crystallize. Some bureaucrat got worried, and they put many of these samples together into biscuit tins, poured in cement on them, and threw them out. Imagine the idiocy of throwing out these crystals because some fool was afraid of these sugars!

Some years ago we went to Ames, Iowa, where all the Henry Gilman samples were stored in one big room, thousands of them. A company wanted $8000 to remove them. I offered to pay for the chemicals and send a truck to pick them up. The only mistake we made was that the truck we sent wasn't big enough so we had to send it twice. There were 20,000 bottles. We immediately discarded about 12,000 because these were chemicals which Aldrich listed. There were another few thousand where the labels had fallen off or the materials had decomposed. We were still left with 3000 chemicals and we published a little yellow catalog of the Gilman samples, just as we published a blue catalog of Woodward's samples.

About collecting paintings, when I was a kid, I lived in a home surrounded by paintings, but they were modern Austrian, and I didn't like them. In time, I realized that I liked Dutch historical paintings and Dutch portraits best. In the early 1950s, I started buying pictures that I liked and that I could afford. Today I'm a dealer as well as a collector. I buy about 200 paintings a year. Last month I bought a very fine Rembrandt and yesterday I bought a beautiful Ter Borch, probably the best outside of a museum. But I'm trying to have a nonelitist gallery. I have many pictures here from a hundred dollars up. On the other hand, the Rembrandt cost several million dollars, but we never bring such expensive paintings to

Milwaukee. I spend about a third of my time buying and selling paintings, another third writing, and the remaining third working with chemical companies. I invest in them, consult and advise them. I'm trying to find customers for them and suggesting new products. I just bought a 10% interest in a large but very ailing English company, called Anglo United, which owns Coalite. I know Coalite Chemical very well, and I think I may be able to help them.

Do you now do 100% what you like to do?

Yes. I'm far happier today doing what I'm doing than I was four years ago. Then I had to work with the top management of Sigma-Aldrich in St. Louis. In the hundreds of meetings in hundreds of days in St. Louis, there was never a day that I really enjoyed myself there.

What do you miss from those days?

The most enjoyable part of my work was the 50% when my wife, Isabel, and I simply walked from lab to lab, talking to students, asking them, "What are you making that Aldrich should be offering? What do you need? What should we add to our Catalog?" My last such visit was early in 1992, before I was thrown out of the company, and I really miss those visits. The most productive visits were to the top schools. We realized early on that 90% of the best research is done in 10% of the universities.

We would spend a most enjoyable day on the sixth floor of Chandler at Columbia University, for example, talking to each of the Gilbert Stork students, learning what they were doing and what they needed. Then a year or two later, I'd meet the same students as post-docs at the ETH or Cambridge, and two years later, they'd have their first Assistant Professorship at an American university. It was a wonderful give-and-take. I was far better at that than I ever was as research chemist.

You give a great emphasis in your autobiography on your being Jewish. But you are Jewish by choice. You might have become Catholic like your mother's side of the family.

I was adopted by my father's sister, a Jewess. I also studied Judaism. At the time of the Anschluß, I was a boy of 14 in Vienna and saw all the propaganda of how terrible Jews were. I asked myself the question, "What if these people are right?" So I felt I had to study Judaism, and it became clear to me that the Nazis weren't right. I'm a convinced Jew and my two wives, Danny and Isabel, who both came from religious Protestant backgrounds, became convinced Jewesses.

Coming back to the question about my Catholic mother, I hardly knew her. She would come once a month to the house, and occasionally she would say, "Bobby [as I was called then], anyone who could be a Catholic and isn't is going to go to Hell." I didn't really know her as a person.

In front of a painting of his "Mother," his father's sister, Gisela Reich, née Bader, painted by Tom von Dreger in 1917 (photograph by I. Hargittai).

I very much wanted to visit her in October of 1947. She was then very sick and I asked Louis Fieser for three weeks off. "No," he said, "if you go, you'll lose your fellowship. Go next Summer." Sadly, my mother died in April of 1948.

Referring to your two interests, what do you find common to chemistry and art?

First of all, there is the restoration of paintings. This relies heavily on the knowledge of chemistry. Second, for hundreds of years, artists have tried to create something beautiful. Many artists today are desperately trying to create something new. They often use chemistry in their search for new materials.

People often ask me whether I restore my own paintings. I never do. When I buy a dirty old painting, I determine whether there is much nicotine on it. Often just saliva will take a lot of dirt off. But I stop at that point and work with two very competent conservators. I have just acquired a painting which may be a late Rembrandt, with a terrible varnish. When it is cleaned, we'll know better.

The conservators have to choose the right solvents, which usually take off the varnishes easily. The cleaning has to be done very carefully, and it may take a very long time. Often there is overpaint that has to be removed. Today we realize that all restoration should be reversible, but it wasn't so 200 years ago. Today competent inpainting is done over an easily removed varnish. Years ago, restorers often painted oil on oil. Almost all paints were based on linseed oil, which takes 300 or 400 years to polymerize completely by oxidation. If a painting from the 16th century was overpainted in the 17th century, that overpaint has also polymerized and is very difficult to remove.

When I say overpaint, I'm not speaking of painting a new picture over another one. In the past, paintings were not so valuable, and possible damages and paint losses may have been corrected in an unprofessional manner. The restorers may have just been house painters, who overpainted much more than was needed. I have seen a little book, published in London in 1752, entitled *How to Be a Butler.* It advises that if the master has a dirty painting, the butler should take a bucket of wood alcohol (i.e., methanol) and a sponge. This would take off the dirt, but also the varnish and the top of the paint layer. There are so many paintings that have been skinned because the top layers have been removed.

I like nothing better than to find a 17th century painting to which nothing has been done. After 400 years, the polymerized paint film is practically indestructible, so the paint film holds the canvas together. But most paintings have been relined because over the years the canvas became as brittle as old paper. Restorers came along and backed the original canvas with a new canvas, putting glue in between. They'd put the painting face down, add a layer of glue, and the new canvas. Then they would use a hot iron to make sure that the new canvas adhered to the old. But when you do that to a thickly painted picture, the paint film gets flattened out, and this is terrible. Today, of course, competent restorers, if they have to reline, do it on a vacuum table so that the paint film doesn't get flattened out, but it is preferable to avoid relining altogether.

What was the biggest mistake you made in the building of Aldrich and Sigma-Aldrich?

We had some great chemists, but I didn't treat them well enough financially, and they left. We started the company in 1951 with a capital of $500, and I had to be penny-pinching. This was nothing new to me. I had never had any money, and even now I am still careful on a personal level. I realize that I should have been more generous with our best people.

Was there any question in our conversation that you'd like to return to?

Yes, concerning Aldrich. Clearly, my heart is with the company. I'm very proud of having founded Aldrich. Today research is different from the way it was in 1950 because Sigma-Aldrich supplies so many thousands of chemicals. We've made research very much easier and saved chemists of the world millions of hours that they can now devote to their own research rather than to making starting materials. I'm very proud of what I've done and of the many friends I've made around the world.

Added on February 23, 1998:

It is great fun — and instructive — to read what I said two-and-a-half years ago. Fundamentally, nothing much has changed. Isabel and I now divide our time four ways: working with small chemical companies — and actually being involved in the start-up of two — takes about one-third of our time.

Of course I am proud of the service that Sigma-Aldrich has provided to chemists, but I had really no idea quite how many friends we had among chemists worldwide. So I was staggered when I learned that I was elected one of the top 75 distinguished contributors to the Chemical Enterprise (*C&EN*, January 12, 1998), a surprise sweetened by my friends at Aldrich congratulating me in an ad — as daring as kind.

My fine arts business has expanded considerably — the Rembrandt I bought in 1995 now hangs in a museum in Aachen, and just last month I bought another beautiful Rembrandt portrait and also an exciting Rubens, both at Sotheby's in New York. An excellent art historian and dealer in New York, Dr. Otto Naumann, handles most of the important sales, so I am more involved in finding than in selling great art. And I still find three or four great paintings a year for our own collection.

There are still many invitations to give lectures, both on the history of chemistry and art. I am particularly proud of my work on Josef Loschmidt, who is finally being recognized as a great chemist.

We now also spend more time helping people, both the ablest and the neediest through bursaries, scholarships, fellowships and awards. We are trying to improve our inner cities and to help the traumatized, for instance, in Bosnia. Our biggest effort was to acquire Herstmonceux Castle for Queen's University. At first it created many problems for Queen's, but the International Study Centre is now running smoothly, and meeting hundreds of students there gives us such pleasure.

Our first granddaughter, Helena, is now three, and we are eagerly awaiting the arrival of more grandchildren shortly. Time flies — but what a time it is when you watch your grandchildren. There is such a foolish saying: time is money — time is life that none of us can buy.

Jacquelin K. Barton, 1999 (photograph by I. Hargittai).

11

Jacquelin K. Barton

Jacquelin K. Barton (b. 1952) is Arthur and Marian Hanisch Memorial Professor at the Division of Chemistry and Chemical Engineering of the California Institute of Technology. She graduated in 1974 and got her Ph.D. from Columbia University in 1978. Following a postdoctoral fellowship at Bell Labs and Yale and an Assistant Professorship at Hunter College, she was at Columbia University in 1983–1989 as Professor of Chemistry and Biological Sciences during the last three years there. She has been at Caltech since 1989. Her many distinctions include membership in the American Academy of Arts and Sciences and the American Philosophical Society, the American Chemical Society Award in Pure Chemistry, many other awards, medals, honorary degrees, and a large number of committee memberships in various organizations. In 2002, Dr. Barton was elected to the National Academy of Sciences of the U.S.A. In our conversation at Caltech on May 20, 1999,* my first question was about Dr. Barton's family background and initial interest in chemistry.

My family lived in New York City, and there were no scientists in the family. My father was born in New York City, and was a Supreme Court judge in New York State. I was very close to him. My mother was born and raised in Belgium, where my grandfather was in the diamond business. The family is Jewish. On the day the Germans invaded Belgium, my grandfather was in London on a business trip. My mother and her family

*This interview was originally published in *Chemical Heritage* **2002**, *20*(3), 6–9.

first went to the office to get the diamonds and then got into their car and traveled from Belgium through France to Spain, trading diamonds for gasoline, and there they got on an Indian ship, which took them to England.

Since I was very young I was always very interested in math and in art. I've always been interested in shapes, in structures. In my day young girls didn't go into science. I went to a girls' preparatory school in New York City. There young ladies didn't take chemistry but they took lots of courses in art, music, and languages. So I never took chemistry in high school. But I took a lot of math. It was at the very beginning of some co-education between the girls and boys school and so I was permitted to go to the boys' math classes. Then I went to Barnard College at Columbia University and I took my first chemistry course, which had also a lab, and I loved this. It was my feeling that chemistry gave me a chance to bring together abstract mathematical ideas and shapes and structure, and also have a touch of reality to it. In 1974, I went to graduate school at Columbia University and worked on metal complexes, particularly blue platinum complexes, with Steve Lippard. It was there that I became interested in metal-DNA chemistry.

What did you do after you got your Ph.D.?

I went to Bell Labs and worked with Bob Shulman, an NMR spectroscopist. He then moved to Yale. In this period his group was using NMR to examine cellular metabolism. As part of that effort, I used NMR to study how spores germinate. It was quite different from my graduate studies. But I really wanted to have my own lab. So I went to Hunter College, which is part of City University of New York. I wanted to be a teacher and to do research. And I was very much a New Yorker. So Hunter seemed like a perfect place for me. Of course, it was not a typical path to go as an Assistant Professor from Hunter to Columbia, where I went two years later, but you might notice that the paths of woman professors tend to be non-traditional. For me it was a wonderful thing being at Hunter. I firmly believe that no matter where you are, Hunter College or Caltech, the first couple of years you have to learn how to be a professor, how to teach courses, how to write grants, how to interact with your colleagues. I actually got my first NIH grant while I was at Hunter College.

Of course, Hunter College is not the same as it was decades before, but my colleagues were wonderful. There is a very wide distribution of students there, and some are very motivated. But the focus at Hunter has to be on teaching, with research being only a part of that effort. While I was at Hunter, in an environment where I didn't feel a lot of pressure, I learned a lot of the things one needs to learn at the beginning of one's academic career. But the other important thing I learned at Hunter was that I really did want to do top flight research. And that was difficult at Hunter. You had to teach a lot of courses, didn't have as many graduate students, and the graduate students weren't of the caliber that they are at Columbia or at Caltech. But it was being at Hunter that made me realize that that's what I wanted. So it was very valuable to me.

Then I moved to Columbia where I was an Assistant Professor, and as far as my colleagues were concerned, I was a new Assistant Professor starting out. The faculty was very supportive. And many of the faculty earlier had been my teachers, and my heroes. I got tenure in two years and became a full professor the year after. I moved out to Caltech in 1989.

While I was at Columbia on the faculty, Peter Dervan, later to be my husband, was at Caltech. He has been at Caltech for over 20 years. We met each other as members of the chemistry community. We sat on the same committees, and even at the beginning reviewed each other's papers. To some extent, in the nucleic acid field, he was probably originally a competitor. But really he was someone I looked up to because of his beautiful work. In 1989, I came to Caltech for a sabbatical. Then and now Caltech was an outstanding place to do chemistry. In 1990, Peter and I were married.

By then we both had well established, independent careers. There aren't many examples of spouses with independent careers in chemistry today at major research universities. But increasingly, there have been couples who both want to go into academic research. Some graduate students and postdocs look to us for some insight about how to do it, but we had the advantage of having first developed our independent careers, before we became a couple, which made it much easier.

I would like to ask you about your research.

When I was still at Columbia I was working on chiral metal complexes bound to DNA and I was also doing some studies in collaboration with Nick Turro, my wonderful colleague at Columbia, on the photophysical properties of ruthenium complexes bound to DNA. The complexes I was using in DNA binding studies could be considered as derivatives of those that had been used by Henry Taube in his classic studies of electron transfer reactions. Turro and I had a joint postdoc Vijay Kumar who did an interesting experiment one day of taking several of our metal complexes that bound to DNA and putting them together in a test tube and asking whether or not the electron transfer was enhanced in the presence of DNA? The observation was that in effect it was. That was our first paper. There are various ways you can think about it, and I thought of three possibilities that could lead to this enhancement. It could be a local concentration effect, namely that the molecules, because of their association with DNA, would increase their local concentration and that would give you the rate enhancements. A second possibility was that not only would they associate with DNA, increasing their concentration, but that they would diffuse more along the DNA helix than in the three-dimensional space of the bulk solvent. So the three-dimensional search would reduce, effectively, to a one-dimensional search, and this facilitated diffusion mechanism might account for the rate enhancement. In this respect you could think of DNA being some sort of a railroad track; that's actually what I expected to be

Jacquelin Barton with a model of DNA (photograph by I. Hargittai).

the explanation. The third possibility was instead thinking of the DNA helix as some sort of a wire, that is, that the metal complexes weren't moving towards each other and colliding but, instead, that the electron moved through the base-pair stack. We began to carry out experiments to look at those different possibilities. It looked more and more like the third possibility could not be easily eliminated. These experiments were in the late 1980s. During the same time our ability to synthesize oligonucleotides and to modify oligonucleotides with metal complexes became feasible in our lab. Then we carried out an experiment, which is heavily cited, where we took two metal complexes, intercalators which stack in DNA, as the donor and acceptor and appended them to either end of the oligonucleotide duplex. One intercalator, a luminescent ruthenium complex, was the excited state donor, and the other, a rhodium complex, served as the electron acceptor which could quench the ruthenium luminescence. We found that the ruthenium-modified DNA glowed, but with rhodium intercalated at the other end, the luminescence was actually quenched. So the result was consistent with long range electron transfer over 40 Å through DNA. Now we weren't the first by any stretch of imagination to be asking whether DNA was an efficient medium for electron transfer. If you just look at double helical DNA in terms of its chemical structure, the question is apparent. And in fact physicists, biologists as well as chemists had been debating, essentially since Watson and Crick, whether this π-stack array would facilitate electron transfer? If you look at solid state materials, similarly π-stacked, they tend to be conductive along that direction. But double helical DNA is a π-stack in solution, and the only one in solution known so far. DNA itself is stabilized to a great extent by this π-stacking, about 2 kilocalories per base-step. The stacking, or electronic interaction among the bases, is a major source of stability for the DNA duplex.

In the context of the ongoing protein electron transfer studies, at that point a far more mature field, our observation of fast electron transfer in DNA over 40 Å was probably outrageous, certainly controversial. Many did not focus on the distinction between protein and DNA as biopolymers, that the protein medium was largely sigma bonded while the DNA base pairs were π-stacked. There was also no precedence in terms of solution experiments for our results, yet I didn't think we were proposing anything outrageous because the characteristics of solid state π-stacked materials were similar to those of DNA, albeit dynamically.

To me the key to DNA electron transfer is this base-pair stacking. Our intercalators were also well stacked and so represented effective probes of this π-stack. In any case, it became important for us to do more kinds of experiments on more different systems, and so we made a lot of different DNA assemblies. We also wanted to examine whether stacking was a key important parameter. In one experiment, for example, we used an organic intercalator, an ethidium derivative on one end of the DNA duplex and attached one of our metal intercalators to the other end. We synthesized larger and larger duplexes to vary the distance, and again, the result was fast quenching, too fast for us to measure, and a very shallow distance dependence in the yield of quenching. With this same system, we carried out what I thought was a very useful experiment. We introduced a base mismatch in the center of the DNA duplex. The idea was that if electron transfer was sensitive to stacking, let's introduce a small and local perturbation in the stacked structure and see its effect. What we found was that, in the presence of the intervening mismatch, we lost the quenching. So this was an indication that stacking was critically important. Long-range electron transfer required a well-stacked assembly. The result also demonstrated that the path for electron transfer was through the base stack, which I think was an important point.

In what form did the controversy take shape?

Attacks at conferences, many articles written in *C&E News*, articles titled "DNA is not a wire." There were lots of papers with donors and acceptors unstacked where no efficient electron transfer was observed. And there were theoretical papers stating that what we were observing couldn't possibly be. At the same time we kept getting different results using a range of techniques where we were seeing this fast electron transfer.

Did this convince your opponents?

No. But that's OK because that kept on pushing us towards better and better experiments. I think that the last experiment that we published in *Science* this year may have made clear how exquisitely sensitive the electron transfer was to base-pair stacking. In this experiment we no longer appended metal complex intercalators but we simply looked at electron transfer from one modified base to another modified base. In this experiment we used two modified adenines as our fluorescent excited electron acceptor to oxidize guanines. The modified adenines were very similar in structure, very similar in redox characteristics, very similar in energetics. But when they were incorporated into DNA, one was well-stacked in the helix and one was poorly stacked. The difference in electron transfer rates and the dependence of the rates on distance for these modified adenines was very large. Just a small perturbation in stacking would allow or destroy coupling along the helix. This stacking perturbation was enough to turn on or turn off the electron transfer through DNA. Frankly, if I had known when we had done those first experiments with Vijay Kumar how exquisitely sensitive this was going to be, I'm not sure if I would've gone down that road, that is, to think in advance that we would have to be able to come up with experiments which would depend upon such subtleties in structure. So I think that all the great debate really did push us towards better and better experiments to try to get to this issue. The fact that my training and my perspective was in nucleic acid chemistry also allowed me to think about the structure and dynamics of DNA as the critical element, which might be governing and modulating the electron transfer. This is where we are now.

What are the possible consequences of your findings?

There are two things to think about. The first is biosensors, utilizing the sensitivity of electron transfer in DNA to mismatches in its base-pair stacking suggests a way to detect mutations in DNA. Maybe DNA might actually be exploited as a nanowire. The other aspect lies in this whole field of radiation biology. DNA is not an unimportant polymer; it is our genetic material. If we have radicals that impinge upon DNA at a given site, then those radicals could migrate through the DNA and cause damage at a distant site. We have started investigating this issue, which I think is an important one. In fact a few years ago, using one of our chemical assemblies we were able to demonstrate such "damage from a distance" on the DNA helix. We have now seen damage as far as 200 Å away from the site where the electron hole is injected. And whether the damage at a distant site arises depends upon DNA sequence, structure, and can even be modulated by DNA-binding proteins.

When we eat free radicals can this damage our DNA?

Our DNA is constantly being damaged. But we also have quite effective DNA repair machinery. It turns out, however, that we don't actually understand, in 1999, how this repair machinery operates, how the damage is sensed. Maybe nature exploits DNA electron transfer chemistry. How these repair proteins find the damage in our DNA, how they recognize it, we don't yet understand. A lot of cancer, the genetic predisposition to some colon cancer and breast cancer in particular, appears to be associated with problems in correcting DNA damage that is mutations in the mismatch repair machinery. If this damage in DNA is not fixed it can be the beginning of cancer. The damage can arise even from a distant site. We did recently another experiment that showed that. The experiment was on the cover of *Science* and it showed that not only could you damage DNA from a distance but that you can also repair DNA from a distance. I'm an optimist and I see in all this not only the problems nature has but the problems for which nature finds solutions.

You are also a wife and mother and you showed me a drawing by your daughter. She is eight and a half. You live in a high-pressured environment ...

But I have a wonderful husband* who understands.

How do you manage things?

I'm incredibly lucky. I have a great family. Actually, we have a five-year calendar and need to be well organized. We also try to minimize the amount of traveling that we do. We say "no" a lot.

Do you discuss chemistry over dinner?

We don't discuss chemistry over dinner and you absolutely must put this in this interview. We do not bring our work home with us unless one of us does an absolutely terrific experiment. People imagine that we talk about chemistry all the time but we don't at all. We like to play, we like to go swimming, we like to go to the beach, we like to travel as a family. We separate our life at work from our life at home. Every once in a while Peter and I go to a meeting together (when my mother comes from New York to take care of our daughter) and then we hear each other talk about our work. And we're interested in it because we don't usually talk about it at home. Sometimes we ask each other a question at the meeting and people look at us and think we are a little crazy.

---

*The bioorganic chemist Peter Dervan at Caltech.

You have had great visibility. You must be in great demand. There are not many woman professors of chemistry in places like Caltech.

I am often asked about the situation of women in science, but I am not a sociologist, I'm just one woman-scientist, and I try to do my best. I have had disadvantages and advantages associated with being a woman scientist. I really believe that the best thing that I can do for women scientists is to do good research. Really even more importantly what I can do is to train outstanding women in my laboratory. Then they too can become woman professors and they will be on all the committees and they will be more visible, and then I can get the work done that I have to do. Actually I am very proud that there are now many women chemists who have come from my laboratory who have become professors around the country, Anna Marie Pyle at Columbia, Cathy Murphy at the University of South Carolina, Sheila David at Utah are just a few; there is beginning to be quite a bunch. Really, when I am in a situation where I am asked about being a woman in science, what I try to get across is that I have a great job. You can be a woman scientist in a major research institution and also be a person with a family and be happy.

Ad Bax, 1998 (photograph by I. Hargittai).

12

Ad Bax

Ad Bax (b. 1956, in Holland) is Chief of the Section on Biophysical NMR Spectroscopy in the Laboratory of Chemical Physics, National Institutes of Diabetes and Digestive and Kidney Diseases, National Institutes of Health (NIH), in Bethesda, Maryland. He received his undergraduate degree in 1978 and Ph.D. in 1981, both at Delft University of Technology in The Netherlands. He spent much of his time during his doctoral studies at the Physical Chemistry Laboratory of Oxford University in England, working under Ray Freeman. He was a Postdoctoral Associate in the National Solid State NMR Facility at the Department of Chemistry, Colorado State University at Fort Collins before moving to NIH in 1983.

Dr. Bax has received a number of distinctions, including Maryland's Outstanding Young Scientist, the Young Investigator Award of the Protein Society, the Gold Medal of the Royal Dutch Chemical Society, the EAS Award in Magnetic Resonance, and the Bijvoet Medal. He was elected Corresponding Member of the Royal Netherlands Academy of Sciences in 1994.

With 21,655 citations, Ad Bax was ranked first on a list compiled by the Institute for Scientific Information, which surveyed the number of citations received for work published in chemistry over the period 1981–1997. He was followed by J. A. Pople (14,044), P. v. R. Schleyer (13,559), R. R. Ernst (13,069), and G. M. Whitesides (12,310). The beginning of the time period examined coincided with the beginning of Dr. Bax's scientific career.

I recorded our conversation in Ad Bax's office at NIH on April 25, 1998, and the transcripts were subsequently augmented with references by Dr. Bax.*

You are the most cited chemist today and by a large margin.

It's very nice that people cite my work but the number of citations is only one marker of how successful a scientist is. Magnetic resonance has really taken off over the last 20 years, and it's used by thousands of chemists every day for a wide range of purposes.

It's really remarkable how the field keeps on evolving. First there was the discovery of the chemical shift and J couplings. Then came Fourier transform NMR, which made ^{13}C accessible. Then there was two-dimensional (2D) NMR, which together with superconducting magnets made it possible to look at much more complex molecules. At the same time, magnetic resonance imaging took off. Now we have three- and even four-dimensional NMR and the possibility to easily enrich proteins and nucleic acids with ^{13}C and ^{15}N. This makes it possible to study the three-dimensional structure of such macromolecules. Some of the very recent work on very weakly aligned molecules may turn into yet another mini-revolution [Tjandra, N.; Bax, A. *Science* **1997**, *278*, 1111].

So, NMR has been an incredibly fertile field in which many people have made critical contributions since its original discovery, foremost of all Richard Ernst. They developed NMR from an interesting technique for physicists into an extremely powerful tool, applicable in all areas of chemistry.

The Bijvoet Medal (photograph by I. Hargittai).

*This interview was originally published in *The Chemical Intelligencer* **1999**, 5(1), 6–11

I've been fortunate enough to have made some interesting contributions in the later stages of this development. Several of these have become very popular, and the citations reflect that others find these experiments useful. In retrospect, many of these popular experiments really were rather straightforward, whereas some of our more challenging work has not become very popular yet.

Would you care to mention an example of each?

One of these popular experiments, for example, is the homonuclear Hartmann-Hahn experiment. It is used widely in both small molecule and protein NMR for proton resonance assignment purposes. We call it HOHAHA [Bax, A.; Davis, D. G. *J. Magn. Reson.* **1985**, *63*, 207]. It turned out that Richard Ernst had developed a similar experiment called TOCSY, well before us [Braunschweiler, L.; Ernst, R. R. *J. Magn. Reson.* **1983**, *53*, 521]. His theory for TOCSY is fully applicable to our experiment. The original TOCSY experiment didn't work very well for technical reasons and the technique had not caught on.

I must admit that I was only remotely familiar with this work when my associate, Donald Davis, accidentally discovered some persistent artifacts in an experiment now known as rotating frame Overhauser spectroscopy. These artifacts showed magnetization transfer between J-coupled spins and to me the explanation for this transfer looked very similar to the so-called Hartmann-Hahn magnetization transfer, used in solid state NMR to enhance ^{13}C sensitivity. Once the analogy was clear, it was quite straightforward to improve the magnetization transfer process and to make it work really well.

Then, as an example of an "unpopular" experiment, which I like a lot, I could mention AMNESIA. This stands for Audio-Modulated Nutation for Enhanced Spin InterAction [Grzesiek, S.; Bax, A. *J. Am. Chem. Soc.* **1995**, *117*, 6527]. This is a rather unusual type of magnetic resonance experiment where the spins are quantized along a radio-frequency field, and audio-frequency pulses are applied to move the spins around. It's like doing NMR at 10 kHz, but using a 500-MHz magnet for it. We used it to get more efficient HOHAHA magnetization transfer between aromatic and aliphatic ^{13}C nuclei in proteins. I think it's a really neat experiment and it actually works quite well. Unfortunately, it's a real pain to set it up correctly with commercial spectrometers. Also, we probably chose the wrong acronym and nobody remembers it.

Another interesting set of experiments which have not caught on yet are aimed at measuring ^{15}N and ^{13}C chemical shift anisotropy (CSA). We can make measurements on the CSA tensor in isotropic solution and we're trying to correlate our results with values of adjacent torsion angles. Of course, we're also calculating the CSA tensor by *ab initio* methods, which appears to show remarkable agreement with the experiment. Unfortunately, the relation between CSA and structure turns out to be rather complex, which may prevent it from becoming widely used.

How do you change the angles of torsion in the experiment?

The calculations are done on a small fragment of the protein molecule. In the experiment, the whole protein molecule is present. If there are a hundred amino acids, we have one hundred sets of torsion angles. So, one experiment on a protein gives us a hundred different data points. Then we compare these data with results of the *ab initio* calculations. We can monitor the effects of hydrogen bonding, side chains, nearby charges, and see whether we get an agreement with the calculated predictions. The results are very promising for the backbone α-carbons in polypeptides but much less so for carbonyls, for example. This may have to do with the level of theory used in our calculations.

How much interaction do you have with Richard Ernst?

We meet maybe twice a year at various meetings and I try to visit his lab whenever I make it to Switzerland. It's really sad that the Swiss academic system forces him to retire, very prematurely I'd say. He's been doing fantastic novel experiments for well over 30 years now and there are no signs of him slowing down. His enthusiasm really rubs off on the whole NMR community, so I hope he will stay active for a long time to come.

NIH is different from a university setting. Are you happy with it?

Very. I am extremely well supported here at NIH. I'm left to do research with virtually no other obligations and have a great group of colleagues.

Why did you leave Holland?

It was hard to find a decent job in The Netherlands. I kept applying and they kept turning me down, so I came to the U.S. for a postdoctoral position. Then, while I was a post-doc, I got several job offers, all in the U.S., and accepted the one from NIH. I originally thought of it as a temporary position and that I would one day return to Europe. But more recently, when we had the option to go back, we first went for an extended visit with our two small children. They went into culture shock, having to adapt to European rules and regulations. Besides that, it rained virtually every day. Although these were really only very minor considerations, they dampened the enthusiasm and made me think harder about leaving a place where my work has gone so well.

My research requires a very strong infrastructure. It would be hard for me to work just by myself without other groups around me. What's ideal about my setup here at NIH is that there are four other NMR groups and we're interacting all the time. I might never have started working on protein NMR had it not been for Dennis Torchia here at NIH. He was doing protein NMR long before I was and he convinced me to apply my small molecule experiments to proteins. Since then, NIH has expanded its protein NMR program and we now also have sections headed by Marius Clore, Angela Gronenborn, and Robert Tycko. These are all very smart and successful scientists and terrific colleagues to interact with. So we openly discuss our ideas and new experiments and get immediate feedback. For example, if I develop a new experiment, the other groups immediately try it out on the proteins they are working on, which helps me, and them. At a university such intense collaborations are probably not easily possible. Besides, one would never have five groups with such an extensive overlap in their interests.

Can you keep up this level of production in the long run? Does your number one status make you nervous?

I don't look at the *Citation Index*, although in the back of my mind it's of course a nice feeling. But what I'd really like to see is that NMR becomes even more useful. I'd like to do things for the field, for chemistry, and for biomedical research in particular.

What are your more immediate aspirations?

I would like to improve the quality of the protein structures that we determine, increase the resolution and accuracy. We're primarily interested in the shape and surface of these molecules. In order to understand the function of a protein, in order to be able to inhibit its function, we must have a good idea about its shape and surface. At present, an NMR structure is frequently of lower resolution than an X-ray structure. But the NMR information is important and is often unavailable from X-ray crystallographic analysis, especially concerning the dynamic aspects of the structure. It's also an important advantage that we work in solution in the native state of the protein, in its natural environment, with the right amount of salts and other components, just as in the real organism. For small globular proteins the crystalline and solution structures are virtually identical. For many other proteins this is not the case.

An interesting example is calmodulin, which consists of two domains. The two domains are connected by a long α-helical linker. When the structure was first determined by X-ray crystallography in 1984, it showed an intriguing dumbbell shape. It was a controversial result because many questioned whether the linker helix really existed in solution too or was only an artifact of crystallization. We used calmodulin to develop many of what is known today as triple-resonance NMR techniques [Ikura, M.; Kay, L. E.; Bax, A. *Biochemistry* **1990**, *29*, 4659]. These techniques have allowed us to assign all the resonances in the two domains and determine their structure. We found that the individual domains had the same structures as determined by X-ray diffraction in the crystal. However, for the α-helical linker, we didn't find the helical character close to the center of what was supposed to be this helix. Moreover, we found that the rotational diffusion of the protein in the solution did not correspond to a rigid dumbbell shape. The rotational diffusion of each of the two domains was almost isotropic, as if they did not feel each other's presence. This told us that the linker must be extremely flexible [Barbato, G.; Ikura, M.; Kay, L. E.; Pastor, R. W.; Bax, A. *Biochemistry* **1992**, *31*, 5269].

We studied this effect as a function of temperature and found that the flexibility of this helix was independent of temperature. This flexibility turns out to be critical to the function of this protein. Calmodulin's role is to activate over two dozen different enzymes in a calcium-dependent manner. It does so by binding with very high affinity and specificity to a region on the target enzyme. So the question is, how can one protein bind specifically to two dozen different targets and leave all other proteins alone? Calmodulin can do this because it consists of the two lobes, like having two hands that are connected by this flexible linker. A single "hand" binds the target with relatively low affinity, and only if the second hand finds something "to grab on to" will there be tight binding and activation of the target enzyme. The two "hands" can change their relative positions, and they don't even have very high site specificity when acting separately. But together they are exceptionally specific. Thus, the flexibility of the linker is key to the function of this protein. Crystallographers still think of this protein that the α-helix has to unwind before the protein can bind the target, but the real story is that there is no α-helix in solution. Or if there is, it has a very low population.

Another example of why NMR is a powerful complementary technique to X-ray crystallography is that calmodulin can only bind its target in the presence of calcium. The question was how calcium can change the conformation of the protein. For the comparison you need both structures, one in the presence of calcium and the other in its absence. However, it proved impossible to crystallize this protein in the absence of calcium. By NMR, it was rather straightforward to determine the conformation of the protein in the apo state [Kuboniwa, H; Tjandra, N; Grzesiek, H; Klee, C. B.; Bax, A. *Nature Struct. Biol.* **1995**, *2*, 768]. What happens is that calmodulin's hands are closed in the absence of calcium, they are wide open in the presence of calcium. And when they are open, that is when they can grab the target peptide. It's a beautiful mechanism, which gives us insight into how the metal acts as a messenger which regulates proteins inside the cell. It's also a nice example of what NMR can do.

Have you ever considered working for a biotechnology company?

I do some consulting for such companies, and I probably could make more money working there full time. But it would be very difficult to maintain my independence in such a situation. Because of the pressure to perform and produce in a given project, the scientific freedom is compromised in such an environment. It's especially difficult to work on more basic, long-term projects. And as I have already mentioned, this is where one of my main interests lies, to make protein structure determination more robust and faster.

Another of such long-term interests is the study of the dynamic aspects of the protein molecules, the amplitude of internal motions and the asymmetry of the motion. The dynamic properties of residues near the active site of an enzyme, for example, are absolutely critical for it to carry out its function. NMR is in a unique position to study these problems at an exquisite level of detail. So far, we have not really reached this ultimate goal yet, but we're working on new techniques that will hopefully provide us detailed insight into what exactly is going on within the protein during a reaction.

Yet another project is aimed at extending the size limits of structures that we can determine by magnetic resonance. Originally, before the mid-1980s, when I was getting into the field, the limit was about 10,000 Daltons. Today we're able to go up to 40 kilodaltons. Ultimately, we would like to apply NMR to integral membrane proteins. They're anchored in the cell membrane. These proteins are involved in signal transduction, transferring the signal from the outside of a cell to the machinery inside it. All of them are important drug targets. This is how one can control what the cell does.

These membrane proteins are very large, and since they are not soluble in water, we can't study them in solution. At the same time they are also very difficult to crystallize. The first one to be crystallized was the photosynthetic reaction center by Deisenhofer, Huber, and Michel, and they got the Nobel Prize for it. Since then, there have been only about half a dozen more membrane proteins that have been crystallized and whose structure has been solved. Considering how many there are and how important they are, it's a real pity there is so little detailed structural information about them. That's one of the ultimate goals we're after. We have several ideas that need to be explored, but this will be another one of those long-term efforts if it is ever to be successful. However, if we get it to work it would be a fantastic breakthrough. It could help in designing novel, more specific drugs with less side effects, for example.

Please, tell us about your family background.

I grew up on a farm in Holland. We were six children. My father was a reluctant farmer. He inherited the farm and since his elder brother had studied law, my father had to become a farmer. He was very unhappy with it because he had always wanted to study. He taught himself Latin and won a contest in creative writing; the prize was a trip to Canada. He encouraged all his children to study whatever they wanted to study. He was a very positive influence, but he died nearly 25 years ago.

Your present family?

My wife is Ingrid Pufahl. We met back in Oxford during my doctoral studies. She completed her Ph.D. in sociolinguistics here in the U.S.A. Her expertise is analysis of language, the relationship between sociology and language. She can explain how and why the same phrase may mean different things to different people, how your social background may change the meaning of what you are saying. Ingrid's thesis was on the comparison of television news in the United States and Germany. One appears to be a show, the other tries to look more like a respectable objective news program, but it turns out not to be so objective either. We have two children; Nicolas is 8 and Christina is 6.

Any heroes?

There are many people whom I admire enormously. I've already mentioned Richard Ernst. Then there is Nicolaas Bloembergen, born in The Netherlands. This man is also phenomenal. Not only did he get the Nobel Prize for nonlinear optics, but as a graduate student he also developed NMR relaxation theory, pretty much from scratch. Considering how little was known at

The Bax family in 1998 (photograph by I. Hargittai).

the time, it's incredible how this man worked out this complex theory so thoroughly. I could also mention Richard Zare. Besides obviously being a brilliant scientist, he has sacrificed a lot of this time to help advance science and engineering. He's a tremendous advocate for science.

Then, outside of science, I still have my sports heroes, particularly in rowing. They are the German Peter Kolbe, the most famous sculler in the mid-1970s, and his main rival the Finn Petti Karpinnen. Both of them have retired from rowing but I used to idolize them and they've both had a big influence on my non-scientific life. I'm an active sculler myself. This is rowing in a boat, which is very narrow, about 10 inches wide, and 27 feet long. Every morning at 5:15, I'm out on the river here, the Potomac, with my friends. We don't talk, we just row. Crazy as it may sound, it's a very enjoyable thing to do.

Do you read?

Very little, besides the scientific literature.

TV?

About 10 minutes of the news, if I'm lucky. Maybe a movie on Saturday night. I try to do first-rate science and to do sports at a competitive level, and with two little children there's simply no time left for anything else at the end of the day.

Donald J. Cram, 1995 (photograph by I. Hargittai).

13

Donald J. Cram

Donald J. Cram (1919–2001) was Professor Emeritus at the Department of Chemistry and Biochemistry of the University of California at Los Angeles (UCLA) at the time of the interview. Donald Cram, Jean-Marie Lehn (see, interview in this Volume), and Charles J. Pedersen (1904–1989) received jointly the 1987 chemistry Nobel Prize "for their development and use of molecules with structure-specific interactions with high selectivity." I visited Professor Cram and his wife, Dr. Jane M. Cram, on April 1, 1995, in their home in Desert Palm, California, which is about a two and a half hour drive from Los Angeles. Their home is beautiful and spacious in a well-guarded and walled community of about a thousand homes. We were sitting on a terrace between the house and the swimming pool, and my Dictaphone recorded not only our conversation but a lot of birds singing in the background. It was a beautiful day, and we proceeded unhurriedly from topic to topic.*

You lost your father at a very early age, and this had a profound influence on you, according to your autobiographical notes.

My father died when I was not quite four years old. He had been a successful lawyer in Canada, who emigrated with his family of my mother and three girls to the U.S. to manage a citrus venture he and friends had invested in, in Florida. The venture failed and the family moved to Vermont, where

*This interview was originally published in *The Chemical Intelligencer* **1996**, *2*(1), 6–16

my father tried to become a farmer, about the time I was born. Shortly thereafter he died, leaving my mother with four girls and me, the oldest child being 13 years old. The absence of any male in or close to the family led me from an early age to seek a father image, first from the many books I read, and later in the teachers and scientists I encountered. I was in my mid-thirties before an inner core of values was in place, and my character was formed. It was a composite of what I had admired in people I had read about or had come to know. Possibly even of greater importance were my many encounters with people who illustrated what I did not want to resemble. In retrospect, I judge that what are ordinarily considered to be misfortunes provided me with the advantages of being able to test myself against circumstances and to grow in confidence, skill, and judgment with each encounter.

Donald Cram's mother with her children (courtesy of Donald Cram).

How about your mother?

She was a very determined person, who taught me the right things when I was little. She was a rebel. She was raised in a strict Mennonite faith, which she rejected and escaped from by marriage at an early age. She warmly embraced my father's values and culture. She kept our family together until I was 16 after which it dissolved and I was self-supporting. She introduced us to literature and music early on. She and my older sisters would read to me the first parts of books, but as soon as they became exciting, they would stop. For me to find out what happened in the stories, I had to learn to read, and did so very early. She arranged to barter my talents at mowing lawns and emptying ashes for music lessons (violin and piano). She bartered my labor for everything from food to dental care. These experiences introduced me to the marketplace and provided me with lessons in homespun economics.

At some point in your studies, you took your first course in chemistry and taught yourself solid geometry.

My high school chemistry teacher, a Mr. Poor, fully lived up to his name, both as a teacher and in his knowledge of chemistry. The textbook was good and had many good problems in it. After working all of them, I decided to become a chemist of some kind. No course in solid geometry was available, but my math teacher (the soccer coach) bought me a fine textbook on the subject whose many problems I battled with and solved without help, which wasn't available. Then I took the New York State Regents Examination in the subject and obtained a grade of 94. Throughout my career, I found textbooks to be much clearer than teachers. Books, experiences, and bad examples have been my best instructors. These early experiences coupled with my mother's telling me that if I tried hard enough I could accomplish anything, became center pieces in my attitude toward life. As I developed, almost everything became a challenge. The educational system first set goals, but later I set them for myself.

You did your Ph.D. work in 18 months.

There were special circumstances that made this possible. Rollins College in Florida generously gave me a four-year full scholarship in 1937, in the midst of the Great Depression. My most important professor there, Guy Waddington, after my first chemistry course with him, told me I might make a good chemist for industry, but he doubted I could become a research professor in a leading university. The word "research" had a magic ring for me. It stood for every day being different, the antithesis of repetition, which had been common to the many jobs I had held up to that time. I hated and still hate repetition. The idea of being a research professor sounded as if you could do your own research and be your own boss, an idea I treasured. I resolved to become a research professor of chemistry at a major university. The prerequisite for this was to get my Ph.D. (my job-hunting license) from Harvard. After my 1941 graduation from Rollins, I applied for a teaching assistantship at 17 universities and was accepted at only 3, the best of which was the University of Nebraska. After 10 months there, I left in 1942 with an M.S. degree and a good thesis, which resulted in my first two research publications. I had planned to go directly from there to Harvard, but instead, in 1942, I joined Merck and Co. at the behest of my draft board.

How did it happen that you were not inducted into the armed services?

At Rollins, I took the Civilian Pilot Training Program offered by the government and obtained an airplane pilot's license in 1939. I had just landed a job for the summer in New York City when a government offer arrived for me to return to Florida to take an advanced course that would have led to service as a flyer in World War II. Badly needing the money, I decided to keep the job, which consisted of selling National Biscuit Company crackers on the east side of Manhattan from about 78th Street to 130th Street, right through the middle of Harlem. I got a very good course in ethnics, the differences between city folk and country folk, the differences between populations from various countries. I lost 40 pounds, and I probably learned more sociology then than all the rest of my life put together. That was very good for me. In fact, a good portion of my early life was pretty much learning what I didn't want to do, what I didn't want to be. During this time I learned the difference between being "street smart" and "school smart." During summers, I lived with my mother on Morningside Heights overlooking Harlem, the result of which was that my draft board was in Harlem. My draft board encouraged me to continue my education through my master's degree and then to take a job in chemistry to aid the war effort. Accordingly, I joined Merck and worked on the penicillin program. During my three years there, I not only learned laboratory technique, but spent the little spare time available solving textbook problems in all fields of chemistry to prepare myself for a Ph.D. program. I also conducted process research on the riboflavin (vitamin B2) total synthesis.

One of the most exciting experiences at Merck involved the scaling-up of the reductive amination of ribose with an aromatic amine, hydrogen gas, and Raney nickel catalyst. The key to getting a good yield was to get the reaction over within four to five minutes, which meant the catalyst had to be both plentiful and very active, and the stirring very rapid. The test for active catalyst was to see if it *sparked* when dried on a piece of filter paper. I headed a team of six workmen who carried out the reaction involving 50 pounds of ribose worth about $150,000 (a lot of money in 1943), about 50 pounds of superactive catalyst, an equivalent of 3,4-dimethylaniline, a large open stainless steel kettle, two lightning stirrers, six large high pressure tanks of hydrogen gas, and about 400 liters of methanol solvent. To avoid blowing up a pilot plant in case of accident (air plus hydrogen plus catalyst spells trouble), we ran the reaction in an open shed between buildings on a windy winter day. With both stirrers going, one workman added the ribose, another added the amine, and I added the catalyst (under methanol) at the same time as the other four workmen fully opened the hydrogen tank valves feeding the gases through hoses to the bottom of the kettle. The reaction was over in four minutes, and we obtained an excellent yield, and a fine mixed feeling of relief and accomplishment.

Max Tishler was my boss from 1941–1944. He was tough-minded, a fine scientist, and an admirable person. After the war was over, at my request, he arranged for me to get a teaching assistantship at Harvard.

How did Harvard compare with Nebraska and Rollins?

I was better prepared for Harvard than Harvard was prepared for me. Some of their postwar staff were missing, and they were just getting organized. My first two days there, I took four qualifying examinations and passed them, which allowed me to plunge into research and a minimum of advanced, stimulating courses. Bob Woodward's and Paul Bartlett's courses were particularly exciting. My year at Nebraska and three years at Merck made Harvard easy. I outworked everyone else and had a good thesis completed in 18 months. Perhaps the thing that most impressed my fellow students at Harvard was my setting up a carbon-hydrogen combustion analysis train for my over 70 new compounds, because it was faster and cheaper to do them myself. My most negative impressions of Harvard were my $800 personally paid storeroom bill and the administration's attempt to charge me rent for use of the laboratory during vacations and intervals between semesters.

What brought you to UCLA?

Having lived in New England, Florida, New York and the mid-west, I wanted to sample the west coast, but needed the money to get there. While at Harvard, I had become friends with John D. Roberts, who was one of the first people to get a Ph.D. from UCLA. He had accepted a teaching position at MIT after his postdoc year at Harvard. Upon my questioning him about the differences between UCLA, USC (University of Southern California) and Caltech and their relative proximity to the ocean, he suggested and agreed to help me get a temporary and part-time teaching position at UCLA. He also offered me a four-month postdoctoral stay at MIT so I could finance my move westward. By the middle of 1948, UCLA had offered me a tenure track position, and I have been at UCLA ever since.

You moved from a New England upbringing to California and have become very Californian.

I came from Vermont which, because of its lack of resources, was shielded from the large European migrations and therefore was like nineteenth century New England. The contrast was stark. In California, everybody was selling something to someone else, but nobody seemed to be making anything. Vermont's economy to me was a mixture of barter and spending as little as possible. Yes, I have become a Californian in my nearly 50 years out here. I was recently asked to give a one-minute address to the combined Assembly and Senate of California. I thanked them for their sizable contribution to the estimated total costs of about $40,000,000 (today's dollars) of my research program over a 45-year period. In return, I supervised 120 Ph.D.'s and supported about 100 postdoctoral coworkers from 43 different countries and taught an estimated 10,000 undergraduates. I also told them it was impossible for me to imagine a more satisfying career.

Please, describe your early years at UCLA.

In 1947, UCLA was a relatively small and young institution with superb prospects for growth. The chemistry department without exception was forward-looking and congenial. Saul Winstein, a physical organic chemist, seven years my senior, was its leading scientist. He was well established for his beautiful research on neighboring-group effects. He was a bright, vigorous, and enterprising person with an unusually critical mind. His standard questions in seminars, many given by world leaders, were directed toward determining how the speaker knew what he claimed to have established through his research. He was very challenging, sometimes a little brutal, but fundamentally a fine colleague. We naturally became intellectual competitors, to my immense profit. Struggling with him over interpretations of research results during my first 10 years at UCLA was probably the most important educational experience in my life. It was far from comforting — just the opposite — but it was very stimulating. After the first five years, I decided our mutual interests and those of the department would be better served if we became more symbiotic in our relationship. Accordingly, I started nominating him for various awards, ultimately for the Nobel Prize, which I was told he would have gotten had he not died at the age of 57. We enjoyed a warm friendship during the last 15 years of his life.

You seem to admire tough-minded people. Were you tough too?

My early experiences in Vermont with my employers and in New York City with my food-store managers made professoring seem pretty soft. My search for excellence required that I set high standards for my coworkers. I tried to find ways around being tough. Out of my 220 coworkers over the years, maybe five or six I should have fired early, but didn't. For one fellow I tried the following. He went home for two weeks at Christmas time. I worked at his bench on his research problem using his compounds, recording my results in his notebook, and advancing his project substantively. I hoped to shame him into working harder by setting a good example.

Did it work?

It worked for about a week. It took him over five years to get his Ph.D. I have seen about five people who fought their way from being losers to winners in research, and this I admire. More often, losers do not change even in the best of environments.

So you made 215 correct choices out of 220 coworkers.

The choices as time went on were made more by the people who worked for me than by me. Once I had established a reputation for working hard myself and expecting a maximum effort from my coworkers, few weak people tried to join my group. Quite a number of my former students told me later they had chosen my group because I expected so much from my co-workers. In particular, I tried to stimulate them without doing all their thinking for them. I told them time and again that we were not only in the business of training good laboratory workers, but also imaginative, creative leaders. I am gratified that more than sixty of my former coworkers have become Professors of Chemistry, several have been elected to the U.S. National Academy of Sciences, and at least one will probably get a Nobel Prize. I am very proud to have been associated with them.

How do you compare the conditions in your beginning research career to the conditions for a beginning Assistant Professor today?

When I started at UCLA, there was little infrastructure for research, there were no such things as research grants or "start-up" money, the teaching loads were heavy, and coworkers were all teaching assistants who did research in their left-over time. The instruments were dull, and the results obtained per unit time and effort were miniscule compared to what they are now. Our society is now organized to identify and educate the most talented people, those who possess the intelligence, vitality, ambition, and good health it takes to have a career as a scientist. When I passed through the educational system, perhaps only 20–30% of the gifted people even obtained a higher education. Now our society is seeking creativity and originality and new solutions to old and current problems. The world is an entirely different place than it was when I was a boy.

On the other hand, my early years taught me how to handle adversity and stress, how to respond positively to challenge, to be self-reliant, self-determining, and individualistic. As importantly, my early years fostered enterprise and creativity and closely linked hard work to reward. I believe today's youth compared to my generation are deprived in one very important respect — they are not challenged, tested and graded enough; life has been too easy for them, they too often have to learn to handle adversity too late in life; they understand too much the "carrot" but not enough the "stick" side of incentive. The distillate of my remarks here is that my early environment stimulated self-discipline, whereas the current environment is too rich in self-indulgence.

Tell me about your personal, as distinct from your academic, family.

By the time I entered college, I had decided not to have children, a decision that was never regretted. Accordingly, I was careful to court only girls who wanted to have professional careers. My first wife of 27 years, Jean, was a social worker. She is now married to my favorite colleague at UCLA,

Kenneth Trueblood. My second wife, Jane, was a Professor of Chemistry at Mount Holyoke College, who in 1952 had used my research on the phenonium ion as the subject for a seminar she gave at Emory University as part of her Ph.D. program. We met and married some 17 years later. We just celebrated our 25th wedding anniversary. We have co-authored an elementary textbook, many research review articles, and the Royal Society of Chemistry published recently our monograph *Container Molecules and Their Guests*, the fourth volume in their series on *Supramolecular Chemistry.* I am greatly indebted to Jean and Jane for their contributions to my career.

Did chemical research dominate your life to the exclusion of other interests?

Not really. There were several things learned early on: that there were boundaries with regard to how hard I could profitably work. I need 7–8 hours of sleep — for me sleep is enjoyable — and vigorous exercise, which is indispensable; another need was to wipe my mental slate clean once in a while. By that I mean my science had to be set aside periodically if I was to maintain creativity, optimism, objectivity, and devotion to long range objectives. Besides skiing, long-board surfing, and tennis, I sing and play guitar folk songs. Reading biography, history, novels, and of advances made in our knowledge of cultural and physical evolution keeps me intellectually engaged in matters other than my research.

What books have you recently enjoyed reading?

J. C. Beaglehole's *The Life of Captain Cook*; Peter Hoeg's *Smila's Sense of Snow*; Kristof and WuDunn's *China Wakes*; Herrnstein and Murray's *The Bell Curve.* I have friends who advise me as to what books are likely to give me a high yield for my reading time. Getting a high yield has always been one of my objectives. At San Onofre Beach where I surf, my friends ask me when I come in from the ocean, "Did you get a high yield?", by which they mean, did I catch several good challenging waves.

You have referred to having been raised poor in Vermont, and now you live in luxury in California.

I had a friend at Harvard who sagely said, "I have never found an adequate substitute for money." My earliest job paid 15 cents an hour, and a recent consulting job paid me several thousand dollars a day. I was raised to be frugal, later learned how to be prudent, but as time went on, I found it easier to make money than to spend it wisely. Now in my mature years, I have finally had time to adapt my life style to my means. Society has been very good to me. Every year it paid me more to do what I wanted to do. I was lucky to have had my career during a period of optimism and expansion, and in a country and a state which valued chemical research and higher education. The most important result of my getting a Nobel Prize was that it extended my research career. In 1989, I faced mandatory retirement after 42 years at UCLA. Yet in 1995, I still have a small research group at UCLA which is doing things undreamed of a few years ago.

When I was entering your community, the guard asked me at the gate whom I was coming to see and when I told him, he apparently did not know you.

There are a thousand homes here.

Of course. But only in one of them lives a Nobel laureate. Is it almost like living incognito?

That doesn't bother me. I prize the lack of attention. I surf on a beach where all kinds of people surf. I like to be known and liked not for some title or prize but for the way I am. I like to be liked for reasons other than the fact that I've been successful in my profession. Again, that's a challenge and to some extent I deliberately make use of the challenge. I also like a variety of different kinds of people.

Then you don't feel isolated here.

Not at all. I always did get most of my stimulation from reading the journals, from my coworkers' research results, and from asking myself absurd questions about where research treasure might be buried. I am almost daily in touch with my research group at UCLA by fax and phone, and we have research group meetings at UCLA often enough to maintain a critical mass of cohesive spirit. My five coworkers are all carefully selected professionals. I do all my literature reading and research paper writing at home — I just finished my seventh research paper written in the last 12 months. We are working in a field we invented and have no competition yet that needs to be monitored. This is an ideal place to do my writing, reading, and thinking.

You originally coined the expression host-and-guest chemistry, but now you are referring to it mostly as container chemistry.

Not really. The words container chemistry we apply mainly to capsular complexes such as our spheraplexes and carceplexes, in which decomplexation is mechanically inhibited. The term molecular container is more suggestive of geometry than the more general term host. Good nomenclature elicits images and aids reasoning by analogy, the organic chemist's "best friend."

The evolution of our thinking that led to container compounds was very simple-minded. We asked ourselves what geometric feature was common to nature's active sites, particularly the enzyme systems and RNA. The answer was concavity. The simplest way to get cooperativity between catalytic or binding sites is to imbed functional groups in an enforced concave surface, so that these groups converge. Multiple and simultaneous contacts between substrates and receptor sites are required to provide the binding free energies required for the molecular machinery created by evolution. The crystal structures of nature's complexes provided this information. A literature search in the 1960s revealed that only a few simple organic compounds containing enforced concave surfaces were known, examples being cholic acid and the cyclodextrins. Both were known to complex organic compounds. Accordingly, we established a research program of designing, synthesizing and studying the binding properties of organic compounds containing enforced concave surfaces, shaped like saucers, bowls, and vases. We called these compounds *cavitands*, and their complexes *caviplexes*. The suffix *-and* was derived from the word *ligand*. The family name *host* was applied to the complexing partner with the *concave* surface, and *guest* to the partner with a *convex surface*. Once cavitands were in hand, it was inevitable that they would be attached together at their rims to compose closed-surface compounds with large enough enforced interiors to imprison guest molecules. We call such hosts *carcerands*, from the Latin word *carcer*, meaning prison.

What are your chief tools?

Our design of hosts involves Corey-Pauling-Koltun (CPK) molecular models, which are based on innumerable crystal structures of simple organic compounds. They employ bond angles and diameters of bonded atoms adopted by a committee sponsored by the NSF and NIH, who chose values of about 90% of van der Waals radii with strong elastomer bonds that are somewhat deformable. The committee was composed of biochemists, organic and physical chemists, and crystallographers. These models store a vast amount of information and were beautifully designed, at least for our purposes. Predictions based on them in our hands have turned out to be over 90% right. We have over 200 crystal structures of our hosts and host-guest complexes, most of which were predicted in advance by CPK model examination. Our crystal-structure determinations have proved to be an invaluable tool and, to me, still have magical and satisfying qualities. Of course, NMR spectroscopy has been our chief criterion for structure in solution. Such spectra have been indispensable in obtaining kinetic and thermodynamic data. Our most thrilling and stimulating tool has involved finding out if the host molecules actually perform the tasks we have set for them.

Did your solid geometry contribute to these studies?

I was attracted to solid geometry and to host-guest chemistry because of my life long intrigue with the shapes of objects. Visualization is perhaps the most substantial part of the imagination. I have never felt satisfied that I understood anything thoroughly unless it could be reduced to an image. My biggest strength is my creativity which grows out of comparisons of not totally unlike images in the imagination. Mechanical modeling was a "natural" for me.

Did you ever tell Linus Pauling how valuable molecular models became to you?

Had I seen him after about 1960, when he essentially quit chemistry, I certainly would have acknowledged my debt to his early use of molecular models.

How about computational molecular modeling?

I knew that this approach would ultimately supercede mechanical modeling, particularly when the host's mechanical models became so large and heavy that their physical support and manipulation became a problem. We like mechanical models because of their transportability, their occupation of three dimensions, and their use as a teaching aid, and because handling them exercises one's tactile proclivities.

Were you anticipating a Nobel Prize?

I always measured myself against people I admired. Prelog is someone I have long admired. He and I competed in some early work on asymmetric induction and then later on host-guest chemistry. When he got the Nobel

Prize, I first started to think about my own candidacy, particularly when people started to approach me for reprints of my publications to be used as the basis for nominations. Among others whom I hardly knew, Prelog and Barton told me they had nominated me. The only thing one can do to deliberately get a Nobel Prize is to do exceptional research and bring it to the *scientific marketplace*, which means publishing your results and giving seminars on it all over the world. This is what I had always done anyway. Another important condition for getting the Nobel Prize is longevity — particularly if one gets it for creating a cohesive body of science, as distinct from inventing a new important instrument or technique, which can happen early in one's career. By 1987, when it came, I was not very sanguine about my chances of being so honored.

Did you have any interaction with Pedersen and Lehn, prior to sharing the Nobel Prize with them?

I had met Pedersen only once in person, but I was much stimulated by his important 1967–69 series of papers on crown ethers. Our only personal contact was a three- or four-hour meeting I arranged in my DuPont Hotel

Donald Cram with a CPK model (photograph by I. Hargittai).

room about 1974. I showed him CPK models of hosts we were making. He was a charming, modest man, admirable in all respects. He had never obtained his Ph.D. degree, never run a large research group, and had published his finest research results just as he retired. He was delighted that we used his results as our starting point. Lehn and I had met many times at meetings and had many stimulating conversations. He, like Prelog, is an international kind of person, very creative, and possesses wide interests and knowledge. His approach to molecular recognition in complexation has been entirely different from mine, although the underlying intellectual framework is the same. We have always picked different systems to study, and our results have been complementary rather than competitive. I was very lucky to have shared the prize with such fine people. Interestingly, each of us belonged to widely differing age groups. It would be less pleasant to share the prize with someone you did not like.

Tell me about how you entered this field of host-guest complexation chemistry.

To retain my fascination with chemistry, I have had to change my research fields about every 10 years. Since about 1948, when we initiated our work on cyclophanes, we started to think about the fact that the same physical laws governed biological chemistry and synthetic organic chemistry. We had the long range ambition of modeling some aspects of the biological system with designed, simpler synthetic organic systems. In the late 1950s, I came to know Fritz Cramer when he was at Heidelberg. He was studying the complexation between the cyclodextrins and various organic compounds and learning some very interesting things about complementarity between the complexing partners. In the early 1970s, I borrowed from one of my biochemistry colleagues a large set of CPK molecular models. During my first three days playing with these models, I never got dressed. I constructed a large variety of potential host systems with the models and graded them as to their probable ease of synthesis, the importance of the potential questions their study might answer, and their utility as vehicles to teach graduate students and postdocs how to think about and do research. The envisioned hosts ranged from enzyme mimicking catalysts to synthetic ionophores. I filled a large notebook with these structures and my expectations for the compounds they represented. With such a record, I could avoid duplication. This took about two months of work spread over about four months. The next thing I did was to try to sell my best ideas to everyone in sight, with the molecular models in hand. In fact, before we had any results, I traveled all over the U.S. and Europe giving seminars *on what marvelous things we were going to do* with this field. My friend at Caltech, J. D. Roberts, told me this was stupid, because most of my ideas wouldn't work — that is clearly the nature of research. He was right! In retrospect, I think this was my way of raising the challenge and my courage to turn my whole research group in a new direction. Among others, I showed Albert Eschenmoser in Zurich and Salo Gronowitz in Lund my ideas and my models. Later, Albert told me that he thought at the time that I should have my head examined. Even worse, I could not persuade any graduate student to make his Ph.D. thesis work depend on such tenuous ideas. I had to force one of my postdoctoral fellows to do our initial work, followed quickly by four others. Within a year we had encouraging results, and within three years my whole group of 17 coworkers were working in the field, and we had solid financial support. This was about 1971–1974.

The central idea of the research was that the exquisite chemical activities of biological processes depended largely on complexation involving large numbers of weak but additive attractions and that enzymic catalysis, immune responses, and genetic information storage, retrieval, and replication might all be modeled. We would use simpler host frameworks than those composed of amino acids, sugars, phosphate esters, and the heterocycles of the natural world. The art lay in developing rigid frameworks containing binding or catalytic sites preorganized to act cooperatively on particular guest compounds. In effect, by complexation we hoped to turn ordinarily remote, uncollected, and nonoriented functional groups into arrays of preorganized, neighboring, and cooperating groups. This was a field ideal for me. I have a tolerance for ambiguity, for failure, for groping, coping, and gambling that this field certainly exercised. This year's gamble in research is next year's conventional wisdom.

You did work much earlier on carbanions. How about carbocations?

We studied both carbanions and carbocations using the stereochemical fate of stereogenic centers as a probe of reaction mechanism and solvation phenomena. These studies involved substitution, elimination, and rearrangement reactions, most of which involved carbanions and carbocations as short-lived reaction intermediates. The chiral systems were designed and synthesized, the kinetics of the reactions were examined, and the configurational changes leading to product were determined. From the changes in symmetry properties in going from the starting materials to the products, we could infer the symmetry properties of the short-lived reaction intermediates.

Did this early work help you to develop the field of host-guest chemistry?

This work on organic reaction mechanisms and our development of cyclophane chemistry were of great use to us in our later work. We did not shy away from tackling either multistep syntheses (up to 30 reactions) or highly asymmetric, designed systems needed in our studies of enzyme-mimicking systems. We needed both equilibria and kinetic techniques and an understanding of the importance of solvent effects in our more recent studies.

Your complexing partners are normally held together by attractions much weaker than those of covalent bonds.

This is true except when *constrictive binding* (mechanical encumbrance) is involved, as in our carceplexes. These complexes have to be heated to temperatures that crack open the hosts (~ 370–400°C) before the guests escape their prisons.

Are there appreciable geometric changes in host and guest upon complexation?

Our carcerands and hemicarcerands are rigid enough not to reorganize conformationally to fill their own cavities. They more closely resemble a suit of armor than a suit of clothes. This fact brings up interesting questions about what kinds of driving forces are involved for dissolved guest molecules entering and filling the interior vacuum in hemicarcerands. These hosts possess portals connecting the inner and outer solvent phases large enough at 100–200°C for properly sized guests to enter and exit the interiors of these shell-shaped compounds. We believe one of the driving forces for complexation is the entropy of dilution associated with dividing one large vacuum into many small ones and dispersing them throughout the bulk phase. We are also running into new kinds of isomeric complexes due to restricted rotations of guests relative to hosts. We have carried out a variety of chemical reactions on incarcerated guests, one of which produced cyclobutadiene, which is stable at ambient temperatures, protected from predator molecules, like a turtle in its shell. Oxidations and reductions of incarcerated guests have been run with reagents dissolved in the bulk medium, indicating that electrons and protons readily pass through the chinks in the host.

How do you get empty carcerands?

We have not been able to make empty carcerands, because carcerands do not form when the shell closures are run in solvents too large to be incarcerated. When run in smaller solvents, one or two molecules of solvent template the shell closures. Shell closures leading to hemicarcerands are also templated, but by heating hemicarcerands to 200–250°C in solvents too large to fit into the interiors of empty hosts, the guest departs the inner phase, driven out by mass law. Then new guests can be introduced by heating the empty host in the presence of large excesses of guests, the complexation being driven by mass law.

How many different carcerands, hemicarcerands, and guests have you studied, and what are the extremes in the guest sizes?

About eight different carcerand systems, fifteen different hemicarcerands, and more than seventy different guests have led to several hundred fully characterized complexes. Guests have been as small as water, oxygen and nitrogen, and as large as [3.3]paracyclophane (18-carbon atoms) and 1,3,5-trisisopropylbenzene.

How did you work jointly with your wife on your books?

This was a real challenge! For co-authors to be of use to one another, each has to provide criticism to the other. Yet Jane and I love each other. Reconciling these two roles led to some real battles. I am blessed by having a very forthright, analytically minded wife. Everyone needs a critic and she is mine. She is probably more intelligent than I am, but she is not as creative and daring in her thinking. We complement one another. Our science is so demanding that one person seldom has all of the qualities needed for success. It is better to have the interactions of several people, with input from each. Highly imaginative people have a tendency to be wild, and critical people less so. We have a sort of Gilbert and Sullivan relationship, one supplies the lyrics and the other the music. It is hard to sing solo.

Did she have an appointment at UCLA?

No — her only appointment is to keep me reasonable.

Does it work?

Yes! I would never have received the Nobel Prize without her. That is very clear.

Does she know this?

Oh, sure.

Does she agree with your judgment on this?

Sure. She agrees, and more than that, she got more pleasure out of it than I did, and I am just delighted. If you're raised in Vermont in the 1920s in my family's circumstances, you become acquainted with a balance of value given for value taken. Vermont was a semi-agricultural state then with very little circulation of money and obligation. There was a lot of generosity within the family, but not outside because there was not much to be generous with. All this I am saying in reference to my wife and the Nobel Prize. Every new idea that I have had since 1970 has enjoyed the benefits of her criticism.

She also has to appreciate that you are so much occupied by your work.

Actually, she has worked very hard recently to deoccupy me with my work! I seem to have to work intensively at times to remain happy. I have to demonstrate to myself, now and then, that I can do it.

How do you organize your life out in this beautiful environment but one which can hardly be enough for you intellectually?

There are many temptations here that are fun not to resist. Partly I am out here because the dry, warm climate raises my pain threshold. Although I am generally healthy, 11 operations of one kind or another left a lot of sore places. When my Nobel Prize was announced, I had a painful case of the shingles, and even wearing a shirt was very uncomfortable. Incidentally, George Olah was the first to inform me of the Nobel Prize. He called and woke me at 6:30 a.m. and said "Get up Don — you have a big day ahead of you. I just heard the announcement of your Nobel Prize on the radio."

To get back to your question about the intellectual climate out here — we have made friends with some of the most interesting and knowledgeable people we have ever known. As importantly, they are retired and have the time to discuss all sorts of matters, ideas, and events in a civilized, unhurried, and intelligent manner. One of the central benefits of getting older is that contexts become larger, judgments lose their sharp edges, values are in place, and responses to challenges are more tempered.

Jean-Marie Lehn, 1995 (photograph by I. Hargittai).

14

JEAN-MARIE LEHN

Jean-Marie Lehn has a joint appointment as Professor at the Institut Le Bel, Université Louis Pasteur in Strasbourg and at the Collège de France in Paris. Donald J. Cram (1919–2001), Jean-Marie Lehn (b. 1939), and Charles J. Pedersen (1904–1989) received the Chemistry Nobel Prize in 1987 "for their development and use of molecules with structure-specific interactions with high selectivity." The conversation with Professor Lehn took place during a brief Budapest visit of his on May 8, 1995. The interview was squeezed in between a press conference and a lecture, and the schedule was running late.*

You started as an organic chemist.

Yes, an organic chemist and, in fact, a natural products chemist.

Now you are also a very conceptual chemist. Not every organic chemist develops general concepts the way you do.

Over the years I have become interested in a number of other areas. After the more natural organic products type of chemistry, I was interested in more physical chemistry and in the late 1960s, early 1970s, we had done really very physical chemistry. This included molecular physics of liquids and motion

*This interview was originally published in *The Chemical Intelligencer* **1996**, *2*(1), 6–16

in liquids. We'd also studied a number of molecules by *ab initio* computations. At that time I became interested in molecular recognition. Things evolve often from initially rather simple ideas which touch a general problem and which, if dealt with successfully, can develop into more general views. The selective binding of metal ions led first to molecular recognition and then expanded into the general concept of supramolecular chemistry.

Was there a specific turning point? You could have become one of many natural products chemists.

We are all different, of course. What happened may be linked, perhaps more to a frame of mind. At one stage I wanted to study philosophy. I always like to put what I am doing in a more general framework. I also feel that having a concept is inspiring by itself. It makes you think more broadly and makes you look at the objects you're studying in a more general fashion.

Initially, the reason why I was interested in sodium and potassium recognition was because I was interested in the nervous system. Maybe that was the outcome of my studies of philosophy. One of the basic functions is the propagation of nervous influx, which rests on sodium/potassium levels and their changes across membranes. As a result of that, I became interested in how to bind sodium/potassium and how to transport it. At that very time, there were these very important papers on ion binding antibiotics by Shemyakin and later by Ovchinnikov in the Soviet Union and by Pressman in the U.S. For me they were an inspiration as they showed that it was chemically possible to bind sodium/potassium ions, and thus, also, to recognize them. This was the beginning of my interest in binding sodium/potassium. Important related work had also been performed at the ETH Zurich at about the same time. This was around 1964–66, and my own interest started about 1966 after I'd returned from the U.S. I started developing my own lab in Strasbourg in late 1965. We had mainly NMR and other spectroscopies but I was also interested in the binding of these metal ions. Then, of course, I learned from Pedersen's work that crown ethers bind metal ions but even earlier there were some papers which had indications about that. One of them was by Jeff Wilkinson reporting work in which alkali metals were dissolved in ethers. Another was by Herb Brown and co-workers about the effects of glimes in borohydride reactions linked to the binding of the sodium ion.

What distinguished your work from the traditional molecular chemistry?

The main difference is that I was looking for the more general concept. It's quite clear that chemists have been developing molecular chemistry for many years. By making and breaking bonds you can create molecules. This field has by now reached fantastic complexity. Chemists are extremely knowledgeable in doing that, with very high precision and efficiency. Not everything is known yet, of course, and there is still a lot of things to be done in synthetic chemistry. Molecular chemistry has a fantastic future. However, once you've reached a certain level of development, at which you can make and break covalent bonds to build molecules almost at will, the question you may ask is, can you do the same at the level of weak interactions? That is, at the level of intermolecular weak bonding? There are forces between molecules which can bind them together weakly. The question is how can you manipulate these intermolecular interactions to generate supramolecular structures?

I consider supramolecular chemistry as the extension of molecular chemistry to the more complex systems. The task is to enable the chemist to handle non-covalent structures as he or she had learned to handle covalent structures. Because the bonds are weak, a number of new properties appear. I like to stress the analogy with information science. The reading of information is subtle and must be reversible. There is need for reading, and this requires weak interactions, in which you can bind and dissociate again. Also, supramolecular structures have a property which you may call healing. When they associate wrongly, they can dissociate and recombine.

Ernest Eliel, Jean-Marie Lehn, and Jack Dunitz in 1970 (courtesy of Ernest Eliel).

In other words, they have a self-healing mechanism. One major characteristics of supra-molecular structures is this capability for self-processes. There are processes that occur spontaneously, but they don't occur for covalent bonds at normal temperatures and pressures because they are too stable for that. You may in this respect consider molecular chemistry as stable chemistry or fixed-structure chemistry, and supramolecular chemistry as fluid chemistry. The term fluid here means that the structures can be arranged and rearranged, assembled and disassembled depending on the surroundings. Adaptation is another important characteristics of supramolecular systems. A molecular structure, as far as connectivity and gross change is concerned, is very stable, not influenced so much by the medium. A supramolecular structure can, in principle, adapt to the medium.

How about intramolecular bond lengths and bond angles in the supramolecular structure?

Covalent bond lengths are rather well defined. Solid state structures provide a wealth of information about what happens between molecules, about intermolecular distances and angles. A number of groups have been interested in the design and properties of molecular solids.

Theoretical modeling investigations have also been much developed. They help us understand intermolecular interactions and predict the ways molecules fit together. This is the theoretical part of supramolecular chemistry.

You have a very high visibility. Books, new journals, meetings, organizations. How do you cope with all this?

I'm trying to do what I like to do, and as actively as I can. With time you begin to think more broadly. As you get older, there is less and less time left. There are two ways of doing it. One is that you take more time to look at things. The alternative is, since there is less time left, one has to do more! For the moment I'm rather of the second opinion.

You came from a humble family background.

My father was a baker but also an organist at the same time.

How difficult has it been to handle various changes in your life, to cope with increasing responsibilities?

One learns with time. Before accepting various functions, especially those that take me away from chemistry, I have to think hard to convince myself to do it. One has to be careful accepting things in which one may be less and less competent. On the other hand, as you think more about what you are doing and about the place science occupies in society, you feel more commitments to many things, to your fellow scientists and to science itself. Science is the strongest force in transforming society. The progress of mankind depends on knowledge, that is, on science and its applications. By saying this, I don't mean that, for instance, philosophers and composers are useless, not at all! I only mean that the objective transformation of society is based on the knowledge we have gained about nature. We need not be arrogant scientists. At the same time, just because we may be accused of "scientism," we should not shy away from stressing the importance of science. Science has nothing to do with any dogma. Science ceases to exist when there is a dogma. For me science is an approach, a way of taking things in a rational fashion, thinking, and trying to do the best with what we know, trying to get solutions in complicated cases to the best approximation, knowing that this is not the perfect solution, and that there is no absolute solution. This is just the antithesis to many very careless and emotional approaches that we see in the world around us these days. Science can merely offer some options for rational solutions, and it's then up to the decision-makers whether they want to use these options or not. Of course, one has to be careful about the deeds of these so-called decision-makers or politicians. I'd only cite what is happening in the south of Europe. I am sure that in many places at war, the people would be quite happy to live together. However, because of political ambitions, artificial problems are created in cases where they did not exist. Then, of course, when you raise antagonism, people finally think that they do in fact have enemies. The people involved often didn't even know, or did not care, that their present enemies were of other race or culture. They only were made to be aware of this. That's crazy. I believe that the scientific spirit is the antidote of this.

In your youth you did chemical experiments at home. How did it all end?

The first one ended in an explosion! So there is some danger in it. On the other hand, we have to learn to live. I became very interested in chemistry in high school. I first studied philosophy and at the same time experimental science. I became interested in chemistry because it appeared to me very logical. When I was 19, I started buying chemicals and doing experiments at home.

You love music. Is there any interaction between chemistry and music?

Music is something organized. Perhaps there is some relationship between chemistry and music in their organization. What I like in music is to see the full picture, to see how it has been constructed. I also like tense music, Béla Bartók, for instance, and Beethoven. I like music when there is a stream, when it is flowing. Gabriel Fauré, also. I like its intensity. I also like Bach but more for its construction.

Do you ever talk chemistry to your musician friends?

There are two composers whom I know quite well. One is Pierre Poulez, who was my colleague at the Collège de France and retired just this year from his Chair. The other one is György Ligeti, who is Hungarian. I had the occasion to talk with Ligeti about what we're doing and he was very interested. Then we met a year later, and not only did he remember what I'd told him but asked me also about what had happened in the meantime.

I think that if you are open and are willing to talk about your science in an understandable fashion, you can also penetrate the world of Ligeti. It's different, of course, from a Mozart symphony, which you can like without knowing how it is constructed. It is more difficult to like and appreciate the more modern works on first approach. You may also want to know what's in it, what its structure is and so on, and it will be a different kind of liking.

Take Paul Klee, for example. You can like Klee's paintings without any preparation, and I liked Klee. Then I read the book that Paul Klee had written on his paintings, and I got a different perspective of his works. When you know what it means when an arrow in one of his paintings is thin or thick and so on, the painting takes on a different meaning. So, again, it's love at first approach and love when you know better.

Do you have children?

Two.

Are they following your footsteps?

They are not scientists. David, the older one did mathematics at the beginning. Then he changed and now he's a writer. The younger one also started in science first and then he switched to musicology. I must say I envy him. He's now concerned mainly with opera.

How about your interest in music?

In some ways I am a frustrated musician. You can't do everything you like. I like to improvise. Composing is different and, again, you have to learn it, and I didn't learn it enough.

My father played the piano and the organ, and I took lessons until the age of 11. Then I had to go to high school in a different town, five kilometers from our own, so I didn't have much time. I continued to play though but I didn't take lessons any more.

Do you remember any teacher in your career who was a very important influence on you?

One has many influences in life. Just three years before the end of high school, I had a friend who was in his last year of school at that time. He was taking philosophy, and he used to tell me about it. I became very interested and began to read Freud, Kant, and others at the age of 15. It wasn't so easy. All this looked fantastic to me. My friend's name was Claude Koenig and he died years ago. I don't think he ever knew how strong an influence our discussions had on me. At that time I just felt the excitement. When I read Freud for the first time it was a big shock to me. When you're brought up in a conventional way, it's something which makes you think very hard. I remember that I tried to contradict Freud but he always won!

Another influence, this one in chemistry, was Guy Ourisson. I liked chemistry before that but he was probably the spark that lighted everything up. He gave excellent lectures, and I decided to do my Ph.D. with him. Then, of course, Woodward had an important influence on me at Harvard.

At a much earlier age, there was another important person for my development. This was my primary school teacher Pierre Charlier. He was one of those typical products of the French Third Republic ideal of democratic education. He was a Republican teacher, a very dedicated teacher, and wanted to motivate the young children. I remember that he worked with us overtime, many times, to bring us from the primary school to high school. I owe him a lot.

In high school I wasn't sure whether or not I liked what I was doing, but when I began to read philosophy, things started to change. At 15 or 16, you begin to ask yourself questions, and you question things that you had accepted before. As Paul Valery said, "On pense comme on se heurte." You think as you feel resistance. You bump into something, and you begin to think. This is exactly what happened to me.

Bruce Merrifield, 1996 (photograph by I. Hargittai).

15

Bruce Merrifield

Bruce Merrifield (b. 1921) is Professor Emeritus at the Rockefeller University in New York City. He received his B.S. degree from the University of California at Los Angeles (UCLA) in 1943. He received his Ph.D. from UCLA in biochemistry in 1949. Professor M. S. Dunn was his Research Director. He has been with the Rockefeller University (originally, The Rockefeller Institute for Medical Research) since 1949, concluding his active service as John D. Rockefeller Jr. Professor (1984–1992). He was awarded the Nobel Prize in Chemistry in 1984 "for his development of methodology for chemical synthesis on a solid matrix." He first used this methodology for the synthesis of the enzyme ribonuclease. This methodology is credited with having brought about a revolution in peptide and protein chemistry. We recorded our conversation in Dr. Merrifield's office at the Rockefeller University on May 8, 1996.*

What dictated the choice of the enzyme ribonuclease, containing 124 amino acid residues, for synthesis? This work was mentioned in particular in the introduction by the Swedish Academy of Sciences at the Nobel ceremony.

I selected ribonuclease A in part because it was a "Rockefeller enzyme." First of all, the activity of this enzyme was discovered here at Rockefeller University by René Dubos. Then Moses Kunitz isolated and crystallized

*This interview was originally published in *The Chemical Intelligencer* **1998**, *4*(2), 12–19

the protein and developed a convenient assay for it. Alexander Rothen studied its molecular weight. Next, Stanford Moore and William Stein (Nobel Prize in Chemistry in 1972), again at this university, determined the sequence of the amino acids using the automated analyzer that they had developed. I knew all of these men and something about the enzyme. It was a nice choice for us. It's big, but it's not too big. It was hard to find a good protein at that time in the range of a hundred residues. It is also quite stable, and it can be denatured and renatured to regain full activity as shown by Christian Anfinsen.

Would you please tell us something about your background?

Both of my parents grew up on farms in Texas, and then they moved to the city of Fort Worth when they were married. My dad really did not like the farm life. After serving in World War I, he went to a business school and then into sales. He was a very nice looking man and personable, and everybody liked him. He made a good low-key salesman. He was a combination between furniture salesman and interior decorator. But when the Great Depression came, nobody needed furniture. By then, we were already in California. He was struggling to make ends meet, but he always had a job. So we scraped along, and that affected me all my life. I can't bear to waste things, I'm not extravagant, never buy anything on credit, and that came directly from the Depression. I was 7 years old when the crash came.

Did your parents live long enough to witness your success?

Not quite. They knew things went well in the lab, but they both died before I received the Nobel Prize. I would have liked them to have lived to see it, and I would have liked my high school teacher to see it, and my two mentors too. But all the people who really mattered were gone by then.

Both my mentors had a big effect on my life. Dr. Max Dunn at UCLA, whose picture is here in my office, knew everything there was to know about amino acids. Dr. D. W. Woolley's picture is also up here; he was here at Rockefeller when I came. He was a fine biochemist, a brilliant man. He'd lost his eyesight in his mid-twenties. It was diabetes. But he carried on and maintained his lab, with a technician, of course. He only had to read something once — it was read to him — and he'd remember everything, and he could correlate information from many areas into a single idea.

Max Dunn around 1950 (courtesy of Bruce Merrifield).

Wayne Woolley (courtesy of Bruce Merrifield).

I worked with him for several years on his projects until I got the idea of solid phase synthesis. I told him the idea. It was kind of funny; we were riding up the elevator and he got off without a word. Next day he came in to see me and said, "that may be a good idea, why don't you work on it?" That made all the difference. He could have said, no, that's not what I want to do, I don't want you to work on it. The Rockefeller Institute, later Rockefeller University, was set up more like the European system. You had a lab head, and people worked under him. So when I was in his lab, I worked on what he was doing. He directed it all. Then, after a while, he'd let you do a little bit on your own. That was standard, there wasn't anything unusual about it. It's changing a lot now. Young people who come in now are appointed as Assistant Professor and independent Head of Laboratory.

Why was it changed?

Some people thought it was better.

Was it?

No, I thought the old mechanism was all right. First of all, we were unique. There was almost no place in the United States that operated that way. Since the Institute had been highly successful for many years, I thought it was worth keeping it special. However, the people in control thought that it ought to be the other way, and that's what happened.

You said one day you had this new idea. How would you summarize it?

I was working on a peptide growth factor with Dr. Woolley, and we isolated it and I sequenced it and synthesized it by standard methods. It was only a simple peptide of 5 amino acids, but it took me 10 or 12 months before I could make it and a couple of its analogues. I realized that peptide synthesis was quite hard to do. People who had a lot of experience could have done it faster than that, but it was still slow business, and the yields were not very good. So I had in the back of my mind that there must be a better way to do it. Then one day or one night, I don't know which, I just had this idea; couldn't we anchor the peptide onto a solid support? If you could put the first amino acid onto a substance that was insoluble in all the solvents you used and then you added the next amino acid and the next one and the next one, you would build up a polymer chain. After each reaction, you could filter and wash, and your peptide would stay attached to the insoluble support. You could wash it thoroughly and get rid of all the excess reagents, and you were ready to do the next step. You wouldn't have to take each peptide intermediate and purify it and crystallize it, which was the standard way to proceed. Then, when you had your sequence assembled, at the end, you could break the bond that was holding the peptide to the support. Of course, you had to have the right conditions under which this bond would break and the rest of the bonds would remain stable. Many ways have now been found to make the attachment and to break it.

You make it sound incredibly simple.

It is a simple idea. You can say it all in one sentence. However, it took a long time to develop it, partly because I didn't have a lot of experience in the field; I had a little, but not a lot. All the variables had to come together at the same time. You had to find a support with the right properties, you had to find some way to attach the peptide, and then you had to find some way to increase the peptide length and to protect your amino acids with something that wouldn't fall off while you were extending the chain. You had to pick the right solvent and reagents, and so on. They all had to be correct at the beginning. It took me more than three years to find a workable combination.

Were you patient?

I was patient but, what was more important, Dr. Woolley was patient and the university system was patient. If I had been an independent Assistant Professor, the head of my own lab, I couldn't have survived. No way that I could have gone three years without a publication. For me, the Rockefeller system was a good one. You had to have the right head of lab, of course, but Dr. Woolley believed that it was a good idea, and he let me go ahead. Then, when I got through, and it did work, and I wrote the paper, he said, "my criterion of whether I'm a co-author is whether I would've done the work anyway, without you," and he said, "I wouldn't have in this case, so you can publish it by yourself." That made an enormous difference in my career. He was so well-known that people would have considered him to be senior author, and even if I would have been first author they would have associated this work with him and not with me. So I owe a whole lot to him.

Originally you came from California to New York. This is not the usual direction.

I came for one year, after I'd got my doctorate at UCLA, but then I stayed.

Did you become a New Yorker?

No, I never have. I didn't grow up one, and it's very hard to become one. There are many important features of the city, but you don't feel like you really belong, and you're not accepted by the native New Yorkers. Eventually we moved out to New Jersey, but it's not much different. I stayed because I thought Rockefeller was an excellent place and a good fit for me. It took 14 years, though, to get tenure.

You had an excellent teacher in Dr. Woolley, and he had an excellent coworker in you. How about your pupils?

Dr. Woolley had several people here, I was just one of the group. We were not that exceptional. He was. With my students, I've been lucky. I haven't had a lot of them, maybe 10 or so. Rockefeller is not a regular university, it's a graduate university, and they take about 20 students a year, for all areas, and that's very few. The number entering chemistry or biochemistry is only one or two. I never had more than one student at a time, and some years I didn't have any, and I don't have any now because I'm getting older.

Garland Marshall, Arnold Marglin, George Barany, Svetlana Mojsov, Wesley Cosand, Mark Riemen, and Bill Heath were particularly good students, and there have been others who were good. Some, though not all, stayed in the field. A couple of them went into industry. Mainly, I have worked with postdocs, and several of them have been really good. There was a variety but always a small group. Right now, I only have three research associates and three technicians.

Where does your support come from?

It's always been NIH. I have had two grants, and we're in year 40 of one of them, which is unusual. When I first came it was different. You were not allowed to get outside support, and everyone was supported by the Institute. As outside support has increased, it has been possible to reduce university support to the laboratories, although the university still provides significant assistance.

You write that your wife joined your group in 1981.

She was a zoology major in California, at UCLA. She took the biochemistry course for which I was a Teaching Assistant, and that's how we met.

Did you start dating while she was your student?

You were not supposed to, but we broke the rule. She got her Master's degree in zoology and was working on the Ph.D. when we got married and came here. She went back two winters to continue, but her professor died and that ruined everything for her, and she never completed her degree. She also went up to Columbia for a while, but she wasn't satisfied with the work she was doing and wouldn't write it up as a thesis. Then our children started coming along. She enjoyed them very much, but it was hard for her to give up the career she had planned. Finally, when the children were all grown but one who was still in high school, she decided to come into the lab. She picked up the lab work without any problem. She never had a job here and just worked as a guest.

You said somewhere that you had set out to do synthetic work on peptides in order to answer some biological questions. What were those questions?

For any biologically active compound, a peptide hormone or an enzyme, you would like to produce it in a large amount, and you would like to find out how it functions. You can do that by making suitable analogs, and finding out what part of the molecule is necessary for its function. You also would like to make inhibitors or antagonists of a hormone or other active compound because that helps you block the action of the natural material, and you can often deduce how it works. In some cases, you'd like to block its action and produce a drug. These are some of the main things that I could foresee.

Did you file patents?

Originally the University frowned on patents. They'd had a few bad examples, and they felt that was not a good way to use your time. Woolley didn't like it; he tried once or twice. So that was generally discouraged, and I didn't know anything about patents anyway. Then somebody from Merck got interested in what we had been doing, and they talked to the President of the University and asked why we didn't take out a patent. But it was too late; I couldn't patent my basic idea because I'd already published it. But since then, we've had a few patents. It's not all bad but can also take a lot of your time and effort on things that don't really lead to anything. Some people, however, have done well in that way.

What's the relationship between your work and natural product chemistry?

That's interesting. Peptides are certainly natural products, but the classical natural products chemists don't recognize them as such. Peptides are excluded from their repertoire. I got a nice volume from K. C. Nicolaou the other day. He's a natural products chemist, probably the best in the United States, but no peptide was included. There is much more varied chemistry in the synthesis of the relatively small natural products than in the repetitive pattern of peptides, and they just feel that peptides ought to be excluded from the field of natural products.

What about computations?

There are some very sophisticated studies in which people generate the shape of the surface of molecules and how they must interact with their receptor, and from that they can design new analogues. We can do that too and we especially want to do that in one project we have where we're making antagonists to the hormone glucagon. Glucagon and insulin are involved in diabetes, and they work in opposite direction. One raises the blood sugar, and the other lowers it. One of the people in my lab, Cecilia Unson, has managed to isolate a gene for the glucagon receptor. She is able to express it so that the gene produces the receptor. What I want to do is take our antagonist and bind it to the receptor and find out what the hormone looks like when it's bound. You can find out what it looks like when free in solution by nuclear magnetic resonance and you can crystallize it and determine its structure by X-ray diffraction, but that's not necessarily what it looks like at all when it's bound to a receptor and functioning as a hormone. If you know its bound conformation you can see where its contacts are and then you may be able to design modifications of the structure of the hormone that will bind better or block some of the hormone's actions. Both NMR and X-ray crystallography have been used for glucagon, and for the receptor we can use NMR even though it's quite big relative to the ligand. For this, we have to have the receptor in some quantity and relatively pure. One of the problems is that it is naturally in a membrane. Then the question is, when you get it out, does it do the same thing as it does when it is the membrane? That's not clear.

This place has had great tradition in the field of peptides and proteins. Can you comment on that?

Max Bergmann was the foremost peptide chemist in the world for several years. He was a student in Germany of Emil Fischer, who was the father of the field. Bergmann came to America just before World War II. He was recognized to be important, and our director offered him a position here at Rockefeller. I now have his old office. He assembled an excellent group of people. There was Leonidas Zervas and then Joseph Fruton, who came out of Columbia University as Bergmann's first postdoc. He did very important work on synthesizing substrates for enzymes such as trypsin and determining their specificity. Fruton made small tripeptides and could show that the enzyme was specific for one kind of peptide bond and not for others. I always considered that to be a very classic piece of work. Then he moved up to Yale as Professor and Head of department. He also has done extensive writing on biochemistry and the history of science. Stanford Moore and William Stein were also in Bergmann's laboratory. They later gained fame by devising an amino acid analysis method and building an automated machine. Among the others in this remarkable group were, Emil Smith, Klaus Hofmann, Carl Niemann, Conrad Frankel-Conrat, Mark Stahmann and Paul Zamecnik.

Did you know Erwin Chargaff?

Not very well, but he was my big competitor when I was a graduate student. My thesis research at UCLA was to develop methods to analyze purines and pyrimidines, which are the basic constituents of nucleic acids. The old methods were just terrible, not quantitative and very slow. Finally, I realized that I could use microbiological methods and determine the three pyrimidine bases. I did that and it worked. I never could find bacteria that were specific for the purines. If I had gotten that to work, I would have beaten Chargaff, but just after I had developed this methodology, he came out with his chromatographic methods. Then he was able to determine the ratio of the bases and show that they were not all present in equimolar amounts in nucleic acids, but two of them were about equal and the other two were about equal. I missed the opportunity. Chargaff's methods were so much easier, all my work was just lost. Too bad, but that's the way methodology works. Although Watson and Crick say that they did not rely on Chargaff's data, it certainly fit their theory perfectly.

What's your present work about?

We continue our work on synthetic peptides. The group is small, and we work on only two problems. One is a group of peptide antibiotics. These are linear peptides. They're quite active and have some effect on malaria parasites, tuberculosis bacilli, and many other bacteria. There are so many resistant strains developing now, particularly in hospitals, making the major antibiotics useless. Any new compound that works by a different mechanism is important. The peptides we work with are called cecropins because they come from the giant cecropia silk moth. They were discovered by a man in Sweden, Hans Boman, and we have collaborated with him for several years. We synthesized a lot of these peptides, and we also made the longer precursor protein. Soon I became concerned about whether these peptides work by some kind of a receptor, like a hormone does, or not. So we made the all-D-peptide from D-amino acids, which is the mirror image of the natural all-L version. Because it contains unnatural amino acids, it is resistant to enzymes. The normal enzymes do not work on D-amino acids. It turned out to be fully active, however, in a whole group of test organisms and very resistant to enzymes. This says there is no receptor involved in the action of these peptides. I felt we had something that ought to be developed into a useful antibiotic at least as a special purpose antibiotic. It would be a new kind of antibiotic working by a different mechanism than penicillin and other well-known antibiotics.

Why is there so little interest?

The big companies don't see it as a major product. They always say that if it's not going to be $100 million a year in sales, they can't work on it. That's a pity, because, for instance, tuberculosis is coming back, and the normal antibiotics that had been useful in the past are not effective in many cases, and people just die.

When you develop a new antibiotic, how far can you develop it and at what stage does somebody else have to take over?

We can design and synthesize an active structure and measure its activity on various microorganisms and may get some information on its mechanisms of action. However, we can't do extensive *in vivo* animal or human clinical studies. I've talked to some of the big companies, and they showed a little interest but not much, and nothing happened. Then I talked to small companies, and we found one that was interested. They were working on related problems themselves, and I keep hoping that they'll be successful.

I apologize for the next question, but do you have any suggestion why all the amino acids in nature are L?

Oh, people make a career of that, but I don't think anybody knows. There may have been some chance event that selected one isomer a little bit over the other. Apparently, you can show mathematically that if one isomer has just a little advantage over the other, it will dominate as the organism grows, and the other one will disappear. Then the question is, why was it just a chance event? However, it didn't just have to be chance; it could be some optically active silica or clay that can give preference to one over the other. The selection has also been attributed to violation of parity as in radioactive decay. People keep working on this, but whether they are getting any closer to an answer, I'm not sure.

You have had some plastic surgery on your face. What happened?

When I was a teenager, I had an infection on my leg and I went to a dermatologist with a brand new X-ray machine. He cured the leg but than he said, you have some acne on your face, why don't I cure that too? So he gave me X-ray treatment for that. It was fairly early in X-ray treatment, and the doses were very large. Then about 15 years later, tumors began to show up, and they have kept growing, and I have two or three a year that have to be removed. Don't ever let your kids have X-ray treatment unless it's absolutely critical.

You have dealt with biological molecules, and there's so much progress in this field, but also a possibility of tempering with nature. Any comment on this?

Of course, I may only have a layman's comment, but I think there is a lot of good to be done in agriculture, in animal science, and in human treatment. There was a man here at Rockefeller, he was here when I came, in 1949, and he did probably the first such experiment. He was able to introduce a gene into two girls who had a disease in which a key enzyme was missing, and they were essentially cured. That was so early on, at least 40 years ago, that my friend never got any lasting credit for it.

What was his name? At least we can give him some credit here.

Stan Rogers. He was truly the forerunner of genetic treatment of people and I see no reason why it isn't going to be highly effective eventually. This assumes proper caution and oversight.

You received the Nobel Prize in 1984. Were you surprised?

The Nobel Prize is a real lottery. There are so many highly qualified people. I don't know how the committee can possibly weigh everything and make a good judgement. For me it was a terribly big surprise. Some people may have told me something about having nominated me, but they usually don't really mean it. So I didn't know that I specifically had been nominated and I certainly didn't expect to win. There are some people who sit by their telephone every year fully expecting a call. That's bad, however, because they are almost sure to be disappointed.

We had one case here; it was Avery, McCarthy, and MacLeod who discovered the transforming principle and the fact that the genetic material is nucleic acid. Everyone in this and related fields would say that was the most unfortunate oversight the committee ever made. It was a truly major discovery, and yet they didn't honor it. There were some critics at the time; some people would criticize just for the fun of criticizing, but evidently that affected the committee. They had six or seven years to decide, but Avery died and the prize is not given posthumously.

Did the Nobel Prize change your life?

Very little, except I get interviews occasionally and invitations. It didn't change the lab and didn't change my home life. We live in a small town in New Jersey, and the people there knew about it and they wanted to do something for me. They said, we'll name the library for you, but the librarian said, no, not while I'm alive. She said, he never comes in here. So they thought a while and finally decided to name the street we live on for me. It's a little street, just one block long. It's now Merrifield Way. There is also a Merrifield Prize in the local high school that was set up with volunteer donations, so I go there every year for the graduation ceremonies to give out the prize. Apart from that, I'm not very much for public engagements. The Nobel ceremonies were a great event though. They sent us two first class tickets, and we turned them in to the airlines to get tourist class so that the kids could come. After we were sitting on the airplane, a stewardess came back and said, we'd like you to change seats and come up here with us, it was on a 747, so they took us upstairs, all eight of us, and we were all impressed by that. We were also impressed by the ceremonies and the banquets and by the King and Queen. The royal family is very important for the prestige of the Nobel Prize, and I think they contribute much to their country.

Árpád Furka, 1999 (photograph by I. Hargittai).

16

Árpád Furka

Árpád Furka (b. 1931 in Romania) is Professor Emeritus of Organic Chemistry at Eötvös University in Budapest. He received a high school teacher's diploma in chemistry and physics from the University of Szeged in 1955, his Diploma in Chemistry and his Dr. rer. nat. degree from the same university in 1959, and his Cand. Sci. (equivalent to a Ph.D.) and D.Sc. degrees from the Hungarian Academy of Sciences in 1961 and 1971, respectively. He has been at Eötvös University since 1961, with a postdoctoral stint at the University of Alberta in Edmonton, Canada, in 1964-65 and an extended sabbatical at the Advanced ChemTech Company in Louisville, Kentucky in 1995–1999. He is best known for having developed the "portioning-mixing method for the synthesis of combinatorial libraries." Having originally formulated his ideas in 1982, he first communicated them in 1988, and that date may be considered the beginning of combinatorial chemistry. In 1996, he received the Leonardo da Vinci Award of Excellence from the Moet Hennessy-Louis Vuitton Foundation, in 1999, the Academy Award from the Hungarian Academy of Sciences, and in 2002, the Széchenyi State Prize of Hungary. We recorded a long conversation in September 1999 in Budapest and below are some excerpts from his narrative.*

I was born in 1931 of Hungarian parents, in Kristyor, Romania. This village is in Transylvania in a region of gold mines, and my father worked for one

*This interview was originally published in *The Chemical Intelligencer* **2000**, *6*(2), 37–40

of the mining companies. He was originally from this region but my mother was from the Great Hungarian Plain. My parents had six children. Our mother tongue was Hungarian but I went to a Romanian school. In 1942, my mother decided to return to Hungary and she took the children (except one of my married sisters) with her to Kunágota where she had come from originally. My father stayed behind because he didn't want to forfeit his pension. We were poor, my mother was ill, and there was hardship. I started working when I was still a child to contribute to the family in any way I could, and I didn't continue my studies after I completed the mandatory general school, when I was 14. There was unemployment and I had all kinds of odd jobs. For example, I helped in harvesting and the payment was in-kind; the wheat I earned fed us during the winter.

I always regretted that I couldn't study and eagerly read every book that came my way. Then, in 1950, I was among those who were offered the possibility of attending an accelerated course and completing my secondary education in one year instead of the customary four years. It was tough. We lived in the school and could leave only on weekends. We got a good education but everything was stripped to the bare essentials. Our certificate was called something like a special maturation, and this term has had a certain connotation in Hungary. To some, it signifies a deficient education; to others, tough perseverance. What I missed was the cultural environment that others, more fortunate than I, spent their childhood in. In any case, I regained some of the lost years in my education.

In 1951, I was sent to the University of Szeged to become a high school teacher of chemistry and physics. I didn't mind that I wasn't even asked about my preference because I always liked sciences. My dream, though, had been to become an astronomer. At the university my speedy high school education successfully withstood the test and not only did I keep up with my peers, often I could even help them in their studies.

In 1955, I got my teacher's certificate and went to teach high school physics in Makó, in Southeastern Hungary. After one year, however, I was called back to the University of Szeged, where I was charged with teaching in the Department of Chemical Technology. Gábor Fodor, who was Professor of Organic Chemistry, was directing our research. My first project was acyl migration, and I enjoyed it a great deal. The upheaval in 1956 interrupted my university work, which had barely started some weeks before. I used the forced break of several months to study English. I was then happy to resume my research when the conditions permitted it. Professor Fodor soon left Szeged for Budapest and later he immigrated first to Canada and then to the United States. He is now Professor Emeritus at West Virginia University.

In 1961, I was offered a position at Eötvös University and was transferred to Budapest. Again, this was not my own doing. After I had already been in the Department of Organic Chemistry for quite a while, I learned that its head, Professor Victor Bruckner, had wanted to fill the opening with someone else when I was virtually forced down his throat. This let me understand why I never became one of his favorite coworkers. I never felt fully at home in Professor Bruckner's department although there was no open hostility and nobody hindered my work in any way. My alienation was as much my own doing as their lack of trust in me. A small episode will give you a sense of the atmosphere that I contributed as much as anybody else to creating. Professor Bruckner used to have a weekly afternoon tea for his senior coworkers. He must have been somewhat reluctant to invite me, but nevertheless he sent word to me that I would be welcome at these afternoon teas. However, I sent word back that I didn't like tea. That must have sounded awfully rude, and it was, I see that now. But at that time, I didn't drink tea and I took the invitation literally.

Nevertheless, I enjoyed working in the new field of peptide chemistry. Just the other day, I found a notebook of mine from those days in which I had recorded some of my ideas. I used to take my ideas to Professor Bruckner but he never showed any interest in them. One of these ideas, in about 1962, was a solid-state synthesis of peptides.

In 1964, I got a postdoctoral fellowship from the National Research Council of Canada and spent one year at the University of Alberta. That year with Professor L. B. Smillie was very fruitful. Smillie had learned a technique at Cambridge for the determination of the position of disulfide bridges in proteins, using a combination of oxidation and diagonal paper electrophoresis. He also determined the amino acid sequences around the five disulfides of chymotrypsinogen B. Based on his preliminary work, I almost completed the sequence determination of this protein of 245 amino acids during my one-year stay. Smillie's reputation got a boost from this work, and, shortly afterward, he established a peptide research institute with himself as director. Last time we met was in 1993 at a symposium. Upon my return to Budapest, I couldn't continue this kind of work, as I lacked the necessary instrumentation, in particular, an amino acid analyzer. The head of the department made all decisions about what should be purchased. So I turned my attention to working out new methodologies, such as the C-terminal peptide isolation and how to enhance the solubility of enzymatic hydrolysates of proteins. Eventually, I defended my D.Sc. degree and, in 1972, I became full professor. Later on, I cooperated for years with the Chinoin Pharmaceutical Company in Budapest. We isolated natural peptides, determined their sequences, and then synthesized them and their analogues. At that time, I began dreaming about the possibility of preparing full series of small peptides, not only selected sequences. Although this seemed impossible using the available techniques, I kept thinking about it. Finally, and quite unexpectedly, I found the solution.

My initial thoughts about what is known today as combinatorial chemistry were also recorded in my notes. I described a technique for synthesizing mixtures of very large numbers of peptides simultaneously, and it included a methodology for selecting the active peptides among them. Thus, this was a description of combinatorial chemistry. We performed the synthesis part but, lacking partners for the rest of the work, the selection methodology, known today as the iteration method, was not tested in practice. I was looking for partners, but neither the biologists I approached nor Chinoin was interested in my proposal. However, Ms. Éva Somfai, who was in charge of patents at Chinoin, suggested to me that I have my notes notarized, which I did,[1] hence the date of 1982 on my notes. Ms. Somfai didn't think my proposal was suitable for a patent since it was about a research methodology. Later, however, others filed similar patents. At that time, I couldn't even think about filing a patent by myself as our salaries were meager. Since I couldn't do anything without a sponsor, and there wasn't any, only my notarized notes have remained from that time.

Two colleagues of mine, Dr. Ferenc Sebestyén and Dr. József Gulyás, worked on the experimental methodology, but electrophoresis did not clearly show the products. Mamo Asgedom, Dr. Sebestyén's doctoral student from Ethiopia, proved very successful in this project. Originally, we identified the peptides with a two-dimensional high-voltage paper electrophoresis technique. This technique was based on an observation made by R. E. Offord in the early 1960s. Offord worked in Cambridge where many peptides had been isolated and purified by paper electrophoresis. He had collected all the data and found that the charge and molecular mass of the peptides determined their mobility. He had set up a formula,[2] and I incorporated this formula into software to identify the peptides on the two-dimensional electrophoretic maps. This software gave us a computer-predicted peptide map, which could be compared with the experimentally established one. A built-in normalization procedure made the match between the prediction and the experiment perfect.

As for publishing my methodology, I presented it for the first time in poster sessions at two international symposia, in 1988, one in Prague[3] and the other in Budapest.[4] Following the two meetings, I submitted our first article in February 1990, and it appeared, after an initial rejection and a

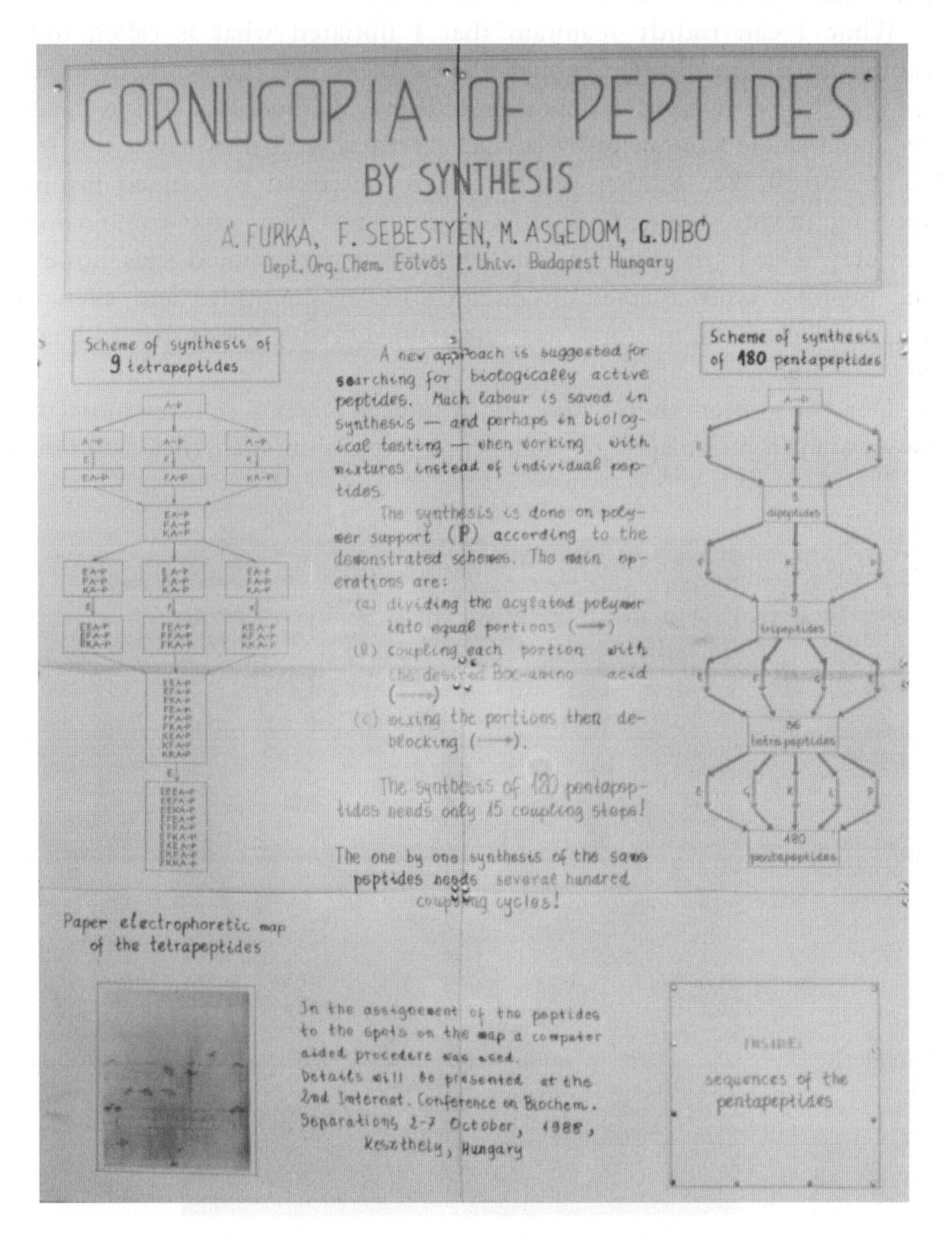

Poster in Prague, 1988 (courtesy of Árpád Furka).

long and painful delay, in 1991.[5] Subsequently, it proved very important that I had presented my methodology to international audiences in 1988 and that the contents of my posters had appeared in print. They have been much cited and today there can be no doubt as to the priority of my discovery, and I don't think there is any doubt either.

While I can rightly maintain that I initiated what is called today combinatorial chemistry, I would like to mention prior work by H. M. Geysen,[6] whose multipin technique preceded my methodology. This technique produced single substances by parallel reactions, and thus it was not truly combinatorial. Yet another technique also described by Geysen involved applying mixtures of amino acids in couplings that gave combinatorial mixtures.[7] Due to the differences in the reactivity of the amino acids, however, the peptides were formed in unequal molar quantities and even the composition of the mixtures was uncertain. For this reason the technique did not find wide application.

Our two posters didn't have much impact initially. It takes more than two posters from Budapest to catch people's attention. Also, they didn't show

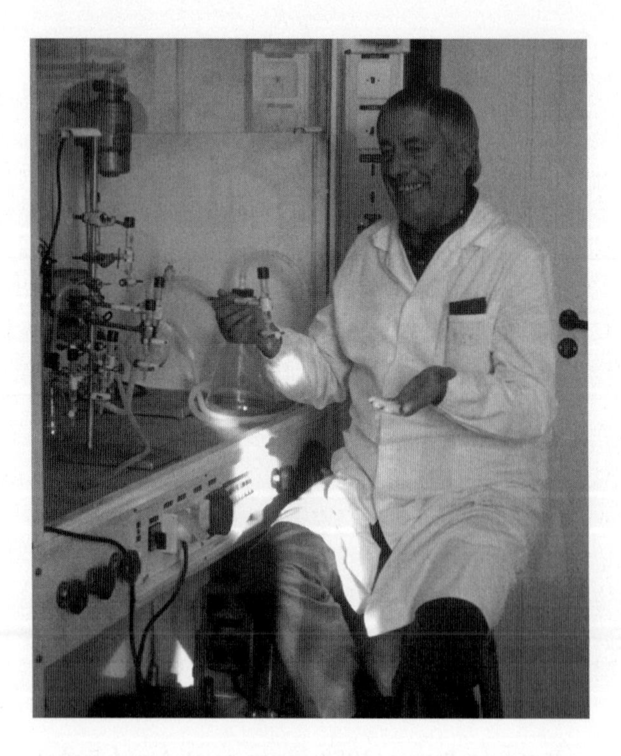

Árpád Furka in the lab in 1993 (courtesy of Ferenc Sebestyén)

how one could select the useful products from among all those present in the mixtures. It was a painful experience when, in my absence, Sebestyén presented our methodology to the Commission of Peptide Chemistry of the Hungarian Academy of Sciences and our peers rudely rejected it. This was between our 1988 posters and the appearance of our paper in 1991. Since then, the situation has changed, and in 1999, I received the Award of the Hungarian Academy of Sciences.

Recently, I've worked out a new approach to improve the applicability of my original methodology for drug research. This improvement is aimed at producing larger quantities of the desired products and at differentiating between the products. The original approach yielded very small quantities on each bead of the solid support, and you had to determine the structure of the substance on each bead in order to identify it. Professor Nicolaou of the Scripps Institute in San Diego has suggested a new approach in which the beads of the solid support are enclosed in a permeable capsule together with an electronic chip. This chip carries information about the identity of the product formed in the capsule. After each synthetic step, the capsules are sorted and regrouped. Those capsules that will be exposed to the same reagent in the next synthetic step are grouped together. This approach preserves the productivity of my original technique but, at the same time, offers the advantages of the parallel synthesis: larger quantities and known products.

My latest methodology, however, would eliminate the need for the chip or any other labeling of the solid-support units. Instead of labeling the support units, we arrange them into spatially ordered groups. The spatial arrangement is preserved during the chemical reactions and, before the next reaction step, the units are rearranged according to a predetermined pattern. With the use of appropriate software, a computer can track the synthetic history of each support unit and can predict the spatial position of each product. In model experiments, we prepared 125 tripeptides using Chiron crowns as solid-support units. We formed spatially ordered groups by stringing the crowns on polyethylene fish line. We used a simple, manually operated device to rearrange the crowns before the second and third coupling steps. After sorting, of course, the crowns were stringed again. Many different patterns can be used for regrouping, and each of them needs different software to predict the final position of the products. For our manual device, the "semiparallel" pattern gave the optimal sorting speed, and the prediction of the positions of the tripeptide sequences on the final three strings proved to be perfect. My present goal is to build an automatic machine, which would enhance the speed of sorting by a factor of 10 to 100. Alas, our grant application has been turned down by NIH because the technical description of our proposed automatic device was considered to be incomplete.

I have recently returned to Budapest after an extended sabbatical in the United States and am continuing my work at Eötvös University. I am 68 and I know my abilities. I'm good at generating one new idea after another, but I'm poor at making contacts with people and securing the necessary funding for my experiments. The lack of proper conditions for creative activities makes my situation almost impossible. But I don't want to ascribe all my difficulties to my surroundings. My own abilities have severe limitations and I just have to live with them.

If I had to single out the most significant of my current ideas that I would like to put into practice, that would be an improvement in the testing of the products of the combinatorial syntheses. My suggested approach wouldn't just test them for a few selected applications, but for all possible targets. The current testing techniques may leave products of great drug potential untapped because of the limitations in their testing. My new ideas would make a significant thrust in the direction of more complete testing. Once we would start the experiments I'm dreaming of, the continuation would be self-propelling. Today my dream may sound just as unrealistic as combinatorial chemistry seemed not such a long time ago.

If I could have three wishes, they would be good health, seeing more of the world, and getting my research funded. I could make annual support of $50,000 go very, very far.

During the past decade or so, a truly new field of chemistry has emerged. According to many, it has also led to a revolution in drug research. All the big pharmaceutical companies have started combinatorial chemistry groups and have invested millions of dollars in the field. The development of new drugs, however, takes time and the results will emerge gradually. There's also secrecy surrounding this kind of work. Beyond its application in drug research, this methodology is beginning to be employed in other fields. Yet the biggest benefit of this new field may not even be the new substances it produces but its impact on our way of thinking.

References

1. See http://szerves.chem.elte.hu/Furka/
2. Offord, R. E. *Nature* **1966**, *211*, 591.

3. Furka, Á.; Sebestyén, F.; Asgedom, M.; Dibó, G. In *Highlights of Modern Biochemistry*, Proceedings of the 14th International Congress of Biochemistry, VSP: Utrecht, The Netherlands, **1988**, Vol. 5, p. 47.
4. Furka, Á.; Sebestyén, F.; Asgedom, M.; Dibó, G. Abstracts, 10th International Symposium of Medicinal Chemistry, Budapest, Hungary, **1988**; p. 288, Abstract P-168.
5. Furka, Á.; Sebestyén, F.; Asgedom, M.; Dibó, G. *Int. J. Peptide Protein Res.* **1991**, *37*, 487.
6. Geysen, H. M.; Meloen, R. H.; Barteling, S. *J. Proc. Natl. Acad. Sci. USA* **1984**, *81*, 3998.
7. Geysen, H. M.; Rodda, S. J.; Mason, T. *J. Mol. Immunol.* **1986**, *23*, 709.

Guy Ourisson, 2000 (photograph by I. Hargittai).

17

Guy Ourisson

Guy Ourisson (b. 1926 in Boulogne-Billancourt, France) is Professor of Chemistry Emeritus of the Louis Pasteur University of Strasbourg and at the time of our recording this conversation was President of the French Academy of Sciences. He did his undergraduate studies at the Ecole Normale Supérieure in Paris, then received his Ph.D. degree at Harvard University in 1952 and a second doctorate, Dr.Sc. in Paris in 1954. He has been at the University of Strasbourg since 1955, first as Lecturer, then as Professor. He was Founding President of the Louis Pasteur University, Strasbourg (1971–1976), General Director of Higher Education and Research at the Ministry of Education (1981–1982), and Director of the CNRS Institute of Natural Products Chemistry (ICSN) in Gif-sur-Yvette (1985–1989), where his predecessors had been Edgar Lederer and Sir Derek Barton. Professor Ourisson's scientific activities have been in organic chemistry, mostly at the interface with biochemistry, biophysics, and geochemistry. Beside the French Academy of Sciences, he is a Member of Academia Europaea and Académie Européenne des Sciences et des Arts, a Foreign Member of the Royal Academies of Sciences of Sweden and Danemark, of the Indian National Academy of Science, of the Akademie deutscher Wissenschaftler Leopoldina, of the Academy of Rhineland-Westphalia, of the American Academy of Arts and Sciences and of the Institute Grand-Ducal of Luxembourg. He is an Honorary Member of the Chemical Societies of Belgium, Switzerland and the U.K., and has obtained many French and international awards and prizes. He has had an exceptionally active public life. We recorded our conversation in Professor Ourisson's office at the French Academy of Sciences in Paris on October 22, 2000.*

First, I would like to ask you to single out something from your research activities.

I am an organic chemist but I have worked much at the interface of organic chemistry and biology. I had had some training in biology and in "natural history" as it was called then. I also had a firm training in organic chemistry and more precisely in natural products chemistry. My first exploration of the possible use of plant tissue cultures for the study of plant biochemistry was in the 1960s, about 20 to 30 years before these tissue cultures became fashionable. My colleague Léon Hirth, a botanist, and I investigated with Pierre Benveniste the biosynthesis of sterols in tobacco tissue cultures. These are difficult to establish and to maintain, but once this is done, they provide the equivalent of microorganisms in culture. We soon found something unexpected, and could confirm it was quite general: the biosynthesis of sterols in tobacco, but also in other plants, does not proceed via lanosterol, like in animals or in fungi. Rather, it goes via an isomer of lanosterol, cycloartenol; this has been isolated in some tropical fruit, but had never been suspected to be a possible essential intermediate in the biosynthesis of the sterols. The difference between the two metabolic pathways, one via lanosterol and the other via cycloartenol, became understandable only decades later thanks to our study on the evolution of polyterpenes and to Benveniste's discovery that it was possible by selective poisoning to obtain maize plants devoid of sterols and containing instead only cycloartenol derivatives. This was my first exploration of the borderline between chemistry and biology.

Simultaneously, we also engaged in developing chemical criteria for the taxonomy of some groups of plants, in using modern tools for chemotaxonomic surveys. One of our two major targets was a family of large Southeast Asian trees, the *Dipterocarpaceae*, *Dipterocarpus*, *Shorea*, *Hopea*, etc., huge trees producing balsams, the gurjuns, which had some importance for perfumery. We managed to organize an expedition in Southeast Asia to collect good samples of fully identified trees.

*Excepts of this interview have been published in *Chemical Heritage* **2002**, *20*(1), 6–9.

Did you go there yourself?

No, unfortunately I did not, a trained tropical botanist, Dr. Norman Bisset, did. The work on *Dipterocarpus* gurjuns led us in fact also to explore the frontiers of chemistry with other fields, like, unexpectedly, geopolitics. We could identify in one species of *Dipterocarpus* growing in Cambodia one particular sesquiterpene, caryophyllene, which was at the time in high demand in the perfumery industry. This high demand had to do with post-war history. Caryophyllene is extracted from clove oil; cloves came mostly from Zanzibar, where they were grown by black slaves working for Arab merchants. When the liberation of Africa was under way, the slaves killed the Arab merchants, and cloves and clove oil production dwindled. Until production was restarted in Indonesia, the substitute we had found in the *Dipterocarpus* from Cambodia could have been used in the perfumery industry. We carefully defined the single species that had to be tapped for its gurjun and a perfume company from Grasse placed an order. In Cambodia, Khmer merchants sent Khmer workers to the forest to collect the oil, Chinese merchants acted as intermediaries, and when everything was ready for the first shipment from the port of Sihanoukville, a last check showed that the oil they had collected was in fact a mixture of gurjuns from various species of *Dipterocarpus*, and thus an awful mixture from which caryophyllene was very hard to isolate. The whole lot was valueless. Before a second attempt was made, the Americans started bombing Cambodia, and the whole operation was killed.

Was it a big investment?

No. It was a small investment made by a small perfumery company in France. I had just hoped even such a minor development could have helped Cambodian economy.

Also in the field of chemotaxonomy, our *Euphorbia* project was also very interesting and I regret that it has not been sufficiently published. *Euphorbias* are very widespread plants on all continents, very similar in that they contain abundant milk, latex, containing polyterpenes: toxic diterpenes and triterpenes. They are close cousins of another well-known genus of Euphorbiaceæ, *Hevea*, which produces another type of polyterpenes, rubber. Some *Euphorbias* are cactus-like trees; some are thorny bushes, or herbs. They are very widespread. Some species can be found in America, but many more in South Africa, North of the Sahara, or in Eurasia. *Euphorbia* is a huge genus, comprising 2000 species, and we studied the composition of the latex of some 400 species. This latex, as I said, always contains triterpenes, which we analyzed by modern methods.

In many of the species, those that the botanists considered being the most primitive, herbs or small non-specialized bushes, we always found cycloartenol in the latex. But the cactus-like species were much more varied, and we managed to correlate the changes in composition of these lattices with the distance of the location of the *Euphorbia* from the center of their distribution, which is in South Africa, and with the morphological criteria familiar to botanists. Plate tectonics became therefore a third component in this correlation: morphology or habitus, location related to plate tectonics, and latex composition. We postulated that the cycloartenol-containing species were the most primitive, and that the other ones derived from them by simple changes in the relative position of active sites in the triterpene-producing enzymes, all leading from the same precursor, squalene epoxide, to the various triterpenes. This implied that the presence of some triterpene in the latex was important enough to be a constant character, but that the nature of this triterpene was irrelevant, and therefore free to evolve from cycloartenol to lanosterol, to euphol, to pentacyclic isomers, etc.

Did this project involve a lot of people?

It did involve a lot of help from individuals, from botanists in Africa, who collected the latex for us, from curators of botanical gardens, from friends who sent us specimen plants. I collected some myself in Jamaica and in the French Indies. It would have had to be followed by a study of the structure of the enzymes producing the various triterpenes isolated, but this was too difficult at the time: it is only very recently that the structures of the first cyclases of this family have been defined.

These two projects should probably have been further pursued, but I gave them up in order to yield to a suggestion made by a geologist in Strasbourg, my friend Georges Millot, who was a specialist of sedimentary geology. All sediments contain organic matter, but exceptionally in the form of deposits like those of coal, petroleum or natural gas. Usually, sediments contain only maybe 2% of organic material, but there are millions of cubic kilometers of sediments, and the total amount of organic matter they entrap is considered to outweigh by a factor of 10,000 the organic matter present in all living organisms. George Millot's question was a simple one: "What is this organic matter? Does it contain interesting substances?" Obviously, it did not look like a simple problem to solve, and I resisted. I resisted until George finally convinced me by bribing me with the offer of a mass spectrometer. This was in the late 1960s and mass spectrometers were still rarely available in chemistry laboratories. With Pierre Albrecht and others, we started analyzing shales, expecting to find intractable mixtures. But soon it became apparent that we were not isolating just complex mixtures of unclear origin in the sediments, but that in every sample, whatever its origin, whatever its age, we always found the signature of one and the same family of substances. At first, it made no sense, that 250-million-year sediments, fresh garden soil, shales, clays, sandstones, should all contain substances giving similar mass spectra. The common feature was the presence in the extracts of substances whose mass spectra showed they fell in the range of the molecular weights of triterpenes, around 300–400, and always gave an extremely intense peak at 191, with a similar pattern of fragmentation. Eventually, we could isolate these substances and define their structures; that was partly a consequence of the dissertation of Jean-Marie Lehn on the structure determination of triterpenes.

These sedimentary substances are indeed representatives of one single family of triterpenes, the hopane family. These hopanoids are pentacyclic triterpenes, a family we knew because they occur also in the secretions of Dipterocarpaceae, in *Hopea*, hence their name. Some hopane derivatives were known, from the resins of tropical trees, and now these exotic substances were turning up in every sample of sediments, whatever its origin, its age, its depth. This finding was not only unexpected, we found it also hard to explain. Moreover, once we could isolate the most complex hopanoids from sediments, the most complex geohopanoids, we found that they displayed another peculiarity, which was very hard to explain. They contained not only 30-carbon atoms, like any normal triterpene, not only fewer carbon atoms, as could have been expected from a degradation in the sediments, but also C_{31-35} substances, and the supplementary carbon atoms were in the form of a straight chain of carbon atoms. No naturally occurring substance was known with such a skeleton. Then, Michel Rohmer could solve the puzzle by showing that these C_{35} geohopanoids came from C_{35} biohopanoids, present in a large variety of bacteria. Geohopanoids are the molecular fossils of bacteria. The first bacterial triterpene, a C_{35}-structure, was in fact identified by a Canadian group in a bacterium they were studying in the hope of obtaining a substitute for wood cellulose. We repeated their experiments and found that their substance was indeed a good precursor for our fossil triterpenes.

Ronald Breslow, Guy Ourisson, and Feodor Lynen in 1978 (Photograph by Roussel Uglaf, courtesy of Guy Ourisson).

The final state of the affair is that we have geological hopane derivatives; we call them geohopanoids; and we have established more than 250 structures, from C_{25} up to C_{35}. We have also the microbiologically produced hopane derivatives, the biohopanoids, with more than 50 structures established. These biohopanoids play an essential role in bacteria: they are membrane reinforcers. Plants and animals have cells containing cholesterol or closely related derivatives to reinforce their membranes. Bacteria do not contain cholesterol but their membranes contain the biohopanoids, which are similar in shape and polarity to cholesterol. The geohopanoids are excellent tracers of maturation of sediments: they are universally present; they are most abundant natural substances overall, although they are never present in large concentrations. The biohopanoids are essential for the life of many bacteria. Michel Rohmer has been working on their biosynthesis and has discovered that it follows a new pathway, different from the classical one described by Fedor Lynen and Konrad Bloch who received the Nobel Prize for it. Rohmer's discovery has been extremely rewarding, and it has been used in petroleum exploration and for creating large new families of substances. I owe a great debt to Georges Millot for having pushed me into this project.

You have served as a consultant for or on the board of a large number of companies. Is this usual?

No, it is not usual, at least to that extent. I was almost born and spent my youth within the walls of a chemical factory. Maybe for that reason, I am at ease with industrial problems. I think that I have played a useful role in several companies. Of course, I can't be sure and the only positive sign is that my contracts have been renewed.

How does it work?

Usually, a few times a year I would spend a couple of days in the company. I was also available for advice between those visits. This concerned the consultantships. Another company asked me to run an audit of their research activities. I set up a small group of colleagues and we visited all their research groups and reported critically on their research organization. From this a permanent Scientific Council of the company grew out, and the same then happened with another company. In the chairmanship of one of these Scientific Councils, Jean-Marie Lehn succeeded me and I was happy to see this position to remain in the family. Both Jean-Marie and Pierre-Gilles de Gennes had become members of this Scientific Committee before they received their Nobel Prizes.

As you said, the extent of my involvement has been unusual and one of the questions to be solved for me was what to do with the money. Instead of simply putting it into my pocket, I started using the money in the lab. I gave out fellowships, for example. The money made me partially independent; it has given me a degree of freedom. I have had the same secretary for 40 years, and the Ministry or the University does not pay her.

Your main area of research has been natural products chemistry. How do you view the slowness by which chemists considered biological macromolecules as part of their trade?

The first organic chemist to really take the chemistry of nucleic acids seriously as part of organic chemistry was Alexander Todd, and from Todd came Har Gobind Khorana. I remember well when Khorana published the first synthesis of a gene: many organic chemists considered it as a betrayal of organic chemistry. Todd and Khorana knew that nucleic acids had a structure to be defined or synthesized as precisely as any organic substances, and not just some approximate structure. Proteins also used to be something outside organic chemistry. Fortunately, this is no longer the case.

You had an early excursion into computational chemistry at the beginning of your career.

Indeed, I was one of the firsts to use molecular dynamics for a problem of kinetics, to explain a rate acceleration, which was not understood. That was in my dissertation, back in 1953. The problem was in the strained structure of the sesquiterpene I was studying, longifolene, and I had found strange rate acceleration in a rearrangement of that structure. I tried to explain it by strain release. I had molecular models of the ground state and of the transition state. These were ball and stick models in which I measured the distances with a ruler. I used a logarithmic table for the computations. Fortunately, it gave clear results, and I received my Ph.D. This was for my second doctorate, for which I had no mentor. This demonstrates that I was an early believer in computational chemistry, even though it was a Stone Age form of molecular dynamics.

How do you feel about combinatorial chemistry?

I can only quote Pierre Potier, who is also a natural products chemist, who said about combinatorial chemistry that it is what God did too. The natural products in living organisms are just the products of combinatorial chemistry. As for myself, though, I have never been tempted to go into it. The real intellectual challenge is in the deconvolution of the mixture and in the identification of its components.

What is your current research interest?

When we performed our work on the bacterial hopanoids, and found their importance as membrane stabilizers, I became aware that we had a possible answer to two paradoxes, which had bothered me for years. In all textbooks of biochemistry, you can always find the statement that membranes are very important because all living organisms have membranes around their cells. When the building materials of the cell membranes are described, the phospholipids, in the best textbooks they add that cholesterol must also be present to reinforce the membrane; otherwise it's not stable. Cholesterol requires molecular oxygen to be present, and molecular oxygen is supposed to have been formed late, after the formation of the first photosynthetic organisms. Besides, straight chain fatty acids, about C_{16}–C_{20}, required to obtain the present phospholipids, are extremely difficult to produce. If the Fischer–Tropsch reaction, a reaction producing long chain organic substances, had worked, the Germans would have had arms *and* butter during World War II. From this we deduced that neither cholesterol nor the classical phospholipids could have played a role in ancestral membranes. When we found out that cholesterol is replaced by hopanoids in bacteria, it changed the picture. By then we had already found fossil hydrocarbons the origin of which we could not understand: C_{15} – C_{40} branched-chain hydrocarbons. These new microorganisms have been found in recent years, the so-called Archea, hyperthermophilic bacteria, hyper-acidophilic, and hyperalkaliphilic bacteria. They contain original phospholipids formed from polypropanyl chains, C_{15}–C_{20}, C_{40}. They contain no sterols. We started working on these phospholipids and established a putative "genealogical tree" of all existing polyterpenes, ending with cholesterol as the most evolved one. Then we could establish a stepwise succession in the genealogy, by observing that a new enzymatic reaction was involved with each step. Going backwards, we could then deduce what should have been the most primitive members of the family, which could have been primitive membrane constituents: they appeared to be simple phosphate esters with one or two polyprenyl chains. We synthesized them and found that they are indeed capable of forming very good membranes. We are now studying the hypothesis that these primitive membranes could have given rise to primitive "protocols," and we are studying how they could have become spontaneously more and more complex. This is a completely new way of looking for mechanisms of the origin of life.

Can we now go back to the origin of <u>your</u> life?

My family background is as mixed as you can imagine it. My father was born in Poland, in Lodz, and came to France to study mathematics in 1905 after a short bout in prison for his modest role in the Revolution of that year. In 1982, I met a cousin of his who had last seen him when they were both in prison ... His sister had also come to France at the same time to study medicine.

Was Ourisson his name?

Yes. Read it as "Uri's son." And read *Exodus* to see who was the son of Uri, Betsaleel. The son of Ur was the artisan called by Moses to build the Ark: he was versed in all kinds of manual skills, and, in addition, he had the skill of teaching. A good patron!

Religion?

Not at all. I am a very dry atheist. I am, however, very interested in the fact that Man is usually religious. My father had had a rather superficial religious training in the Torah, which was, however, sufficient for him to like endless theological discussions with members of the catholic clergy in Thann. I witnessed some. He was a dry atheist, and so was my mother, who had had a classical catholic education. My parents wanted my sister and me to have a non-sectarian religious education, and asked the Lutheran minister in Thann to give us private lessons in religion. Later, we followed "catechism" with friends, taught by the very old catholic priest of our village in the Dordogne. I have also read a lot on other religions and I am not anti-religious, even though I do not understand and hate sectarism. On the whole, I think that neither God nor the Devil have resented that I do not believe in them: the second has blessed me with all kinds of privileges, which could have led me to Hell, and the first has given me the wisdom not to take them seriously.

Back to the family, a great-great-uncle of mine, Pavel, was a rather well known mathematician, who died in France in the 1920s, having drowned in Brittany where his tomb is maintained by the township of Batz. The history of Pavel was written in a small booklet for Russian children to incite them to be good students. This was of course back in Soviet times. My father studied mathematics from 1905 until World War I. He gave lessons in mathematics to a friend, Joseph Blumenfeld, the brother-in-law of Chaim Weizmann, who taught him some chemistry in exchange. In 1914, at the outbreak of World War I, my father was not called in the army for medical reasons but was sent to a chemical factory to work for the army. He worked on pyrotechnical projects, and became a chemist. After the war, he was hired by Joseph Blumenfeld, who had become an influential industrialist, to work in a small very old factory, in Thann, in Alsace. The factory was small, but quite well known: it was there that racemic acid had been discovered and that Pasteur had gone to investigate it. Blumenfeld and my father launched there original productions, first the production of titanium dioxide as a white pigment, then the electrolysis of potassium chloride from the neighboring potash mines, then various other productions. My father worked first as a chemist, then as the technical director, and finally as the General Director of the factory. During World War I, the factory had been completely destroyed because the front-line had been stabilized at mid-factory. When World War II came, we, of course, left, and the factory was, again, destroyed. Rebuilding the factory after World War II completely exhausted my father and he died in 1947 when he was 55 years old.

Where did you spend the war years?

In southwest France. This is the other part of my background. My mother was from the Périgord, not far from Bordeaux. Her father was from the region, and her mother was born from the marriage of a local noble man and a Russian princess whose parents had left Russia because they were Catholics. The Russian family was the Ermolovs, and the most famous hero of the family was the famous Ermolov who conquered Caucasus for the Tsar, and had to fight in the mountains of Chechnya. They came to France with plenty of money, which they managed to spend in two generations. My grandmother was born in a mansion and died in poverty. We have a small house in Périgord, and we spent the war years there.

How did you become interested in science?

I was born in Paris but as I said we moved soon in Thann, in Alsace and we lived in a villa within the walls of the factory. When the wind was blowing from the east, it smelled chlorine. From the west, it smelled sulfur dioxide. My friends and I played in the wooden frame of lead chambers that had been built by M. Gay-Lussac and were still used to produce sulfuric acid. I was always interested in plants and insects, which were abundant in the small woods and the meadows adjoining the villa. My father was very generous and absurdly confident in letting my sister and me, and all our friends, do all kinds of things that would be strictly forbidden now. We played with sulfuric acid or zirconium bars, with mercury droplets, I visited all the workshops, there were not even gates to hinder entrance and the workers and the engineers were our friends. Today, this would be inconceivable.

In 1940, when I was 14, we left, and in 1944, my mother returned to live in Paris while my sister and I continued our studies there, while my father had to rebuild the factory and spent much time in trains between Paris and Thann. For two years, I went to Lycée Saint-Louis in Paris to prepare the entrance competition for the Ecole Normale, and then I was at the Ecole Normale. The final degree there was the Agrégation de Physique et Chimie. This degree authorized me to teach in lycées. It would be a kind of master's degree, which I received in 1950 when I was 24 years old. Then I went to Harvard.

How did you decide to go there?

Blumenfeld again played an important role. He insisted that I should go. Simultaneously, my Professor at Ecole Normale was probably the only French professor who, as soon as the war was over, had insisted that students from the Ecole Normale should go out. The first student, Philippe Traynard, was sent to Sweden as early as in 1944; Marc Julia went to England in 1945 and Serge David in 1946, and all three have become leading scientists in France. Two others went to the United States, the first, Germain, to Ipatieff in Chicago, and then it was my turn. By correspondence, I had been accepted for a doctorate in chemical oceanography in La Jolla and

Guy Ourisson and R. B. Woodward among other participants of a meeting in New Delhi in 1972 (courtesy of Guy Ourisson).

I had also considered studying chemical paleontology with Barghoorn at Harvard. The decisive moment was Professor Fieser's visit at Ecole Normale, which made everything easier, and I became his graduate student. Looking back, it would have been better to study with Woodward or Stork, but Fieser came to Paris, and he wanted anyway to get a few French students for his collection of nationalities. My closest friend in Fieser's group was Koji Nakanishi and I had very good friends also in Woodward's group. It was at Harvard that I discovered that I was a European. It was the decisive period of my life intellectually.

I spent two years at Harvard, during which I took 21 exams. The fact that I spent only two years at Harvard was due to a misunderstanding. The pamphlet that I had received from the Admissions Office at Harvard said that the "minimum requirement for a Ph.D." was two years, so I went for the minimum. I had a job waiting for me after these two years at the Ecole Normale. So, I took my Ph.D. in two years. I only discovered many, many years later that it had been totally abnormally short, and I think there are extremely few students who have managed to do it so quickly. I know one — no, two. There are sometimes advantages not to be too well informed. But I had to work hard.

My wife and I had married before going to America. She worked in pharmacology with Professor Otto Krayer. He and Mrs. Krayer were from Freiburg/Breisgau. They also helped us to understand we were Europeans. During our stay there, our son was born. He was therefore an American citizen by birth and he lives now in America, where he works in a small company offering advice for quality certification (ISO, etc.). I prefer to say that he sells quality. We also have had two daughters, who are living in Strasbourg; one is a very successful pediatrician, the other a very successful literary translator from English and from German. I have six grandchildren.

When did you move to Strasbourg?

I have been here since 1955. I am at home in Alsace from my childhood. Some of my childhood friends were killed during World War II, either in the French army or in the German army. Strasbourg is a border region between France and Germany.

Can you single out one person who had the most important influence on your career?

Probably Derek Barton, indirectly as I never worked with him. He had already left Harvard when I was there but he visited us and I had a nice discussion with him. When I returned to France, he invited me for a visit at Birkbeck College in London. Then, when I started my Ph.D. in France, I worked on the structure of a sesquiterpene, longifolene, which one of Barton's students was also working on. They used X-ray crystallography on one of its derivatives and obtained its structure, but they had rather poor chemical data. I, on the other hand, had good chemical data but no complete structure. When Barton was ready to publish their findings on the basis of the structure of the crystalline derivative, he almost published a wrong structure. However, we were in contact and I had told him what we knew, so that we could publish simultaneously, and both published the right structure. Barton helped me acquire a stronger scientific ambition than I would have had otherwise. He was always friendly and always ironic and difficult to please. I was finally his successor as Director of the Institute of Natural Products Chemistry in Gif-sur-Yvette, when he left. I owe him a lot.

What are you most proud of?

I am most proud of not having missed Jean-Marie Lehn and a few others. I sent Jean-Marie to Woodward and this gave him an enormous boost. Jean-Marie and I have remained good friends. I am also proud of having been the founding President of Université Louis Pasteur 30 years ago and to see that it has remained one of the best if not the best university in France. I am also pleased by my experience in numerous contacts with industry and in various administrative positions, which I could dispatch without giving up my scientific activities. I have had a very good life.

France has not been doing very well as regards the Nobel Prize.

You are right, and it is by contrast amazing how well Hungary has done, as is amply demonstrated by the busts in the corridor of your Academy of Science. We can take some comfort in the fact that there has been no similarly successful family in this respect as the Curie family. However, the level of French research had gone down terribly between the beginning of the 20th century and the 1990s. The consequences of the two World Wars have been disastrous, not so much for the material destruction, but for the loss of many of the best scientists who, especially in World War

I, were sent to the front. This did not happen in England and did not happen to the same extent in Germany. This was an important factor. Then, the university system was totally centralized, and what could be called research was done only in Paris between the two wars. One exception was Strasbourg where the take-over of the German University made it possible to do groundbreaking work. For instance, Louis Néel, the physics Nobel Prize winner, had begun his work on magnetism there. Apart also from other exceptions, including some medical schools, it was extremely difficult to have anything done outside Paris. The exceptions included Grignard and Sabatier. Another negative factor was chauvinism, leading to isolation. Contacts with German science were obviously excluded, but contacts were also minimal with American and British science! Contacts were normal only with other French-speaking scientists. Even after I had returned from Harvard, I was constantly called to help with translations. French pride hindered learning foreign languages. The only good years for French science were the "years of the General." The decade of de Gaulle's reign between 1956 and 1968 was the time of real development.

Things have changed completely now. Contacts with the outside are extremely close. What we have not yet learned is to work together. The famous saying states that "*Gallia omnia divida est in X partes.*" France is divided into 60 million individuals. Frenchmen will not easily work together and they continue to lose *Bella Gallica.* And to return to your question, there are two independent factors needed for the Nobel Prize. One is the quality of the potential candidates, and the other is to accept to make a slightly collective effort to push them a little.

How concerned are you, as the President of the French Academy of Sciences, with this question?

I am very much concerned, of course. The Nobel Prize is so much more prestigious than any other prize, and it is considered, rightly or wrongly, as a measure of national success, nearly as much as a football championship. We do give some good prizes, which are not as heavy in monetary value, but which will never reach the fame coming from one hundred years of excellence. Also, our prizes may one day reach the same amount as the Nobel Prize but the amount we spend on the search for worthy laureates would still be far lower than what the Swedes spend on it. My reaction to this year's chemistry Prize was, like that of many of my colleagues, "*Who?*" followed by the notion that I have to look it up.

So, did you?

Not yet. You are very inquisitive.

Are you making any efforts to produce Nobel laureates?

In several ways. The French Embassy in Stockholm is on the lookout for possibilities to bring brilliant French scientists to Sweden to give lectures early enough, and repeatedly. In France, we try to be polite with our Swedish colleagues. People of equal scientific standing, if they are Swedes, have a slight advantage, and this is not a unique French idiosyncrasy.

Are you also on the lookout for talent in France, to provide special support?

That we certainly do in many rather modest ways: the Institut Universitaire de France gives "junior" appointments with lower teaching duties and higher automatic funding, the CNRS or INSERM, or the Ministry of Research, give special grants for carefully selected young group leaders. The Academy tries to identify young talented scientists and to give them Prizes. We have also launched European Symposia, *Scientia Europæa*, reserved to scientists younger than 40, nearly 350 Europeans, from 35 countries, have already been invited there. But all this is really insufficient. Jean-Marie Lehn or myself have been scientifically independent at 28; I do not see how even somebody much more gifted could manage to do it now.

Is there anything higher than being President of the French Academy of Sciences?

In the scientific establishment, in a formal sense no. But from the point of view of real power, it is more important for instance to be head of CNRS. There are two kinds of clout. The Academy has the honor, and we have many small prizes to give. Such a prize may be the first sign for a young scientist that he is recognized. However, the Academy runs no laboratory. We mobilize our members and many non-members when we prepare our reports on specific topics. On the other hand, the CNRS runs French science, it has the money, and its head has a lot of power. In the biomedical field, the same could be said for INSERM.

What is the structure of the Institut de France?

The *Institut de France* is the parent body of five academies:

— **Académie Française**, 40 members, not one more, and they are in charge of the purity of the French language. They are not scientists although scientists may also become members for their literary merits. Thus, François Jacob and Jean Bernard are members.
— Then there are **Académie des Inscriptions et Belles-Lettres**, for archeologists, Egyptologists, etc, **Académie des Sciences**, **Académie des Sciences Morales et Politiques**, for economists, lawyers, political scientists, and **Académie des Beaux Arts** for the fine arts. The five Academies are independent, but have a few activities together, and they share the buildings. The *Institut* is the overarching structure, the only legal entity.

How many members are there in the Académie des Sciences?

Right now we have 150 members plus around 250 corresponding members. We are working hard to try to obtain from the Ministry the right to increase this number.

What is the ratio of women?

About ten percent and this reflects a considerable recent increase. A major signal was the election six years ago of the first woman as President, Marianne Grunberg-Manago, and another election last year, that of Nicole Le Douarin as the first woman Secrétaire perpétuel.

What does the Science Academy do?

We provide solicited and unsolicited advice to the government on science and technology and we try to foster science through symposia, colloquiums, lectures, meetings in Paris and in the other scientific sites; we maintain close relations with the other national Academies. We provide in fact more and more frequently some of our members to other national academies to take part in international evaluation committees.

Where does your budget come from?

From the government and from donations. We try to court rich widows. We are not very successful. We have little recognition to offer to our donors: we can name a Prize in their name, and they are usually not really excited at the idea that we shall just post a marble plaque in a corridor to remind generations of their generosity.

Do the members get paid?

They get a sitting fee in proportion to their attending the meetings; the total acquired by a scrupulous attendance does not exceed a one-month salary, annually.

Can you explain that François Jacob was elected to the Academy 11 years after his Nobel Prize.

I can only guess why. In his time traditional biology dominated the scene and molecular biology was crushed. Later there was a backlash and now there is a lack of traditional biologists.

English is an advantage for American and British scientists and Hungarian and Japanese are a disadvantage for Hungarian and Japanese scientists, respectively. How is it with the French?

The French scientific community has finally understood that it was necessary to overcome the barrier. We have a French-Italian symposium in chemistry where everything is in English. We speak English as well as the Italians do. On the other hand, when I am with Italian colleagues, I insist that they speak Italian and I speak French. I refuse to speak English to an Italian who had had French in school and I had learned enough Italian to participate in a bilingual conversation although I am not able to speak Italian. However, between two Latin-speaking people it is indecent to speak English. Now, when you say that Hungarian and Japanese are disadvantageous, I cannot agree: you can nearly always hide behind Hungarian or Japanese, but if you speak French, you can never be sure not to be overheard and understood by somebody.

Imagine, a French scientist and an English scientist meet or a French and a German. What will they speak?

Guess! Bad English. I speak German with Germans (not too badly), but this is an exception.

Do the French journals accept papers in English?

Yes. The important thing is to have good papers. On the other hand, when something is in French, it should be good, and in good French, unambiguous. I have often claimed that if I ever managed to do something really exceptionally good, like making an artificial living organism, I would publish it in French.

What field would you enter today if you could start it all over?

I would continue doing chemistry or I would go into paleontology from the chemical side. But I am doing right now what I like to do, and I will continue doing it as long as I am in good health.

Mildred Cohn, 2002 (photograph by M. Hargittai).

18

Mildred Cohn

Mildred Cohn (b. 1913 in New York City) is Benjamin Rush Professor of Biochemistry and Biophysics, Emerita, at the University of Pennsylvania School of Medicine. She took her undergraduate degree from Hunter College and her Ph.D. degree in chemistry from Columbia University under the mentorship of Harold Urey. She worked with Vincent du Vigneaud at Cornell Medical College. Following 14 years at Washington University School of Medicine, she has been at the University of Pennsylvania since 1960. During her career she has co-authored papers with many other famous scientists, such as Nobel laureates H. C. Urey, V. Du Vigneaud, F. Lipmann, G. T. Cori, C. F. Cori, and E. W. Sutherland. Paul Boyer, another Nobel laureate talked about Dr. Cohn's pioneering research on the use of ^{18}O with great appreciation (see, Boyer interview in this volume). In Boyer's words, "Her work led to the discovery that mitochondria catalyzed a rapid exchange of phosphate oxygens with water. Her measurement of the ^{18}O exchange reactions played a crucial role in the development of the binding change mechanism for oxidative phosphorylation." She is a Member of the National Academy of Sciences of the U.S.A. (1971), the American Academy of Arts and Sciences (1968), and the American Philosophical Society (1972) among other societies and organizations. She has received many honorary doctorates and other recognitions, including the Cresson Medal of the Franklin Institute (1975), a Senior Scientist Humboldt Award (Germany, 1980), and the National Medal of Science (1982). She is past President of the American Society of Biological Chemistry. Her husband was the late famous theoretical physicist, Henry Primakoff. We recorded our conversation in her home in Philadelphia on March 20, 2002.*

First I would like to ask you about your family background.

My parents were immigrants from Russia. My father was studying to be a rabbi, but he decided not to be ordained and he became a militant atheist. His family was rather wealthy in Russia and he was the only member of his family who came to this country when he was 21 years old. My mother, on the other hand, came with her family to America when she was 17 years old. It was soon after the unsuccessful Russian revolution of 1905. I never found out exactly why they left Russia, because they never talked about it, but I know there were pogroms against the Jews in Russia at that time. They were middle class, but their financial situation was deteriorating. One of my aunts was very angry with my grandmother because under financial stress, she kept the maid, but sold the cow, and there was no milk for her. My parents had known each other in Europe, but got married in America, and I was born in New York City.

Which part of Russia did your parents come from?

It was Russia when they left, then it became Poland, then it was Russia again, and today it is Belarus. They spoke Russian rather than Polish, in addition to Yiddish. I learned Yiddish as a child and I was bilingual. My father was very interested in Yiddish culture, he was a "Yiddishist." From the time I was 13 to 19, we lived in a cooperative that was dedicated to preserving Yiddish Kultur.

How did you become interested in science?

When I was in high school, we had a very good chemistry teacher. We had two years of chemistry instead of the customary one year in most high schools. My best friend's father owned a couple of beauty salons and manufactured cosmetics. He told both of us that cosmetic chemistry was a great opportunity for women. By the time I went to college, it was Hunter College, I was convinced that I wanted to study chemistry.

*This interview was prepared by Magdolna Hargittai.

Then in the first year I became interested in everything else I studied, and I wasn't sure anymore. I was arrogant and I thought that other things, such as the humanities and social sciences, I could learn by myself, but science I had to study in school. I became interested in physics, but Hunter did not offer a major in physics at the time. Women were not supposed to be interested in physics. So I stuck with chemistry and minored in physics. The standards at Hunter were very high. Of the three women Nobel laureates in science who had got their education in the United States, two had graduated from Hunter College. This is not to say that the training in science was strong at Hunter, it was weak, but in spite of that, its graduates have done very well.

Were your parents supportive of your interest in science?

My mother didn't want me to go on to graduate school, she wanted me to become a school teacher. My father encouraged me to become whatever I aspired to become although he was a realist and knew that there would be obstacles, he was aware of all the discrimination. He was disappointed when my brother decided to become a lawyer, because my father's dream for my brother was for him to become a Ph.D. in philosophy. When my future husband started courting me, my father said that although he was not a philosopher, he was the next best

Mildred Cohn in 1927 (courtesy of Mildred Cohn).

thing, he was a theoretical physicist. My father had a great respect for scholarly work. It was in the Jewish tradition too, but it was not only that. If you look at history, the first generation of Jews that were liberated, they were outstanding achievers. After that they became part of the mainstream. There are many Jewish Nobel laureates, and most of them are children of immigrants. Their children will not be Nobel laureates. So it is not only a question of Jewish tradition, which is important, it is also a question of the first liberated generation whose parents wanted them to achieve.

Do you suppose there is a genetic component?

My theory, for what it is worth, is that only the smart ones survived. With all the persecution over the ages, the ones who weren't ingenious and didn't know how to survive, didn't.

How did you continue after Hunter College?

I was determined to become a research chemist. First, I went to Columbia University. I had applied to 20 graduate schools for scholarships but had not received any offer. At Columbia, I could not get a teaching assistantship since only men were permitted to be assistants. The year was 1931. I used my savings from summer jobs and babysitting. At Columbia at that time you could not do research as a graduate student until you had passed your qualifying exam. That took about a year and a half of course work. So that first year I was very disappointed. But my teachers did not disappoint me; there were Hammett and Urey among them. At the end of the first year, after I was awarded my Master's degree, I had to leave school and find a job because I didn't have any money left. It was the Great Depression and jobs were scarce in 1932. Fortunately, in my senior college year I had passed a federal civil service examination for Junior Chemist and that enabled me to land a government job at the National Advisory Committee for Aeronautics. First, I was assigned to do computational work, which I did not like because it was done with manual calculators. Finally, I was transferred to an engine research department, which needed a chemist. After two years, I was able to return to graduate school at Columbia and became Professor Urey's Ph.D. student. The first three months I spent learning how to blow glass, because Urey thought that every one of his students should know how to do that. Though we had a glassblower at Columbia, Urey thought that we might end up in a place where there wasn't one. Then I started on a research problem separating carbon-12 and carbon-13. Urey had a method of using chemical equilibrium of the isotopic exchange reactions for separating stable isotopes. I spent a month or two doing calculations on various equilibria to see which one and at what temperature you got the largest fractionation. Although at that time it was not possible to determine individual partition functions, it was possible to calculate the ratios of partition functions of isotopic species from spectroscopic data. I found that the best reaction, on paper at any rate, was between $^{12}CO_2$ and ^{13}CO, that is, between the isotopic species of carbon dioxide and carbon monoxide. If you could bring those into equilibrium, the fractionation factor was quite good. However, our mass spectrometer was under construction and it was not working properly and I was working for a whole year without knowing whether I was getting anywhere or not. Finally, Urey sent me down to Princeton where Bleakney, who was at that time the outstanding expert of mass spectroscopy in this country, was working and he analyzed my samples for me. The results were all negative. At that point I was ready to give up graduate work and Urey apologized to me profusely. He said he should've never given a student a problem that depended on an instrument that didn't exist yet. His apologies were all very well but I had wasted a year. At any rate I didn't quit and I did go back. My future husband persuaded me not to give up. Then Urey suggested that I study a problem with ^{18}O because I could measure the ^{18}O water by density methods. Of course, for density determinations we had to be able to prepare and handle very pure samples and there was always the danger that ^{16}O from the reagents would contaminate our sample. Fortunately, our mass spectrometer started working and that made possible accurate ^{18}O analyses.

Urey had just won the Nobel Prize when you joined him.

I had decided to work with him before he got the Nobel Prize and joined him a few months after his award. But I had decided that I wanted to do my graduate work with him when I had attended his lectures, he was so inspiring. The Nobel Prize did not make any difference. He was very enthusiastic about his subject. He had a profound insight in chemistry and physical chemistry in particular. Actually, he did not like the term physical chemistry, he preferred the term chemical physics for his type of research. He was the first editor of the *Journal of Chemical Physics.* He started a new seminar series in chemical physics; he didn't like the content of the physical chemistry seminars. I remember when I started working on the acetone — water ^{18}O exchange, I first investigated it in the gas phase, which is what he wanted me to do. Then I thought, why not try it in solution. I did and I found that the exchange was easily measurable at room temperature. In the gas phase I had to go to 80°C to get an exchange. I went to Urey and told him that I ought to study this reaction in the liquid phase and he said to me, "I know something about the gas phase and I know something about the solid phase, but the liquid phase is a complete mystery to me, and I do not do experiments where I have no theory to guide me." I told him that I thought it would be much more interesting to study the liquid phase because we could study acid and base catalysis in solution. So Urey suggested to me that I talk with professor Hammett who knew all about the liquid state. I did it in solution and it resulted in my thesis.

Urey was a remarkable man. Of all the scientists I ever worked with, he is my favorite. He was very naïve about social questions. It was at a time when Hitler was on the rise; I was in Urey's laboratory from 1934 to 1937. He was great friends with Professor Rabi in the Physics Department and he had suggested to Rabi that he move to New Jersey where Urey lived. But Rabi preferred to stay in New York with his own people. Urey told me that he understood Rabi's position since Rabi was a Jew, whereas he, Urey, could probably accommodate to the fascists if they ever took over. But Urey by then had already signed a petition that Columbia University should not participate in the celebration at Heidelberg, Germany to which Columbia had agreed to send a representative. When Urey signed the petition he was immediately labeled as an anti-fascist. Although Urey was very naïve, he learned fast and later he became very active in political causes. He was the only one on the faculty there who wore a Roosevelt button; all the others were Republicans. He was also very generous. He was the only one who cared about the underdogs, the graduate students. He was worried whether they had enough support and in particular he worried about me. One day he said to me, "Miss Cohn, what are you doing for money?" He had gotten me financial aid. At that time they had something called NYA, National Youth Act, like work study program for students today. He said the only thing he wanted me to work on was my own thesis. The aid was available only during the academic year. In the summer, he said to me, "Miss Cohn, why don't you let me lend you some money? Ever since I got the Nobel Prize, I wanted to use some of it to help my students. Some day when you have a job, you can pay me back." There aren't many professors who would do that and certainly not today. He never forgot that he had been a poor boy himself, a farm boy. He told me once that the only reason he had ever gone to high school — there was no high school where he lived so he had to go and board somewhere — was due to his uncle who died and left him $300, that's how he managed to go to high school.

Were you the only woman student there?

No, he had a woman student before me and while I was there he had two women postdocs. People are under the impression that there were no women around in those days. The statistics show that there was a larger fraction of female graduate students then than there were in the 1950s and 1960s. There was a real drop in the 1950s and 1960s in the percentage of women Ph.D.s awarded in this country in science. My interpretation is that this drop occurred because the men came back from World War II and the women were displaced. During the war years, women took up a lot of non-traditional jobs. But afterwards to get the women out, they were told that women should get married and have children, and if they did not bring up their children themselves until they were at least five years old, the children would become monsters and so on. That had an effect. It took awhile before women returned to attend graduate school again, it did not happen until the women's liberation movement started.

After Urey, you went to work for Vincent du Vigneaud, which was quite a different experience.

First of all, medical schools were generally very different from chemistry departments. The medical schools were structured on the German model, they were wholly hierarchical. They would hire a man as head or chairman of the department and he would come in with a whole group of his own and everyone else left. The man at the top determined everything. In most medical schools there was only one professor in each department and everyone else worked under that professor. Du Vigneaud was very paternalistic, but he wasn't mean; he ran a tight ship. You had to take your vacation in August whether you liked it or not. At one point, after we moved to Cornell, he got very busy and he wanted everyone to write him a note every day of what they had done. I refused to do it and I got away with it. That was because I was the sole physical chemist and I was the only woman too. He read those notes and kept very close control of what was going on. He ran a team with very specific goals and he determined all the problems that everyone was to work on. He was doing big science in a day when not many people were operating quite that way. Not all medical schools were like that because when I got to Cori's department, it was different. There, too, was only one full professor, but he let me do independent research after he had found out that I was determined not to be a member of a team anymore, having done that for 9 years with du Vigneaud. It took only 6 weeks for Cori to realize that I was adamant and after that he let me do what I wanted to do and he supported it.

Please, tell us about the Coris.

I wrote a chapter about them in a book called *Creative Couples in the Sciences.*[1] I quoted Carl Cori relating how complementary he and Gerty were to each other, "Our efforts have been largely complementary, and one without the other would not have gone as far as in combination."[2] They were remarkable in that way. He would start a sentence and she would finish it. They were completely complementary. In personality, they were very different. He was aloof and she was very vivacious, she was outgoing and he was not, although he was very insightful about people. They let me read their Nobel speech before they presented it, for my comments. I showed it to my husband, (Carl had written half of it and Gerty had written half of it) and he could tell exactly where one stopped and the other began. Carl's part, succinct and unadorned, read like a theorem in Euclidean geometry. I also wrote a memoir of Carl Cori for the Biographical Memoirs of the National Academy of Sciences, U.S.A.[3]

Which of your scientific result would you consider most important?

There are conceptual and methodological aspects. I found something in my studies using ^{31}P NMR that is very important for enzymology. Studying enzyme reactions involving ATP, I found that the equilibrium constant of the step between the central complexes, that is, the enzyme with two substrates on it and the enzyme with two products on it, is always close to one even if the overall equilibrium constant is 10^4. That's an important finding as far as understanding how enzymes work. The discovery of the ^{18}O phosphate-water exchange in the mitochondria was important too. Methodologically, I was the first one to show that you could see the three phosphorus atoms of ATP well separated by NMR. That has had clinical applications.

The extent of muscle diseases can be followed by the decrease of ATP concentration. Surprisingly, I have one paper, which is a citation classic and that's the one where I showed the effect of metal ions on the NMR chemical shifts of ^{31}P in ATP. It is now used even *in vivo* to find out how much magnesium there is by measuring the chemical shift of the phosphorus atoms in ATP. One can determine how saturated ATP is with magnesium. Now it's used to determine magnesium concentration in the brain.

Which of your results then do you remember with the greatest pleasure?

I mentioned this in my chapter in *Annual Review*.[4] That was when I found the isotopic shift due to ^{18}O on the NMR phosphorus chemical shifts of ATP because I had come full circle. I was now showing an isotopic mass effect in a spectroscopic parameter and that's where I started in my graduate work. It isn't a very important finding, but it gave me great pleasure to discover yet another effect of isotopic mass of a stable isotope with a spectroscopic manifestation.

I would like to ask you about your husband, Henry Primakoff. When did you get married?

I met him at Columbia in 1934 when we were students in a physics laboratory course. He was a senior in college and I was a graduate student. He already knew that he did not want to do experiments. He wrote long lab reports because for every experiment he discussed the theory at length. He was a very brilliant student, straight A. By the time he got his bachelor's degree, they also gave him a Master's degree because he had taken so many graduate

Family photograph. Mildred Cohn, her husband, Henry Primakoff, and their three children, Laura (1949), Paul (1944), and Nina (1942) (courtesy of Mildred Cohn).

courses. For the first year of his graduate studies, he went to Princeton. It was almost equivalent to getting a scholarship in those days because Princeton only charged a hundred dollars tuition for a year. Admission was difficult because only two or three students per department were accepted each year. The Institute for Advanced Study had just been founded and Einstein was there and Pauli was a visitor. Wigner and Condon taught in the physics department. When he was not awarded a fellowship in his second year, Henry went to New York University where he was awarded one and he got his Ph.D. degree there. When he finished his graduate work, we were married. Henry was always very quick and could come up with sharp, relevant comments in discussions instantly when it was needed and not an hour later as I often do. Once in answer to a question after his lecture at a meeting, he said, "Symmetry is like a bikini, what it reveals is trivial and what it conceals is essential." Henry was wonderful on his feet.

Who followed whom with the jobs?

I always followed my husband around. But I will say that he never took a job in a place where I could not have a job. At that time all state universities had strict nepotism rules; a husband and wife could not both have jobs of any kind. In private universities, each one had its own rules. For example, in Washington University, husband and wife could not work in the same department, but they could work in different departments. They made an exception for Gerty and Carl Cori. No exception could be made in a state university.

So this made it difficult for you.

Of course it did because I always had to ask for a job, I was never offered a job. There was an exception though. Around 1947, I got an offer from Johns Hopkins University. They wanted me to run a mass spectrometry facility. They said that they understood that my husband was a physicist and offered to find a job for him too. He was quite willing to go if I wanted the job. I showed this letter to Gerty Cori and she told me to frame it, that she had never received a letter like that.

Children?

I have three children, six grandchildren, and two great-grandchildren.

How did you manage?

You manage. There were times when I had crises, of course. Occasionally, a caretaker for my children walked out on me without notice, on some occasions I had to fire women, but then in St. Louis, I eventually found an excellent woman who stayed with us for 30 years. In New York, they had professional baby-nurses as they called them. In St. Louis, they never heard of them. At that time, unfortunately, you couldn't put a child in a child-care facility until they were three years old. The first two years were always very difficult, but even later you still had to have someone at home after school hours and if a child wasn't feeling well.

Did you have live-in help?

Occasionally, for short periods I did, but in general, I did not. The women came in early and left after dinner, so they worked long hours. The way I managed was the way men do when they compartmentalize their lives. When I was at work, I didn't think about my children and when I was with the children, I did not think about my work. I had a mathematician friend, she was a Ph.D. She had two children about the same ages as mine. She lasted three weeks in her war job. She worried all the time at work about what was happening with her children. You can't do that.

Did your husband participate?

My husband was very European in this regard. Whenever I asked him to do something, he said, "Hire somebody." He never participated in housework, but often played with the children and invented stories for them. Occasionally he would take a child to the doctor. On the other hand, he was so supportive; he just assumed that I would have a career. That was a given. I was the practical one in the family. If a child broke a toy, I was the one who repaired it. He never did anything practical, not only traditional women things, he never fixed the car, he was not interested in the garden, and so on. He was very cerebral.

How did your children take it that their mother had a job?

My oldest was always complaining. When she started school, she was the only child in the class whose mother worked. In those days it was not common for middle class women to work outside the home. My youngest, on the other hand, who is seven and a half years younger, remarked once when I mentioned possible retirement, "What would you do, play bridge?" When my oldest child was a college student, it still bothered her that I was not always at home when she was a child. She majored in psychology and she submitted a paper on the children of working women versus non-working women. She found, fortunately, that there were no more problems with children whose mothers worked than with children whose mothers did not work. She became a professional woman herself. All three children obtained Ph.D.s, my son is a biochemistry professor and both my daughters are psychotherapists.

Were you ever criticized by family doctors or teacher for being a working woman?

Yes. The chairman of the chemistry department at Hunter College had told us that it wasn't ladylike for women to be chemists. Why was he teaching chemistry in a women's college? He wanted us to be teachers of chemistry. I got a lot of criticism from relatives. When I had saved up money for my education, one aunt said that it would be better for me to spend the money for straightening my teeth. A great-aunt said that I would educate myself out of the marriage market. After I had my first child, my mother-in-law carried on a campaign to get me to quit working, but she didn't succeed either. One of the reasons my oldest daughter reacted as she did was that when she was in second grade, she was 7 years old, she joined the Brownies and when the woman who ran it found out that I worked she told my child that I was a bad mother. There was a lot of social pressure against the mothers working.

What would be your advice today for a young woman who would like to do science and have a family too?

The first thing I would suggest is, marry the right man. That's the most important thing. You have to have a husband who is fully supportive. That he does more than paying lip service to equality. My husband was really a feminist. He liked women and respected them. My second advice is that whatever decision they make they shouldn't feel guilty.

You experienced discrimination.

It took me 21 years to get a faculty position. I was always a research associate. That effected me financially too because since it was not a tenure-track position, there was no pension. And I was paid less than anybody else. On the other hand, there were certain advantages. Since I did not have a tenure-track position, I was not in competition with my male counterparts and that helped me to maintain excellent relations with the men. I was no threat to them. Also, I didn't have to teach, which meant a much more flexible working day. If a child was seriously sick, I simply stayed home. I couldn't have done that easily if I were teaching. The most important thing from the scientific point of view was that I was not under pressure when I was at Washington University, I could pursue long-range problems. I didn't have to publish or perish. It took me two years to publish my first paper after I got there. I had to build a mass spectrometer, I had to work out methodologies, and this took time. If I were on a tenure track and had to worry about getting tenure, I couldn't have afforded to choose a problem that I knew would take at least two years before I had any results from it.

Isn't the tenure system a problem even today for women? The decisive period of time is when you have to establish yourself in your workplace and build your family at the same time.

Some schools make some allowance for this. They offer the option of delaying the tenure decision and so on. But dropping the tenure system would not be the correct solution. I never was concerned for tenure. I am a child of the Great Depression and I learned that nothing is secure. I wonder if there was ever a study made comparing women with children with women with no children from the point of view of attaining tenure. Once I made a study of the women at the National Academy of Sciences. More than half were married and of those a very high percentage had children. These are the women who had achieved. I know that many people talk about the problem of tenure for women and I think it should be more flexible, but I am not at all sure that it is a problem of primary importance. In any case we are far away from the problems of my youth. When I was graduating from Columbia's Ph.D. program, all the big companies sent recruiters to the campus in the spring. It was still the Great Depression period in 1937, jobs were scarce and most graduates sought industrial jobs. When the notices appeared on the bulletin board of the chemistry department, they read, Mr. So-and-so from such-and-such a company will interview all perspective Ph.D.s of this year. Male, Christian. Nobody even objected. It was taken for granted; that's the way it was.

Did you ever experience anti-Semitism?

In chemistry, it was very strong at that time. I knew I could not get a job in industry or in a university. There was a brilliant student at Columbia University and the faculty considered him for a faculty position, but he was a Jew so they didn't take him. There wasn't a single Jew on the faculty of the chemistry department at Columbia when I was there. I went back

35 years later to give an invited seminar and at a luncheon with the faculty, I noted that at least half of the faculty was Jewish.

Are you religious?

No.

Did you or your husband participate in war-related work in World War II?

My husband did, he participated in the sonar project. Urey invited me to join his project at Columbia that was part of the Manhattan Project and as I later learned, it was on the separation of uranium isotopes. I had just borne my first child. I visited Urey and he told me that there was a job for me and that I would have to work until midnight every day. Du Vigneaud, for whom I worked at the time, said that we should not abandon basic research because the war would be over one day and we should not let the whole research infrastructure disappear. He wanted me to stay with basic research. Du Vigneaud was involved in war research on the penicillin project. With all these considerations, I decided not to join Urey's project. Then Robert Oppenheimer called Henry and offered him a job in Los Alamos. The Manhattan project was a very well kept secret. The scientists knew but nobody else did. My husband said that this project was not for this war, but for the next war. He wanted to stick with sonar, which was for this war. So he didn't go. He was almost right.

What do you feel about Heisenberg?

I saw *Copenhagen.* My husband knew Heisenberg. Heisenberg was here in the summer of 1939. Henry spent five weeks at a workshop at Purdue where Heisenberg was the chief attraction. Henry got to know him since it was not a very big group. My husband liked Heisenberg very much. Heisenberg told them that there was going to be a war. He was a German nationalist, there was no question about that. I don't think he was a Nazi, but this is a borderline thing, being a nationalist when the country is governed by Nazis. His claim that he made after the war that he deliberately sabotaged the German bomb project is questionable. There is the other view that Heisenberg did not try to sabotage anything, he just made a mistake in his calculations, so he did not succeed but not for want of trying.

You have worked with a lot of great scientists. Do you have any heroes?

Of those I have known and worked with, certainly Urey comes as number one. He was a man of integrity and had the fastest mind I have ever known.

Any message?

A piece of advice for women scientists. They should stress the scientist more than the woman. I remember in the 1970s being on various committees and I was very unpopular because they wanted women to be on all the committees and have all kinds of administrative responsibilities and so on. I said, let women alone and let women do their science and let them influence others by their example. Even my fellow women scientists did not take it well. One can easily become interested in the power aspect. That's all right for those women who want to have that but they should not be encouraged to take up other activities as a substitute for their science. When the government decided to have a woman on every advisory committee, I was asked to sit on five study sections, three at NIH and two at NSF. There weren't enough women to go around. I have limited such activities. I have been active in my professional society and at my university I agreed to serve on two committees but no more. I read about a famous physicist who was considering going to Israel, but he hesitated. His wife persuaded him to seek Einstein's advice. So he went to Einstein and Einstein said to him, "I'm a scientist first and a Jew second." The same thing is for women. They should be scientists first if that's what they are interested in, and a woman scientist second.

Do you have to be aggressive?

It depends on your definition of aggressive. You certainly should stand up for your rights but you shouldn't be going around with a chip on your shoulder just because you are a woman.

References

1. Cohn, M. "Carl and Gerty Cori: A Personal Recollection." In *Creative Couples in the Sciences.* Edited by Pycior, H. M.; Slack, N. G.; Abir-Am, P. G., Rutgers University Press, New Brunswick, New Jersey, 1996, 72–84.
2. Cori, C. In *Les Prix Nobel en 1947*, Stockholm, 1949, 23.

3. Cohn, M. "Carl Ferdinand Cori 1896–1984." *Biographical Memories*, Vol. 61, 79–109, National Academy Press, Washington, D.C., 1992.
4. Cohn, M. "Atomic and Nuclear Probes of Enzyme Systems." *Annu. Rev. Biophys. Biomol. Struct.* **1992**, *21*, 1–24.

Paul D. Boyer, 1999 (photograph by I. Hargittai).

19

Paul D. Boyer

Paul D. Boyer (b. 1918 in Provo, Utah) is Professor Emeritus of the University of California at Los Angeles (UCLA). He shared half of the Nobel Prize in Chemistry in 1997 with John Walker (b. 1941), MRC Laboratory of Molecular Biology, Cambridge, England, "for their elucidation of the enzymatic mechanism underlying the synthesis of adenosine triphosphate (ATP)." [The other half of the 1997 chemistry Nobel Prize went to Professor Jens C. Skou (b. 1918) of Aarhus University, Denmark, "for the first discovery of an ion-transporting enzyme, Na^+, K^+-ATPase."]

Paul Boyer obtained his B.S. in chemistry in 1939 from Brigham Young University and his M.S. and Ph.D. in biochemistry in 1941 and 1943, both from the University of Wisconsin. Following a professorship at the University of Minnesota, Dr. Boyer moved to UCLA in 1963. He founded the Molecular Biology Institute at UCLA in 1965. To mention but a few of Dr. Boyer's numerous distinctions, he has been a Member of the National Academy of Sciences of the U.S.A., has received the Rose Award of the American Society of Biochemistry, and served as editor of the 18-volume treatise *The Enzymes* [Third Edition, Academic Press, New York, 1971–1990].

We recorded our conversation in Anaheim, California, on March 22, 1999, during the Spring National Meeting of the American Chemical Society.* Before the interview, I attended Dr. Boyer's lecture to high

*This interview was originally published in *The Chemical Intelligencer* **2000**, *6*(2), 16–21

school science teachers about his science and his career. Among other things, he told his enthusiastic audience: "I admire you because you teach science to high school students and because you are producing ATP even as we speak," and "I don't know an enzyme that I don't love. They are my life."

Could you please summarize the research that led to your Nobel Prize.

At the time I was in graduate school, it was recognized that living cells capture energy from oxidation of foodstuffs by making adenosine triphosphate (ATP) from adenosine diphosphate (ADP) and inorganic phosphate. The ATP is then used in a myriad of functions — muscle contraction, nerve, brain, and kidney function, metabolic syntheses, and solute transport. How this oxidative phosphorylation occurrs remained for many years a major unsolved problem in biochemistry.

The solution to the problem required important advances in the study of protein structure and function and in enzymology. In the 1950s and 1960s, my laboratory began studies on how the ATP synthase functioned, using, among other approaches, measurement of the exchange of phosphate oxygens with water oxygens. Our studies gave useful information but left the major problem awaiting clarification.

But in the early 1970s, we recognized an unusual feature, that the energy from oxidations was used primarily not to make the ATP molecule but to release ATP from the enzyme. Soon, we also found unexpected evidence that the catalytic sites on the synthase participated sequentially; each of three sites in turn passed through identical conformations for binding, covalent bond catalysis, and release. As the 1980s approached, we obtained results that could be best explained if an internal core of a smaller protein subunit or subunits rotated with respect to an outer structure with three copies each of two larger subunits, and with catalytic sites at the interface of the larger subunits.

This postulate of a rotational catalysis found welcome support in the 1990s from the structural studies in John Walker's laboratory at the MRC Laboratory of Molecular Biology in Cambridge, England, and from studies in Yoshida's laboratory in Japan that provided a visual demonstration of the rotation. The discovery that the ATP synthase is a remarkable molecular machine depended on researches of many capable investigators. I am fortunate that my group attained the first insights into the unusual mechanism of this catalysis and that this led to the Nobel recognition ...

You have said that the Nobel Prize does not make anybody wiser. But does not it make you feel freer to speak about a broader range of issues than you would otherwise?

It does not make you freer, but it does provide you with more opportunity, sometimes more than you want. If anything, because of these opportunities to speak outside your field, you feel your deficiencies more.

What introduced you to chemistry?

My first real exposure was in my high school chemistry class, but I received a chemistry set when I was about 10. I remember playing with it in the basement of my parents' house, and it may have interested me more than I realized. My interest was maintained by science classes in junior high school.

You went to Brigham Young University. Was its affiliation with the Mormon religion important to you?

I was raised in the Mormon faith. I have left the church, and never joined another one. In my biographical sketch for *Les Prix Nobel*, I quoted from the sketch of a previous Nobelist, Harry Kroto, on this subject. He stated: "I am a devout atheist — nothing else makes sense to me and I must admit to being bewildered by those who in the face of what appears to be so obvious, still believe in a mystical creator." And then I added: "I wonder if in the United States we will ever reach the day when the man-made concept of a God will not appear on our money and no longer must be invoked for political survivor by those who seek to represent us in our democracy." Those of us that feel that the search for truth comes in ways separate from religion and that religion does not have things to contribute to the truth of why we are here and what nature is about are too silent. This is why I decided in my biography to make a statement. We have to fight ignorance. When we make laws like one that says there should be zero level of carcinogens in any foodstuff, it reveals a disappointingly poor understanding of science. That is what our legislature did, and such action reveals the inadequacy of the people that are doing our legislation.

Was it Federal legislation? Did it pass?

It was and it passed. Political and not logical forces prevailed. The legislators should have recognized that we cannot have a zero level, it does not make sense.

You made a statement that you often made lucky choices based on limited information. Of course, we always have limited information, but this statement means more than that.

In the choice of my career, I never had a background to tell me what my career would be like. I had little concept of what biochemistry or what graduate school meant. I never recognized that they would open up such a marvelous vista for me. This was a lucky choice.

When I look at research work, as you plan it, there is a big difference if you wait until you are 99% or over 99% sure before you go to the next stage versus how much progress you make if you wait until you are only about 95% sure and yet go to the next stage. If you start on several paths, you will discover more things if you don't demand that surety. It's a research philosophy that you may accomplish more if you're willing to speculate on what might be next and move to test that before you've absolutely got the base to know you should make the move. That's what I mean when I talk about a lucky choice on limited information. An example of an unlucky or poor choice was when I thought that a protein bound phosphohistidine we had discovered was an intermediate in oxidative phosphorylation. But I should have been more cautious at that stage.

Was then the characterization of phosphorylhistidine as an intermediate of oxidative phosphorylation an exception to your lucky choices? You've said that you'd rather take back the paper that reported it.

Not the papers that reported the discovery and characterization of the phosphohistidine, but the ones that said we thought it was an intermediate in oxidative phosphorylation. That was a learning experience. But I perhaps tend to move ahead to the next possibility more readily than many scientists do. If you want to finish off a research project, make sure that everything seems right in that area, document everything, and so forth, you may never get to the next important aspect.

When did you formulate your research philosophy?

I recognized my characteristics in the 1970s. I always pressed my students and postdocs for the evidence and to see both sides of every question. But when it gets to a stage where it looks like something important may be becoming visible and it feels right, I may move on without having the surety that I am going in the right direction.

How much speculation was involved in your main achievements, the postulation of the binding change mechanism and the rotational catalysis of ATP production?

The rotational aspects remained more speculative and were presented that way in my publications. I only said that the rotational catalysis was the best way I could explain the present data. We also had some experiments that people did not pay much attention to but they added considerably to the evidence that there was rotational catalysis. It was enough for me to think that the speculation was likely to be correct, but it remained speculative until John Walker's results became available. On the other hand, I did say that alternate side participation, the second facet, had been adequately established.

What John Walker determined was a frozen structure. Was it a complete proof?

No, not at all, but it made it much more probable. However, even though it was a frozen structure and had ADP at the tight site, it was still representative of the mechanism of the catalysis. The structure also made possible other experiments to probe the catalysis.

If I want to test myself to see how sure I am of something, I would say, "Would I bet a granddaughter on it?" I would not bet a granddaughter on the rotational catalysis; I have barely reached the stage to bet all the scientific support in this country on the validity of rotational catalysis, but not a granddaughter.

You have mentioned that your seminal manuscript on the binding change mechanism had been rejected by the Journal of Biological Chemistry *before you published it in the* Proceedings of the National Academy of Sciences of the U.S.A. *Would you suggest revision of the peer-review system?*

It would be better if the peer-review system did not make mistakes, but I don't think we should try to change it much. Unusual results are often difficult to accept. The two most original things I ever published, I had a hard time to get published at first. Besides the binding change mechanism, the other dealt with how isotope exchange measurements could be used to probe the mechanism of enzyme catalysis. It was finally published in the *Archives of Biochemistry and Biophysics*, where I had originally submitted my manuscript, but only after an extremely difficult time with the reviewers. It was not such a major breakthrough, but for me it was one of the most original things I have done.

You did your graduate studies at the University of Wisconsin.

At that time, Wisconsin had a much better biochemistry department than most other universities. For me it was a lucky choice. The state and responsible administrators deserve credit for their accomplishment.

Then you went to Minnesota.

Had Wisconsin offered me a job, I would have chosen Wisconsin. In Minnesota, I rose readily to full professor in biochemistry. Then UCLA invited me and I moved to California after having spent 17 years in Minnesota. I was elected to the National Academy of Sciences shortly after I went to UCLA, but it was on work I had done at Minnesota. In Minnesota, my main contact was Rufus Lumry, a Professor of Physical Chemistry. Lumry has never been well recognized, but he understood the conformation of proteins and their ability to capture free energy. This was back in the late 1950s.

When you have the three-dimensional structure of proteins, it is held together by many secondary weak bonds, hydrogen bonds. There is not a fixed unique structure. The structures can be deformed by interactions. The sites can be deformed through this matrix of the scaffolding by another site. Lumry appreciated this earlier than others in the field. This is where I got the feeling that conformational changes could be at the core of catalysis, long before we came to the binding change mechanism. I was appreciative of the beauty of the protein molecule, that it was not a fixed structure. From hemoglobin we learned that free energy ranging up to 10–20 kilocalories might be accommodated within the protein structure to be reused again.

I also interacted with the physics department. If it had not been for the physicist Al Nier in Minnesota, I would never have been able to do my ^{18}O work. But it was Mildred Cohn who first introduced me to the ^{18}O technique, through the literature. I did not meet her until later. Here is an example of a woman who had tremendous ability but who was in secondary positions in academia practically until the time when she was elected to the National Academy of Sciences. Mildred continued to make important contributions to enzymology and oxygen-18 measurements that influenced my research over the years.

Mildred Cohn, 1991. She pioneered research on use of the ^{18}O, which led to the discovery that mitochondria catalyzed a rapid exchange of phosphate oxygens with water. Measurement of the ^{18}O exchange reactions played a crucial role in the development of the binding change mechanism for oxidative phosphorylation (courtesy of Paul Boyer).

Alfred Nier, around 1960. He participated in the development of mass spectrometry, including the instrumentation used in the Mars probes. He provided ^{18}O and mass spectrometers that made possible Boyer's researches with ^{18}O (courtesy of Paul Boyer).

Henry Lardy, biochemist, around 1990 (courtesy of Paul Boyer).

Mildred did not continue to study oxidative phosphorylation, and I believe that she considered some of my approaches as not well based but perhaps thought that they might be on the right track. In many ways she is a better scientist than I, in regard to capability, accuracy, knowing what she publishes, being sure that it is right. I doubt that she has any publications that she would like to disappear.

Southern California is a great scientific center of the world. I would be curious to know whether this concentration of great scientists sustains a sizzling intellectual life here?

It is not an intellectual center in the sense that great scientists get together in any organized way on a routine basis. You have people who seek to talk to people who are highly qualified and you have a lot of highly qualified people in Southern California. It is not that there is an elitist group, but there is depth in many fields. I can get interactions with highly respected colleagues in biochemistry, and this is where most of my interactions occur. I do not interact much with organic chemists, for example, because our fields are different. We tend to be narrow. We don't move outside our science. I am disappointed in the university that we do not get more interactions. Now that I am emeritus, I get more interactions with emeriti of other fields; we get together for that. But this was not the case before I became emeritus. My wife and I were members of a tennis club and we got into more interactions with people from other fields in the tennis club than from academic programs.

I would like to ask you about your family.

My wife used to do editorial work in a computer center at UCLA and helped me edit the series of volumes, *The Enzymes*. We have three children. Our son is a surgeon, one daughter is a lawyer and writer, and the other has an MBA from UCLA.

You have said that we have the remarkable ability to blind ourselves to unpleasant facts.

The world is heading on a potentially disastrous course. I am not even sure that we will continue to inhabit our planet. The facts are very prominent, but society wants to disregard them. Our politicians disregard them. Both the politicians and business people are interested in short-range problems.

Our democracy is set up to respond to immediacy but does not have the mechanism for long-range planning. This concerns population control and the protection of the environment, just to mention two areas. I am afraid that we are ruining our planet for our great and great-great grandchildren. We are headed on such a collision course that I sometimes feel glad that I will not be around to see the results. Other times I am more optimistic — there are some signs that disasters and learning could guide humanity into a way to keep and enjoy a steady-state world.

What do you consider to be your legacy?

I will be known for the ATP synthase. I will be known at UCLA also for founding the interdepartmental Molecular Biology Institute. My legacy otherwise will be my wonderful family relationships. My wife has been their principal architect.

You said that you have closed your research laboratories. But if you could be planning research for a few decades from now, which direction would you choose for research?

I have a tremendous interest in living processes. I would be interested in investigating what determines three-dimensional structure, with the eventual goal of being able to design catalysts. When we become wise enough, we will be able to do that.

Enzymes are large molecules. Isn't it possible that during evolution they may have acquired parts that are not necessary for their actions?

That is not so. If you look at the three-dimensional structure of an enzyme, it is not just the positioning of the catalytic groups that is important. These are not rigid structures. The catalytic groups have the ability to change their positions. In order to have an efficient catalyst, the enzyme has to recognize the substrate, and for this it must be in an appropriate conformation. To position requisite groups properly likely requires a big structure. You need a lot of scaffolding to get the groups in the right specific positions. It has also to do something else. It must also be able to recognize the transition state. That involves again considerable structural change. The enzyme, in order to act as a catalyst, has to go from the structure for substrate binding, then for binding to the transition state, and then to a form that provides a reasonable binding of product. Finally, it has to change conformation back to the initial structure to get ready for the next reaction. This is why catalytic antibodies are never going to be very successful. The molecule must have the capability to take a substrate through a guided tour through the transition state to product. That is part of the reason why enzymes need large structures. Evolution has been taking care of all this. There is then an increasing complexity if you take our ATP synthase and compare it to the *Escherichia coli* ATP synthase. Our ATP synthase is much more complex than the bacterial enzyme because Nature had to put other protein subunits on it for various control functions. There are many parameters that need control, and the enzymes must also be turned on and off.

So if I would be starting a few decades from today and would really like a challenge, and considering the great growth of available information, I would love to work toward making a synthetic enzyme. At the same time, I would look toward rational pharmacology (instead of a combinatorial approach). We are getting closer and closer to accomplishing rational design of drugs.

As for other aspects of the living processes, the field of immunology is at the forefront. We need to know much more about how the immune system functions and how it is controlled. This would be among research areas of interest to me. And those areas I have mentioned might not be the best ones. One way to make almost sure to be wrong is to try to predict the future.

Are you involved in biotechnology? Do you have patents?

No, I'm not and I don't. I did not get into the biotechnology area. Mostly my work did not lead to that. Also, I am not enthusiastic about it. The universities are getting too involved in this. I purposely tried not to get involved. The university's goal is to create knowledge, not to apply knowledge. Other parts of our society do the applications much better. I am a purist.

Do you have heroes?

I know people I greatly admire. Let me first mention Disraeli, the British prime minister who said that when he was young, he admired all the great people that were running the country. But when he reached his high position and he looked around and saw the people who were running the country, he thought they were in trouble. To some extent, this also happens in science. I did not have heroes when I was young but found some as I began research. For example, Fritz Lipmann was a hero I admired early in my career. Later, I learned to admire colleagues such as Henry Lardy, a fellow graduate student at the University of Wisconsin. He is an example of the finest type of individuals that humanity produces.

Any message for budding scientists?

When you start to learn science, find if it has an emotional appeal to you. You should not go into science unless you can feel gratification, not just from your participating in science, but from understanding what has been done by you and others. The satisfaction that comes from helping to create and to be able to recognize the beauty of scientific knowledge should be important to you.

John E. Walker, 1998 (photograph by I. Hargittai).

20

John E. Walker

John E. Walker (b. 1941 in Halifax, England) is Director of the Medical Research Council (MRC) Dunn Nutrition Unit in Cambridge, U.K. He shared half of the Nobel Prize in Chemistry in 1997 with Paul D. Boyer (b. 1918), Professor Emeritus of UCLA. "for their elucidation of the enzymatic mechanism underlying the synthesis of adenosine triphosphate (ATP)." [The other half of the 1997 chemistry Nobel Prize went to Professor Jens C. Skou (b. 1918) of Aarhus University, Denmark, "for the first discovery of an ion-transporting enzyme, Na^+, K^+-ATPase."] John Walker got his B.A. in chemistry from the University of Oxford in 1964 and his M.A. and D.Phil. degrees also from Oxford University. He was Postdoctoral Fellow at the University of Wisconsin, NATO Fellow at the CNRS in Gif-Sur-Yvette (France), and EMBO (European Molecular Biology Organization) Fellow at the Institut Pasteur in Paris before he joined MRC LMB (Laboratory of Molecular Biology) in Cambridge in 1974. He was elected Fellow of the Royal Society (London, 1995), received the CIBA Medal and Prize from The Biochemical Society (London, 1995), received the Peter Mitchell Medal of the European Bioenergetics Congress (1996), among many other distinctions. Until October 1998, Dr. Walker was a staff member of the MRC LMB. We recorded our conversation in Dr. Walker's old office at the Laboratory of Molecular Biology on October 8, 1998, and the following narrative is based on this recording.

John Walker was born in the North of England, in Yorkshire. His grandfather was a local politician who had a stone quarry. They were quite well off.

Then the grandfather died suddenly, he left debts, which meant that the family lost the entire business. Walker's father spent his whole life working for the people who bought the quarry from them. So John and his siblings, two sisters, were brought up in quite poor circumstances. Nonetheless, education was important in his family. His mother was well educated, she spoke French, she was very musical, and so was his father who played the organ in the church. They encouraged John to become educated. They invested their efforts in their son because they wanted to see the family improve in the next generation. John's mother focused her efforts in him and had him reading when he was 2 or 3 years of age, she taught him math, long before he went to school, to quite a high level.

Dr. Walker's wife is a daughter of a parson of the Church of England who had been a chemist. He was at Oxford, and he was brilliant at research and worked with Norman V. Sidgwick. But halfway through his studies he got religious and became a parson, then a missionary, and went to India where his daughter, Christina Westcott, was brought up during World War II. They met when John was a student in Oxford through friends who came from the same part of England as he. He was 21 when they got married. She also studied chemistry and worked in textile chemistry, which was very strong in the North of England. She stopped working when their children came along, two girls, who were born in the late 1970s. They did not become chemists.

John Walker went to a grammar school in the town Brighouse. He has reestablished contact with his old school because he wanted to encourage children to go to Cambridge and Oxford University just as he did. For quite a long period of time, every two years they would bring a bus of promising students down to Cambridge. Walker gave them lunch there and told them about the history of molecular biology, about the great people who had been in Cambridge, and about their great discoveries. Then they would go into Cambridge and visit the colleges. They would get a flavor of the place. According to Walker, children from Northern England tend to view Oxford and Cambridge as being extremely elite and essentially inaccessible to them. Cambridge and Oxford University would like to recruit more pupils from state school background. The government encourages this because 90 per cent of the children in Britain are in state schools and 10 per cent in private schools [called public schools], yet in Oxford and Cambridge about 50 per cent of the students are from public schools. There is a lot of political pressure to harness the talent of a resource, which is essentially not being properly tapped. The most intelligent children are not necessarily making their way to the best universities.

Walker went to Oxford and was in St. Catherine's College, which was in a state of transition in 1960. It had a dynamic head, a historian, Alan Bullock, who wrote famous books. He raised money to convert the St. Catherine's Society into a college. Bullock had been in a grammar school in Northern England and he knew that there were northern grammar school boys who were an untapped resource.

Walker read chemistry at Oxford and was particularly interested in organic chemistry. Sir Cyril Hinshelwood gave the very first lecture he attended as undergraduate. However, during his undergraduate period he became rather disillusioned with chemistry. He thought something was missing. Originally he had to make a decision about his future studies when he was 16. He was interested in modern languages but there was a young chemistry teacher, Maurice Walshaw, who was very stimulating so he decided for math and science.

The element that Walker found missing at Oxford was biology. After he finished chemistry he met Edward Abraham who was part of the penicillin team during World War II. Lord Florey of Florey and Chain had been the Professor of Pathology and that is where the penicillin group was. For Walker, there were a few people who mattered, in retrospect, in his life. Walshaw was one and the second one was Abraham who by then was the Professor at the Pathology School in Oxford. Abraham was a chemist, he had been a pupil of Robert Robinson as a Ph.D. student and then he had gone to work with Boris Chain and remained there, continuing to do chemistry on peptide antibiotics. Walker became Abraham's Ph.D. student. When Walker recently visited the Justus von Liebig Museum in Giessen, Germany, he realized that he belongs to a family tree relating a whole string of Nobel Prize winners including Robinson and Hans Krebs.

Abraham guided him in reading appropriate books and appropriate papers, he told him that he had to study a paper by Jacob and Monod and others. Thus Walker began to learn about biology. The four years Walker spent there was an important foundation for his future, which was laid down without any formal education, simply by opportunity. The only course he attended was one on protein structures by the newly appointed Professor of Molecular Biophysics, David Phillips.

Walker got his Ph.D. in 1969 and went to Madison, Wisconsin, to work on peptide antibiotics with David Perlman. Perlman wanted him to take some graduate courses, so he sat in a lot of courses as a postdoctoral fellow. He started working on proteins and slated a number of enzymes involved in antibiotic biosynthesis. Then he wanted to know more about how to study the structures of enzymes and got interested in sequencing methods. He met a French scientist who was giving a lecture in Madison, Edgar Lederer, one of the fathers of chromatography. His group in France pioneered using electron impact mass spectrometry to sequence peptides. They permethylated the peptide bond with methyl iodide and in so doing they could volatilize much longer peptides, and could deduce the sequence from the mass spectrum. Walker went to Lederer's lab in Gif-sur-Yvette, just outside Paris, with a NATO fellowship.

He spent two years in Madison, one year in Gif-sur-Yvette, and there he met a Czech scientist Keil who told him that they were starting a protein chemistry unit in the Pasteur Institute in Paris with himself as head. Walker went there and helped him start the unit and was responsible for sequencing proteins. That was another two years, 1972–1974. Monod was still director, but he was also very ill by then and he was hardly visible.

In 1974, Walker was on an EMBO fellowship and went to an EMBO Workshop in Cambridge on sequencing proteins. At dinner he met Frederick

John Walker with colleagues at the Laboratory of Molecular Biology, Cambridge (photograph by I. Hargittai).

Sanger, who asked him whether he thought of returning to England. Sanger's question made Walker think of the possibility of returning indeed.

Walker did his Nobel-Prize-winning work at LMB. Its most important part was the understanding of the ATP-synthesizing enzyme, which is found in mitochondria, chloroplast, and bacteria. They studied mostly the one in mitochondria. By carrying out a detailed structure analysis of the catalytic part of the enzyme, they were able to get striking insights of the mechanism of the enzyme action from the atomic model that they built. This revived the idea that the enzyme might be working by rotary mechanism.

This idea had been around before but it was derided and other models had much greater sway. When they had the model, the question was, "How does this fit with Paul Boyer's binding change mechanism?" What Boyer showed was, using kinetic techniques, that each enzyme complex had three catalytic sites but, surprisingly, at any moment in time they had different properties and different affinities. What Walker and his group did was to lock the enzyme in one moment in time by inhibiting it. This way they produced the snap shot, which came from X-ray crystallography with the three catalytic sites of the enzyme. They were structurally different, although they were chemically identical. This could be seen just from looking at them. One of them was incapable of binding nucleotide: although there were bags of nucleotide around in the mother liquor, none was bound. That corresponded to the open state in Boyer's mechanism. Of the other two of the three catalytic subunits, one was effectively ATP bound, the other ADP bound.

They also tried to get intermediate states between those three states over the last four years and until about two months before our conversation, they had not managed to do it but finally they have. They had been looking for the intermediate structures by inhibiting the enzyme with different antibiotics. They thought that it was likely that some of these compounds would lock the enzyme in intermediate state but they did not. They kept coming up with the same structure. Eventually it became frustrating, if not demoralizing, to keep growing new crystals, collecting data, solving the structure, and getting the same results. Finally they hit on the way of locking the enzyme on the right position. They realized this only a few weeks before the interview and this observation was followed by actively building a new model. Walker was visibly excited about the possibility of describing the three intermediate states in the catalytic cycle, which he considers to be another breakthrough.

Continuing his story about his Nobel-Prize-winning research, they wanted to understand the mechanism by which the three states interchange, how does this concerted mechanism work. They looked at the model and the most obvious thing was a long, not quite cylindrical but elongated curved structure and the most obvious way was to rotate it. So they proposed that this was an intrinsic part of the mechanism, that is, the rotation of the central gamma subunit. After that, other people were able to use the structural model to design experiments to test the idea of rotation. The first attempts, though showing no rotation, were consistent with rotation. Then, about two years before the interview a Japanese group did the key experiment. They could visualize the rotation directly in the microscope. They attached chemically a long actin filament, which is visible in a light microscope, to the rotating part, and it became visible. This was not just modeling, this was a real biological experiment. The experiment was filmed, it is on the web, and Walker shows it in his lectures. That experiment really settled the issue.

That experiment had its origins in other experiments carried out in the early 1970s when a lot of studies were going on of motility in bacteria. They move towards food sources or away from some things and they do this by chemoattraction or chemorepulsion. The key thing is the flagella. They have several flagella and when there is no attractant or stimulant present, they randomly swim. When they are swimming, the flagella bundle up and this then propels it forward, like a screw. When then they are tumbling, the direction of rotation is reversed so they spin apart and this makes the bacteria tumble. Under the impact of an attractant, for example, some sugar, the frequency of swimming increases and they end up at the source.

When Walker was a postdoc, people argued whether this was rotation, or whether it was beating in a sinusoidal fashion. The key experiment was that somebody made an antibody that recognized the end of the flagellum and they took a cover slip and coated it with the antibody and added the bacteria to it. This then trapped the bacteria attached by the tip of their tail, and one could look in the light microscope and see the bacteria turning around in this fashion. This was the first demonstration of rotation in these motile bacteria. The ATP experiment mentioned above was a derivative from this bacteria motility experiment, a macroscopic demonstration of a microscopic chemical event.

It is an interesting question whether there is any common evolutionary origin between the two phenomena. According to Walker, it may turn out to be that the rotary motor of the flagellum and the rotary motor that drives this rotation in the catalytic part of ATP-synthesizing enzyme have a common evolutionary origin. However, that is only a hypothesis.

The rotation in the catalytic part of the ATP-synthesizing enzyme is the energy-transducing event, called the coupling event in this enzyme. The rest of the enzyme is also of interest. The catalytic part is the knob sitting outside the membrane and there is another part in the membrane, which is the motor providing the rotation. The motor is driven by the proton-motive force, which is being generated across the membrane by photosynthesis or respiration in the way that Peter Mitchell described in the 1960s and 1970s. There is a proton potential across the biological membrane where the enzyme sits. The protons are transported back through the membrane to the site where catalysis occurs. When this happens, energy is released, so the passage of the protons through the membrane provides the rotation in the motor of the membrane part. The rotation next to the catalytic site provides the energy to force the complex to make ATP from ADP and phosphate. This is the way Walker and others think about this mechanism.

This research demonstrated a novel kind of catalysis. Although many were against the rotary mechanisms in enzyme catalysis, it turns out that it is very likely a general principle. The ATPase is the first of a much bigger family of enzymes, which work by rotary mechanisms. Enzymes involved in DNA replication, which have a related structure, are working via a similar mechanism. They hydrolyze ATP or GTP and they have a DNA strand running through the middle and they rotate themselves along the DNA strand, which forces the two strands apart. The structures involved here are the DNA helicases, an essential part of the replication process pulling the strands apart. This work was done subsequent to Walker's work and it was realized that there was a relationship. So it turns out that rotation is a much more general principle than people were prepared to acknowledge even only a few years ago.

The continuation of the work is aimed at understanding how this rotation is generated. They are looking at the structure of the membrane part, which is even more technically demanding than doing the structure of the catalytic part. It was big and it was the largest asymmetric structure solved when they published it, but it was easier than studying the structure of the part in the membrane. They have to isolate the membrane part, then grow crystals of them.

Originally, Walker was a chemist but he rapidly realized when he was an undergraduate that he was actually interested in biology. At the time of our conversation he had already officially left the LMB and had been appointed director of an independent MRC unit in an adjacent building, involved with nutrition. They want to use molecular techniques to study problems of human nutrition. This is not very far away from his interest in mitochondria and he will be continuing with the research teams that he had developed at the LMB. One of their research projects links directly to the basis of obesity in man. This is a human nutritional problem that they want to apply molecular techniques to.

In his new unit, there will be 8 or 10 research groups, about a hundred people. The unit has existed for a long time. One group will come from the old unit, which studies the effects of components in food in relation with cancer. They collect epidemiological data to find links but they would like to understand the molecular basis of these conditions. For example, there is a clear link between colon cancer and eating red meat in large amounts. This turns out to be related to the way in which we metabolize in our guts, in the colon. The idea is to take the statistical findings and understand the underlying molecular mechanisms. Another statistical link is between other components in food that have estrogen-like properties and they can influence breast cancer. Another major topic in the Western World is the link between nutrition and osteoporosis. People are trying to alleviate the problem, once it is there, but, again, we should understand it how the problem arises in the first place.

Walker's big career move was the initiative of the head of MRC. He clearly wanted to give him more visibility. Many people were surprised that he took on this job, which has more organizational element in it than Walker used to have in the past. He welcomes the new challenge.

When Walker won the Nobel Prize, many people wrote to him, among them the Prime Minister. He took the opportunity to respond to Tony Blair and to suggest that the government should continue support basic research in science, especially biological sciences. He suggested that developing biological sciences in a biotechnological direction would provide help to contribute to the future wealth of the country. The question is about the mechanism whose absence made academic scientists being diffident about pursuing practical extensions of their work. The business of patenting is complicated. Walker notes, however, that there has been a big change in attitude amongst younger people. They realize that there are opportunities to see their discoveries reach the market place. It can be fulfilling and helps them to swell their bank balances quite considerably. This is also important in a climate where people in academia are relatively poorly paid, compared with other professions. MRC has been very active and they have set up a biotechnology transfer group. Walker is planning to be personally very involved in the biotechnology phase of their work in the new unit.

Walker knows that some of his friends thought that he would just carry on the same. The ones who are more idealistic thought a little bit let down when he decided on the change. They saw him as the archetypal guy who is extremely focused on a scientific project, keeps focused for a long period of time, and finally solves it, which he did. Fred Sanger said to him on the day that Walker won the Nobel Prize, "Now, get back to the laboratory and win another one." But Walker never set out to win a Nobel Prize in the first place. He has seen people at LMB over the years, brash young people, talking about their doing a Nobel Prize-winning experiment, and many people clearly became disappointed when they felt they merited the prize and they did not win it. It drives people crazy. A few years ago people started talking to him about the Nobel Prize and he told them just to go away. He did not want to think about that because he knew that it could have a very bad effect on people. He maintains that one can be personally ambitious but one can keep it to himself.

As for his interests outside science, Walker used to be very keen on soccer and cricket, but he does not play them anymore. One of his passions now is long distance walking. He walks a lot on his own along the coast of Britain. He finds it extremely relaxing and it gives him an opportunity just to think. When he was younger he could not bear to be alone, he always had to be with somebody, whereas now he quite likes his own company. Then, he has always been passionate about music. He was a boy soprano and he gave public concerts, and they all sang in the church choir. Although he was brought up religious he rejected religion at a particular age. For him it was either science or religion. He saw there a philosophical incompatibility, and one of them had to go. That was his own resolution of what he found to be a very difficult problem in his late teens, early twenties. For him it had an emotional context as well. His mother was very upset and never accepted that he could reject God in the way that he did. There was strong discussion with his wife as she came from a Christian family but Walker found them very tolerant of him and they never tried to strong-arm him. Christianity has remained a part of him because it was a very integral part of his childhood both philosophically and morally. He does not escape that but that is not something he instilled in his children. They used to go and sing in the choir as children but now they do not espouse religious views as far as Walker knows.

Walker has great concern to see LMB remain at the top in its field. He appreciates the long string of very good leadership from people like Perutz and Sanger, by example rather than by political decision. He values the culture at LMB and the maturity of the people who realize that if the place is going to stay at the top it has to be renewed from the bottom. This is why they spend a lot of time over recruitment. They know they have to replace themselves because otherwise the place goes into a downward spiral from which it would probably never recover. Until recently he was Chairman of the Recruitment Committee at LMB. The policy has been, find good people and let them do what they want to do. This is how Walker himself got into ATPase, which was a very unfashionable topic to work on at that time. But he did receive support from Sanger who thought that it was great because it was so difficult and he told Walker so, "If it is difficult, go for it."

The legendary sizzling intellectual atmosphere of LMB has changed though and the population of the Laboratory has changed. When Walker came, the American postdocs dominated the place. This was the place to go in the 1960s and 1970s, if one was interested in molecular biology. Now, he thinks, they do not do this, now the American postdocs see this as a dangerous career move because it means moving out of the United States. This has become much more a European institute most of whose overseas postdocs are from the EU (European Union) because of the EU fellowships. In Walker's comparison, the American postdocs are a bit older because they may be in the graduate school for six years. They are quite mature and they know a lot. They are extremely ambitious, and this made the place crackle.

One of Walker's idols is Frederick Sanger and another is Max Perutz. Perutz has been supportive behind the scenes, just making sure that everything is fine. He finds them both giants in different ways. He also cites Monod, just for his intellect. By reading Monod and seeing him talking in public, Walker found him a fantastic intellect. The same is true for Francis Crick. The first time Walker met Crick he told himself, "My God, here is another level of intelligence that I have never encountered."

Herbert A. Hauptman, 1985 (courtesy of Herbert Hauptman).

21

Herbert A. Hauptman

Herbert A. Hauptman (b. 1917 in New York City) is President of the Hauptman-Woodward Medical Research Institute, Inc., in Buffalo, New York. He and Dr. Jerome Karle of the Naval Research Laboratory, Washington, DC, received the Nobel Prize in Chemistry in 1985 "for their outstanding achievements in the development of direct methods for the determination of crystal structures." Dr. Hauptman is also a Distinguished Professor of Structural Biology and a Research Professor of Biophysical Sciences and Computer Sciences at the State University of New York at Buffalo. Earlier in his career, he worked as a mathematician and as a physicist and as supervisor in various departments at the Naval Research Laboratory, beginning in 1947. He received his B.S. (1937) from the City College of New York, his M.A. (1939) from Columbia University, and his Ph.D. (1955) from the University of Maryland. All his degrees are in mathematics. The title of his doctoral dissertation is "An n-Dimensional Euclidean Algorithm." Dr. Hauptman is a Member of the National Academy of Sciences of the U.S.A. (since 1988). We recorded two conversations. The first on August 3, 1995, in Dr. Hauptman's office in Buffalo,* and the second, in two sessions, in May-June 2001 in Erice, Italy. We then blended the materials of the two conversations by correspondence.

---

*That part of the interview was originally published in *The Chemical Intelligencer* **1998**, *4*(1), 10–17 © 1998, Springer-Verlag, New York, Inc.

As a result of your activities, the possibilities of X-ray crystallography have tremendously expanded. How would you characterize this development?

Prior to the discovery of the scattering of X-rays by crystals, there were no molecular structures that were unambiguously known. There had been conjectures, many of which were very intelligent, but prior to 1912 only the simplest molecular structures could be said to be known. Even the very concept of a molecule consisting of atoms having a definite arrangement was not really well defined before that time. The major contribution of X-ray crystallography was its ability to determine, unambiguously, molecular structures, that is, the arrangement of atoms in the molecule. This, in my opinion, is a major discovery of the 20th century.

What happened in 1912 in Munich was that Paul Ewald was writing his doctoral dissertation under the supervision of Arnold Sommerfeld. The dissertation was concerned with the propagation of electromagnetic radiation in a medium consisting of a regular arrangement of resonators. As it turns out, this is exactly what crystals are. Each of the atoms in the crystal may be regarded as a source of electromagnetic radiation, once there is light or X-rays incident on the crystal. Thus, without knowing it, Ewald was doing work that had important implications for X-ray crystallography.

Ewald went to Max Laue to get his opinion of his dissertation. Laue had only one question to ask: was Ewald's work valid for arbitrary wavelengths of the electromagnetic radiation? Ewald had been thinking of the propagation of visible light in the regular arrangement of resonators, and Laue realized that X-rays were a form of electromagnetic radiation. They had a very short wavelength, much shorter than the wavelength of the visible light. It was more or less known that crystals consisted of regular arrangements of atoms, and the wavelength of the X-rays was of the same order of magnitude as the distance between atoms in a crystal. Therefore Laue thought that perhaps a crystal would serve as a three-dimensional diffraction grating for X-rays and would scatter an incident beam of X-rays in many different directions and with different intensities. The directions and intensities of the scattered X-rays would be completely determined by the crystal structure, that is to say, by the arrangement of the atoms in the crystal. Once he learned that Ewald's work was valid at all wavelengths, Laue immediately urged two junior physicists, Friedrich and Knipping to do the experiment, and Laue's intuition was dramatically confirmed. They observed that the crystal did indeed scatter X-rays. For this work von Laue received the Nobel Prize in 1914 (by then his father had acquired hereditary nobility).

The work was picked up immediately by the Braggs, father and son, who, by interpreting the scattering of X-rays as reflection from the different planes in the crystal were able to deduce the structures of some very simple crystals in the years 1913 and 1914. And for this they got the Nobel Prize in 1915. That was the beginning of X-ray crystallography.

Certainly, no one could possibly have thought in those years that X-ray crystallography would grow and become of such major importance in the next 80 years.

What happened in the next 40 years or so is that the science of X-ray crystallography was firmly established on theoretical grounds, and more and more crystal structures and molecular structures were determined. Nevertheless, crystallographers were in a strange position. They were determining simple crystal structures of molecules having 10 or 15 atoms. They were able to determine the crystal structure of even more complex molecules, provided that the molecules had one or two or three heavy atoms in their structures. These atoms dominated the structures and once the positions of these atoms were determined, then it was usually much easier to locate the positions of the lighter atoms. But there was no systematic method for doing this. It was largely a trial-and-error method because structural chemists and crystallographers knew, or thought they knew, what the structures of simpler molecules were so they could make an intelligent guess as to what the structure was. From that assumed structure they could calculate the nature of the diffraction pattern, which is the assembly of the directions and intensities of the scattered X-rays. Then if the calculated diffraction pattern agreed reasonably well with what was observed, standard methods would serve to refine the assumed structure. So this was generally what was done prior to, say, 1950.

It is true that in those years there was at least one method that might be called a more direct method. This was the method that Patterson had developed by introducing his famous Patterson function, which was simply the Fourier series with coefficients that were the intensities of the scattered X-rays. With this method, one was able to deduce information about the interatomic vectors, and, if the structure was not too complicated, one could then calculate the actual arrangement of the atoms. Crystallographers were surely getting structures in this way.

On the other hand, they were firmly convinced that the X-ray diffraction pattern was insufficient to determine unique crystal structures. What was needed was not only to measure the intensities of these scattered X-rays but their phases as well. But the phases were lost in the diffraction experiment. The phases were needed because the electron density function is a triply periodic function, which is represented as a three-dimensional Fourier series. The coefficients of that series are complex numbers in general. It was only the magnitudes of these complex numbers that could be observed. They were observable directly from the intensities, which could be measured. But the phases of these complex numbers were lost in the experiment. Therefore, the Fourier series itself couldn't be computed. And crystallographers were firmly convinced, up until 1950 or so, that the missing information — the phase information, which contained half of the information that you could get from the diffraction pattern — could not be supplied, and that therefore it was impossible, even in principle, to deduce crystal structures unambiguously from the actual experiment.

The first major contribution, which I was involved in making, around 1950, was the recognition that, in fact, the intensities alone contained all the information that was needed to determine molecular structures by the technique of X-ray diffraction. That is to say, the missing phases, the phase information, could be deduced from the information that could be measured. If one formulated this question of exactly how the crystal structure and molecular structure depends on the observed diffraction pattern as a problem of pure mathematics alone, it meant simply the solution of a very complicated system of simultaneous equations. The number of these equations was far greater than the number of unknowns, which were simply the position vectors of the atoms in the molecule. One observed maybe 10 or 15 times as many intensities as were needed in a strict algebraic sense to determine the molecular structure. The information was there, the problem was how to extract that information. If one was to try to do the obvious thing, which is to solve the system of maybe a thousand simultaneous equations, or transcendental equations with all the trigonometric functions, it would be impossible to solve that system. It was too complicated, and it was an overdetermined system, and it had to be overdetermined because errors in the measured intensities had to be compensated for by measuring many times more intensities than one needed in a strict algebraic sense to determine the structure. So to attempt a head-on solution to this problem was obviously impossible.

The conclusion, which could be drawn was that although it might not be possible to solve the problem, the problem had a solution. There was a solution there, and the missing phase information was in fact contained in the measured intensity information. I think this was a very important contribution, which we made back in 1951 or 1952. It is interesting from a psychological point of view, if no other, that no one in the crystallographic community accepted this conclusion which today seems almost trivial.

I think this notion that half of the information that was needed was lost in the diffraction experiment was simply accepted as dogma without question by all structural chemists and crystallographers. Furthermore, I think that the problem also was that most crystallographers did not have a strong background in mathematics, and many of them mistrusted mathematics. They simply could not believe what the mathematics was saying.

But you never had any problem with publishing your claims.

We never had any problem there. The then Editor of Acta Crystallographica, Isidor Fankuchen, must be given a lot of credit. Even though he believed little or nothing of what we did, he still went about it in a very professional way, sent our papers out for review, and published everything we ever submitted.

The only problem that we had, and it turned out to be good for us, was when we submitted our first paper, which we very brazenly called "Solution of the Phase Problems, $P\bar{1}$" and it went to Ewald for review. I'm sure Ewald did not completely understand it, and it was a very long paper. It was at this meeting in Ann Arbor, Michigan that Ewald came to us and said that this was a very long paper, and he suggested that we write a brief version of it and, then, expand on it to write a monograph. And this is what we did.

For many years no one accepted our argument at all. Even when Jerry Karle and I developed these methods to the point where we were actually led to a numerical algorithm for determining the phases in terms of the intensities, even this work was not accepted although in the 1950s, we actually solved some half-dozen crystal structures in this way. Several of them were very difficult by the standards of those days. The biggest one was the solution of di-para-methoxybenzophenone, which consisted of 36 carbon atoms plus a number of hydrogen atoms. We didn't have the experimental apparatus to do the experiment, and I remember that the data were given to us by Seymour Geller. He had collected his data and knew that there was no way that he could solve the structure. So he simply gave the data to us. And this method worked in a very straightforward way. Even after that, although Seymour Geller was convinced, no one else was. One can go back to *Acta Crystallographica* and find some papers, which, in effect, claim, that this is a method that cannot work.

In hindsight that was a lucky thing for us because no one could ever have claimed that we did not do this by ourselves. There is a paper by Woolfson and Cochran,[1] which I like to refer to, and another by Vand and Pepinsky,[2] which appeared in *Acta Crystallographica*, in which they tore our work to pieces. The paper by Woolfson and Cochran was called "Have Hauptman & Karle solved the phase problem?" and the whole paper was devoted to showing that the answer to this question was an emphatic no! Publishing that paper was, in my opinion, a big mistake. I believe that Woolfson must have been very disappointed that he did not share the Nobel Prize, and I think that paper may have been the reason. To be fair to him, he improved the methods, and he and Peter Main, who worked with him, made important contributions. About 12 years ago, I got a letter from the University of York in England advising me that Peter Main was applying for promotion and he had given my name as a reference. In the letter that they sent me they also included Peter Main's own justification for his promotion. Peter Main has been a good friend of mine, and I gave a very positive account of his contribution. It amused me that as part of his justification he wrote that he had been considered for the Nobel Prize, which I believe was true enough. Then he referred to our Nobel Prize, which as he put it was for "some early mathematical work in the phase problem." I had really no problem with this. Then he went on to say that he considers that his work in applying these methods was part of the reason that Jerry and I received the Nobel Prize. I found it amusing for a couple of reasons. First, I believe that it was true. As a result of his work he succeeded in showing that these methods were, in fact, useful. As a result of his work and the work of others who have made the applications, we were finally recognized as having done this fundamental work many years before. If one goes back to the record, we were the only ones who were consistently making these contributions back in the 1950s. The one big exception was Dave Sayre (and to a lesser degree W. Zachriasen) whose work, in my opinion, has not been sufficiently recognized; but he then left the field for a period of many years. It should also be mentioned that, despite their reservations, Woolfson and Cochran also made significant contributions in those early years. Nevertheless, I believe that as a result of the universal skepticism with which our work was received no others could legitimately claim that they really should have been considered for the Prize.

Coming back to the reaction of the crystallographic community at the time ...

It wasn't until 10 years later that as a result of the accumulation of structures that were determined by our methods, and with the development of the high-speed computers, our methods started to be used by a growing circle of crystallographers. Some people in Great Britain (primarily Woolfson and Main) prepared programs around the mid-1960s that would automatically solve simpler crystal structures. Between the mid-1950s and the mid-1990s, during these 40 years, the method itself and its theoretical underpinning and the numerical algorithms and computing software have developed to the point that now structures having as many as 400 or 500 atoms in the molecule have been routinely determined.

In a recent meeting in Montreal, there was a description of two structures determined using the methods we had developed here; one had about 320 atoms in the molecule, and the other had about 450. These are the first unknown structures of that complexity solved by the so-called direct methods that I'm talking about. So there has been a remarkable development in the last 80 years or so, which could not have been anticipated by Ewald when he was writing his dissertation.

Of course, I have just emphasized the so-called direct methods, which determine molecular structures *ab initio* from the observed intensities, and I have not even touched upon much larger macromolecular systems, such as proteins, which consist of thousands of atoms in the molecule. These structures also can be determined by special techniques. These include such techniques as multiple isomorphous replacement and anomalous dispersion, which take advantage of our ability to modify a given crystal containing molecules consisting of thousands of atoms by diffusing into the crystal a small number of really heavy atoms without disturbing the crystal structure. One then does the diffraction experiment on these so-called derivatives as well as on the native protein that one is interested in.

Special techniques have been devised which enable crystallographers to determine structures having literally tens of thousands of atoms in the molecule. The direct methods have not yet reached this point, but they may eventually also reach this point. You can see from the list of structures solved in the course of 40 years or so how the direct methods have developed from 20 atoms 40 years ago to 500 atoms today. In particular, during the last three to five years, as a result of the work done right here in our laboratory, I have seen that number go up from 100 or 150 atoms per molecule to 500. Furthermore, the computer software has reached the point where this is done completely automatically. The user of the computer program doesn't require any understanding of what is going into the computer program and why it works. It's completely automatic. I am sufficiently optimistic to foresee a similar advance in the next few years as during the past few years.

The field of X-ray crystallography has opened up this enormous gold mine of information, which has also enormous implications with respect to our ability to understand better the nature of chemical reactions and also to relate the structures of biologically significant molecules with their biological activity. This information has enabled us to better understand the life processes, and this carries with it enormous implications because nowadays we can understand the workings of the human body, and all living organisms, much better than otherwise would be possible. We can develop better drugs to treat all kinds of diseases because we understand what causes the different diseases that people are subject to. There is now a whole new field called structure-based design of drugs, which designs drugs in a very intelligent way by understanding the relationship between structure and biological activity. One can design drugs on the basis of desired structures, and the drugs will do exactly what we want them to do without adverse side effects, which most drugs always carry with them.

I would like to ask you about the most recent development in the direct method of X-ray crystallography.

In the early 1990s, my Buffalo colleagues and I stumbled on what has become known as the "Shake and Bake" procedure. In this we formulate the phase problem as one of constrained global minimization. As a consequence, suddenly, the whole field opened up again. We were solving structures with as many as 500 or 600 atoms, in a straightforward way. In the last two or three years structures having as many as 2000 atoms have been solved by this and related techniques in our laboratory and others. That changed the whole landscape. All of a sudden, there was a quantum jump in the complexity of structures, which could be solved. This is what makes the phase problem so interesting, because this was a completely unexpected development. The lysozyme structure is the most remarkable in this respect. Originally, it was one of the first protein structures solved with isomorphous replacement. David Philips did this work years ago. This protein has about 1000 atoms and there are about 200 solvent atoms, so the total is about 1200 atoms. The symmetry is P1. Shake and Bake starts with a random structure and whether it refines to the correct structure or not, depends on the initial guess. Sometimes it may require a hundred trials before we find one, which leads to the solution. But the method is fast enough that even a thousand trials is acceptable. With lysozyme something very strange happened. No matter where you start, it will converge to solution, 100% of the time. You can start with any random structure you wish, run through with the Shake and Bake process, and it'll converge to the solution with a single trial. We've seen this happen with other structures in P1 as well, and this happens only with P1 structures. These structures may have a few hundred atoms to be positioned. Lysozyme is larger and there is an interesting story about it.

One of the first times I presented the Shake-and-Bake method was at a computing meeting after the Seattle crystallography meeting. I was very optimistic and I made some unjustified remarks that we would soon be solving structures with 1000 atoms although at that time the best what we had done was about 500 atoms. Giacovazzo, who was in the audience, said that with 500 atoms we had reached the limit. Then I made this statement that we would solve structures with 1000 atoms before the decade was out, by the new millennium. Chuck Weeks was there and he wrote a statement on a transparency, which said, "I, Carmello Giacovazzo, state that direct methods will never solve a structure with a 1000 atoms or more in it, without heavy atoms." And Giacovazzo signed it! A few years ago there was a direct methods meeting here in Erice and we had still not solved a thousand-atom structure. Sheldrick was at that meeting and had the data for lysozyme and he used a modification of Shake-and-Bake, which he called Half-Baked, and he solved the structure right here. When Sheldrick gave his paper, Chuck Weeks gave him the transparency with Giacovazzo's statement. The beautiful thing about it is that lysozyme happens to have 1001 atoms plus solvent molecules. Giacovazzo was very good about it and said that he should have never signed that statement. It was a good joke, even he agreed.

From the time of your fundamental publications of the direct method to the Nobel Prize, there was a period of more than 30 years.

That's right. For the first 10 years the reaction from the crystallographic community was skepticism at best, hostility at worst. In the 1950s and 1960s, scientific meetings were much more interesting than today. There was always a lot of criticism from the audience, and a lot of it was very unfavorable too. However, people started using the method in the middle of the 1960s. In addition to Woolfson distributing his programs, Isabella Karle determined the structures of fairly complex molecules of 40–60 atoms, which up until that time would have been unthinkable. That convinced a lot of people that there is something here. The full significance of the method was not understood until the middle or late 1970s when people in the life sciences realized that they were able to relate molecular structures to biological activity. This ability came to full flower in the 1980s and I believe that it was that which persuaded the Nobel Foundation that this was important. Of course, when we did our work back in the early 1950s, we had no idea that it would turn out to be that important. For me it was merely a challenging mathematical problem. Once I understood what it was, I became addicted to it, and I still am.

Did you feel any frustration during those 30 years?

No, I never felt that. I can say in all honesty that when the Nobel Prize came in 1985, I was surprised. I knew that my name had been submitted but by the mid-1980s, the method had been well established, and even my name was not attached to it. For most people, it had just become part of the accepted practice. I was also surprised because my contribution was of a mathematical nature, and the Nobel Foundation had not previously awarded the prize for a mathematical contribution.

Your awards show a strange pattern. In addition to one college award, and the Sigma Xi Scientific Research Society Award (shared with Jerry Karle) in the late 1950s, you received the Patterson Award (also shared with Jerry) of the American Crystallographic Association in 1984, at 67, and then, after 1985, of course, there is a long list of awards and memberships.

You're quite right about that. My award came from the City College of New York in 1936. It was the Belden Medal for excellence in mathematics.

Once I saw a big ad in a New York paper saying that so and so many Nobel laureates came out of City College. It was more like 15 than 5.

Yes, a large number. The year I graduated, 1937, there were three future Nobel laureates in that class.

You were then 20. When did you enter City College?

In 1933, when I was 16. I graduated from a very special high school in New York, which was later killed by Mayor La Guardia, 10 years after I'd graduated. It's now been reborn. The name is still the same, Townsend-Harris. It was named after the American ambassador to Japan who opened up Japan in the mid-1800s to the rest of the world. This was a very special and very small high school with severely restricted entrance requirements. It was a three-year school instead of four years. It was the best you could get and it was free, a part of the public education system. Even City College was free, and you were automatically admitted to City College if you'd graduated from Townsend-Harris. City College was a very difficult college to get into. There was severe competition to get in. The only cost was a 50-cent library fee, and in later years we had to buy our own books. If it had not been for City College, I couldn't have gone to college. My parents were poor. My father was a printer. His salary was 28 dollars per week, which wasn't very much even then. My mother also worked, at least part time. The early 1930s were, of course, the Great Depression, and it was impossible to find work. The unemployment rate in New York City was 25%. Everyone I knew was poor. So if not for City College, and if my parents hadn't been so supportive, I wouldn't have gone to college. They wanted their three sons to go as far as they could go. For all of us, those years, going to school was a serious matter, and we worked very hard.

Who or what was the most important influence in your development?

The earliest influences on my life were not people that I knew but people that I read about. A very early one was Bertrand Russel. This was at the age of 11 or 12. Up until that time, I had been a very religious kid and believed in a supreme being.

Were you taught this at home?

No, my family was not religious. My religion was not in the usual sense of any organized religion. My parents were nominally Jewish.

Then, reading Bertrand Russel, I came to understand that this was not for me. I think I read just about everything that he wrote except for the *Principia Mathematica*, which I only dipped into. It's almost a life-time's work to read those volumes.

Then there were others, today you'd call them role models, Fermat, Archimedes, and I read their works at an early age. Fermat, I read in French translation at 15 and Archimedes in English translation.

Did your being non-religious make your social life difficult?

I don't advertise my beliefs or non-beliefs particularly. Being non-religious in the United States is not something that is acceptable. For example, no politician would speak this way and hope to get elected.

Does it matter which religion you don't believe in?

It's not a serious matter although there was a time when there was serious discrimination against certain groups, Jews particularly and others as well.

Did you experience it?

I shouldn't say I didn't. It did cost me a job at one time. This was when I was a statistician at the Census Bureau in Washington in 1940 or so, and I'd learned about another job with the government, a job as a mathematician in the Department of the Navy, for which I applied. I was granted an interview by some Naval officers, and I was asked to wait outside, and whether this was done deliberately or not I'll never know but I overheard them in their discussion referring to me as "smarty Jew."

Did you get the job?

No.

Any other experience?

Not of that kind, but when I was in the Navy during the War there was some anti-Semitism.

In the Navy? During the War?

Yes, but I observed it only on the part of certain individuals.

But this has changed.

No doubt about it.

By the time of the Holocaust you were a grownup. How much did you know about it?

Surprisingly enough I knew nothing about it. I was overseas from 1944 to 1946. Even before the war, there were some refugees in our big apartment building, but we didn't understand the enormity of the problem or what was going on in Europe. It wasn't a topic of conversation. I remember an older German Jew who moved into our building and he was a refugee from Hitler's Germany and he had been a judge. He told me how one day they dragged him out of his court, where he was presiding as judge, and they made him scrub the sidewalk with a toothbrush. I thought that this was unbelievable that they could do this. In those years we were remarkably ignorant, and I was not particularly involved in political matters at all. I was concerned mostly with going to school and I was apolitical. I was one of the handful of people who went to City College during the 1930s who did not participate in radical political activity. I was interested only in learning everything that I could. I insulated myself from everything else that was going on around me. This was true also later when I was in the Census Bureau and when I was in the Navy.

Back to your studies, after City College, you went to Columbia University and earned a Master's degree there. But then, there was no continuation in your studies for a long time.

When I applied for a fellowship for doctoral studies, halfway through at Columbia, to Columbia and other universities, I was turned down by all. I had had a perfect record at City College, but I never succeeded in getting any kind of assistance to go on for a Ph.D. As a result, I had to stop my education at that point. It never occurred to me to ask why. I simply accepted it.

Did City College have a Jewish image?

No doubt about it; perhaps 95% of the students were Jewish at that time. Almost all students were outstanding, but I knew nothing else. City College was my only experience and I just assumed that that is the way it is. It never occurred to me that these students were particularly outstanding. A number of them became quite famous. Herman Feshbach is an example; he became a physicist. He and Henry Birnbaum were my best friends. Henry died within a couple of years of graduating from City College. He was at Cornell and he drowned trying to rescue another student. He would have become another outstanding physicist, just as Feshbach turned out to be.

After I left Columbia with my degree in mathematics, I could not find a job, and I did some tutoring for a few months. When the 1940 census came around, I landed a civil service job in the Census Bureau in January 1940. First I was just a clerk, then W. Edwards Deming, the chief statistician at the Census Bureau, transferred me to his group. After about two years in the Census Bureau I worked for about a year or so as a civilian instructor in electronics and radar for the Air Force and then enlisted for war service. When the war was over, I found a job at the Naval Research Laboratory (NRL). I went to NRL in 1947 and that is when my scientific career started. I wanted to do scientific research more than anything else.

Why?

Science and mathematics were the only things that interested me from the very beginning. For me, the beginning was when I started to read. I was lucky growing up in the Bronx. There were libraries all over the place, and there was one within a five-minute walk from where we lived. I spent all my time there and it opened up a whole new world for me. I vividly remember the book that, for the first time, showed me how to make Platonic solids. I was intrigued by them; it is hard to describe the feelings I had about them. I made them all, and I became addicted to them. Even in recent years, when I had learned to use stained glass, I made these and many other polyhedra using stained glass. I still feel about them the same way I did as a young kid. I also remember when I first saw a curve in analytic geometry and I was fascinated by the shape of these curves. I wanted to learn more about them and I did. I spent an entire summer just looking up curves and learning about their properties. I didn't always understand what I was reading, in fact, I understood very little. The first thing I saw under each curve was an equation, which made no sense to me at all, and I knew I had to learn more. And I remember the day when I finally understood what the equation of a curve meant. When I understood that, analytic geometry became very easy. Such experiences led me to a strong determination that I wanted to be in mathematics and in the sciences.

Then it was a long time, from your early teens to when you were 30 years old, by the time you did really go into research.

That's right. When the war broke out, I knew that sooner or later I would become involved. For a brief period, I got a job teaching electronics to officers of the army. This was in spite of the fact that I knew nothing about electronics. So they sent me to school first, for a few months. It was about 1942 and I had already had my master's degree. It was basically electric circuitry and I practically became an electrical engineer and could solve any electrical circuit, because it required only the solution of a linear second-order differential equation with constant coefficients. Then, I was teaching what I had learned. After a while I joined the Navy as an ensign doing weather forecasting, and I stayed there for three years. For my career, it was a waste of time, but I wanted to be a part of the war effort, especially being Jewish and knowing what the background of the war was. I got out of the Navy in 1946, and, at 29, I still did not have an advanced degree, and I was very frustrated.

I was looking for a job in Washington because my wife loved living in Washington, but I couldn't get a job. At that time there was a freeze on hiring in the Federal Government, and I was looking for a job at some research laboratory in Washington. So I had to go back to my old job with the Air Force for about a year, and kept looking for a job, and succeeded finally in getting a job at NRL and started my work there in 1947.

One of the things that attracted me to NRL was the possibility of going to the University of Maryland, where I could work for an advanced degree at the same time that I was working. I got my Ph.D. in 1955, so it took me 8 years.

By that time you had already completed your Nobel Prize winning work.

That's right.

Why did it take so long for you to get your Ph.D.?

My job at NRL was a job of a physicist, so I thought to get a Ph.D. in physics. I took all the physics courses that I needed to take. I remember that it was 24 credits. I expected to finish in 1953, but the physics department required that I have a thesis, which would be published under my name alone. I thought that this would not be a problem because by that time Jerry Karle and I had published about 10 papers and it was very easy to get our papers published. I thought that I would submit one of our forthcoming papers under my name alone. However, Jerry refused to let me do that, and he was nominally my boss. We had an implicit agreement to publish jointly and to alternate the order of authors between us, and we did this. Jerry's refusal to let me publish a paper on my own made me understand that there was no way for me to get a Ph.D. in physics. So I switched to mathematics. It was no problem for me and I knew that Jerry could never pretend to have collaborated with me in mathematics. They required that I take three courses, two courses in topological groups and one course in advanced matrix theory. This was because, although I had passed the examination for Ph.D. candidates, I had no math courses at the University of Maryland.

My dissertation was an n-dimensional Euclidean algorithm. In number theory, there is a fundamental algorithm known as the Euclidean algorithm, which is used to determine the greatest common divisor of two integers. I discovered that this algorithm can be interpreted geometrically as a lattice on a line. By extending the geometric interpretation to two and higher dimensions, one can get an analogous algorithm, valid for all dimensions. It has very interesting consequences in the solution of certain diophantine equations. These are equations in which one is looking for integer solutions. Just as the Euclidean algorithm is useful as one way of solving the first-degree linear congruence, or linear equation in two variables, one has an analogous situation in two dimensions or three dimensions, and so on. I simply explored that, and discovered many interesting things. I also found that there were many unsolved problems, which this approach might possibly solve. I always intended to come back to this problem, but I never did because I got involved in so many other things.

When did you finally get your Ph.D. degree?

In 1955. I was 38 years old by then.

Coming back to your work on the phase problem, did you, at that time, have a sense of how important your discovery was?

No. Although we knew that this was important work, we did not have the slightest idea what the long term implications would be. It became important only because of the applications, many years later, when it turned out to be of central importance for the life sciences.

This was not your first project at NRL.

No. My first project was to design and construct an electron diffraction instrument, which would be used for gas diffraction. I did that from 1947 to 1948, when the instrument was constructed by the engineers at NRL.

Starting around 1948, I learned about X-ray crystallography and the phase problem, which, incidentally, had not been of any interest to Jerry at that time. His work was concerned exclusively with electron diffraction. Jerry and Isabella were not at all involved in X-ray diffraction. I had a period of free time for a year or so with not much else to do, but learn about X-ray diffraction and the phase problem. Jerry knew something about this, he had some background in it, but it was not an area in which he was planning to do any research or in which he had done any research.

Why did you turn to X-ray diffraction?

Probably Jerry learned about it from going to the meetings that there was the problem known as the phase problem. He must have understood what it was, but it was not an area of research for him. He spoke to me about it and I became interested in it. The real challenge was that this problem was regarded as unsolvable.

It didn't take long, although I'm a little vague as to the timing now, but this must have occurred in 1948–1949, when I finally got involved and understood what the problem was. It was before 1950 when it became clear to me that this argument that the phase problem can't be solved was simply not valid. Jerry and I both understood that around the year 1950. In order to be absolutely certain about that, we took an easy problem; I think it was a three-atom problem, in one dimension, and set up the equations and found that the solution of the phase problem for such a simple problem required only the solution of an equation of the fourth degree. We could solve it numerically and we found that there were two solutions, two sets of phases, which were determined by three magnitudes. In order to make the answer unique, we simply had to introduce one other intensity and that distinguished between the answer and the non-answer. So we could see that in the case of a three-atom problem there was, in fact, a unique answer. That convinced us that this problem was solvable. This was, of course, in violation of all the dogma, which existed at that time. No one believed that, and we were regarded as crazy even for thinking that, but we knew that it was correct. Throughout the whole decade of the 1950s no one, with a few exceptions, believed that we were on the right track at all.

What can you tell us about the division of work between you and Jerry Karle during the time you worked together in the Naval Research Laboratory on the direct method?

It's a difficult question for me to answer. In the first place, it was a good combination. The work was very strongly mathematical, and this was my contribution to it. Jerry Karle's field was physical chemistry. So what we brought to this problem was his expertise in structural chemistry and mine in mathematics. This worked out very well for the 12 or so years that we worked together, from 1947 to 1960. It was in the Naval Research Laboratory, and then I stayed on for another 10 years in the Naval Research Laboratory, working in a different area as a mathematician.

What happened then?

Until the late 1960s, the Naval Research Laboratory was, from my point of view, a very good place to work because I had a great deal of freedom to work on problems which I selected and which I was interested in. As long as one was a productive scientist, this was acceptable. But in the late 1960s, the Navy and Congress questioned the wisdom of supporting work of research scientists that was not directly related to naval missions. By no stretch of the imagination could my work be considered to be of direct interest to the Navy. So around 1968, I was subject to a great deal of pressure to switch from what I was doing to what was regarded as more relevant to Navy needs. I was unhappy about this for two reasons. In the first place, I was not particularly interested in the kind of work they wanted me to do, and, in the second place, I did not like the idea of my work being used for military purposes. Prior to that time, there had been no way that my work could be used that way. The military establishment had been supporting my work. Therefore, there was an implicit obligation on my part to support military research, which I am philosophically strongly opposed to. When I had to confront this issue at the end of the 1960s, I began to understand that the time had come for me to leave.

Where were you in World War II?

I was in the Navy. When the War started, in 1941, I was working as a statistician in the Census Bureau as a civilian, but I soon transferred to the Army Air Forces. I went to work as a civilian instructor in radar and electronics. That was the only war that I felt comfortable in. In 1943, I submitted an application to the Navy for a commission to become an officer, and on this occasion they accepted my application. (On a previous occasion I had been rejected because I wore glasses.) Instead of active service, however, they sent me to school for another 9 months to learn weather forecasting. After that, they sent me to the Southwest Pacific for 18 months, where I did many things except weather forecasting. When I came out in 1946, I was unable to get a job anyplace so I went back to my old job as civilian instructor in the Army Air Forces. Then in 1947, I got a job in the Naval Research Laboratory. After 23 years at NRL, I left for the reasons I've already indicated to take a job at the Medical Foundation of Buffalo, Inc., here in Buffalo, New York.

How about this Institute?

This institute was founded in 1956 by a medical doctor, Dr. George Koepf, who was interested in basic biomedical research. He was not only a practicing physician but also a research scientist. In 1956, a very wealthy patient of his, Mrs. Woodward, volunteered to assist him in establishing this institute with an initial endowment of 3 million dollars, and this was a lot of money at that time.

The mission of this institute is basic biomedical research, and it provides an environment for individuals to carry out independent research in this area. Our primary support comes from government granting agencies — NIH, NSF, NASA. We have a total of 65 people, 25 Ph.D.'s among them. The bulk of our research is X-ray crystallography.

Does the original donor's family continue its support?

Yes. Mrs. Woodward was the Jell-o heiress. Then her daughter, Connie Stafford, continued to support our Institute until she died five years ago, and her daughter, Connie Constantine, continues to support us.

Please tell us about your family.

We got married in 1940, and we have two daughters. We have Barbara, 54, and a younger daughter, 51. Our older daughter lives here in Buffalo, and we see her often. Our younger daughter has her Ph.D. in psychology and lives in Bethesda and works for the Uniformed Services University of the Health Sciences. Her research is in the field of trauma. She is sometimes called when there are major accidents, and it's important to understand how individuals and communities cope with such a disaster.

You wrote a beautiful article about your hobby making stained glass polyhedra.[3]

When I was 8 or 9, one of my favorite books was on geometry, and it described how to build the Platonic solids. I remember making these polyhedra. This interest has remained with me ever since. I kept making polyhedra from time to time, and some very elaborate ones too.

Then about 15 years ago one of my colleagues was taking a class working with stained glass. I got interested in it and registered for the course. It was once a week for five or six weeks, and I did some simple projects. Eventually I started making polyhedra from stained glass. The first ones were very crude, but then I kept improving. Over the years I made about 40 regular and semi-regular polyhedra. Then in 1983, I got interested in packing spheres inside these polyhedra. One quickly learns that in the regular tetrahedron you can put spheres of equal size and they'll pack in the hexagonal close-packing arrangement, and the octahedron and the cube can be filled up too. Then I wanted to do the icosahedron, that is, to fill it up with equal-size spheres, and I found out that I couldn't do it. I did a lot of experimenting with the icosahedron. I took the icosahedron and put one sphere in it. The next step was to put one sphere in each vertex, having 12 spheres that way, and that leaves a hole in the middle which you can fill with a somewhat smaller sphere, of approximately

Herbert Hauptman with one of his polyhedral packing models (courtesy of Herbert Hauptman).

9/10 the diameter of the outer spheres. Then I placed three spheres along each edge of the icosahedron, and it was easy to figure out the necessary diameter of the sphere, and that made an icosahedral shell. Then I made the next internal layer of the smaller spheres, and this arrangement left a hole in the middle. What really made me excited was that when I calculated the size of the sphere that you'd need to fill that hole, it turned out to be the large-size sphere again. So that only two different-size spheres would suffice for the tight packing of spheres in the icosahedron, at least to this complexity. At this point there was no stopping me, and I found that in all cases two sizes would suffice in filling the icosahedron, and the ratio of their diameters would always be the same; roughly the small sphere has 9/10 of the radius of the large sphere.

Once this was done, what remained was the dodecahedron, and for a long time I thought that it must be possible to fill the dodecahedron with spheres of two different sizes. But there was no way that I could do it. I must have spent many months trying to do that. And it didn't work with three sizes either. I was almost ready to give up when I thought to try four different sizes, and that does work. It was about 10 years ago, and by now I've forgotten how to do it.

What made you stop making these beautiful models?

The Nobel Prize. I just didn't have the time anymore.

How did your wife's life change because of the Nobel Prize?

Many more people seek her out in a social way, people who would not have otherwise been doing that. She has gotten to know more people, mostly women, than she otherwise would get to know. She is happy about it, and this has been good for her. She is considered a more important person because of this, and she knows that and has said so herself. And so am I, and I'm not saying this in any derogatory way, it's just a simple fact.

Concerning the Nobel Prize winning achievements, Jerry and you laid down the theoretical foundations of the direct method. However, for its recognition, it was crucial to have the applications. Who was primarily responsible for them?

The initial applications were made by Jerry and myself, and the people at the Geological Survey. During the 1950s, we solved six crystal structures, difficult or impossible to solve in those years with existing methods. Later on, from the 1960s, Isabella Karle, Michael Woolfson, and Peter Main started making important applications. In the early 1960s, Isabella was almost alone in making the applications in a systematic way. By the end of the 1960s, all three were involved. Woolfson and Main, in particular, not only used our work, but they made their own contributions. They succeeded in strengthening the methods over the next few decades. Isabella, on the other hand, was the first to work on very complex structures, for the time. She did these structures, with 70 or 80 all-light-atoms, before Woolfson and Main got into the act. Woolfson and Main then made theoretical contributions and they made their computer programs available worldwide. I want also to mention the important early contribution (1953) of David Sayre who discovered the famous equation which bears his name and which has played such a prominent role in much of the later work.

If the announcement of the Nobel Prize had included Isabella Karle in addition to the two of you, would you have been surprised?

No. I would not have been surprised at all. I was somewhat surprised by the way it turned out, that she was not included. That surprised me.

Because?

Because of my background as a mathematician and the fact that I was junior to Jerry, and I had never received the recognition from the crystallographic community that Jerry and Isabella had. They were certainly much more active in promoting themselves than I was. Furthermore, I have no chemistry background whereas Jerry and Isabella are well known in the chemistry community.

Are you saying that you might have been left out of the prize?

I believe it could've happened. That would not have surprised me. It had already happened that Jerry and Isabella were members of the National Academy of Sciences and I wasn't. She became a member soon after Jerry had become a member. David Harker who promoted me for membership in the Academy, told me that Jerry had prevented my election by arguing that "Herb is too mathematical." I believe that Jerry wanted to win the Nobel Prize for himself and Isabella alone; but I was a thorn in their side because in the period of the 1950s, Jerry and I had co-authored several dozens of papers, equally co-authored, and Isabella had not been involved, except at the very end. It wasn't until the 1960s, when they got their own diffractometer, that they could co-author their papers without me.

What composition of the awardees did you think most likely?

I thought that the most likely result would be Jerry and Isabella to share the prize; the second most likely, Jerry, Isabella, and Woolfson. So, I was surprised at the way it turned out. Whether Jerry was surprised, of course, I don't know, but I'm certain that he was disappointed. He was in this difficult position of trying to promote himself and Isabella and rule me out; but the Swedes, obviously, did not buy that. I wouldn't be telling you these things if I weren't sure of them, but it would be impossible of course to provide airtight proof that this was Jerry's intention.

Who would you assign the greatest credit for the ultimate triumph of your method apart from Jerry and you?

Woolfson, without any doubt. But he was also the one who, in 1954, together with Cochran, tore us apart.

The same Cochran who, together with Francis Crick, described the diffraction of helices?

The same Cochran, a brilliant physicist. If it weren't for the paper that Woolfson and Cochran co-authored, I believe that Woolfson might have shared the Nobel Prize. But they, Vand, and Pepinsky were the only ones who put their objections to what Jerry and I were doing in print. All the others criticized us mercilessly; you have no idea what it was like at those meetings of the American Crystallographic Association (ACA) in the 1950s. At every single meeting of the ACA they tore us shreds. The criticism was made all verbally and there is no written record of that.

If you were to decide who would you give the Nobel Prize today?

Michael Rossmann and Ada Yonat. Rossmann did important work on virus structures and made major contributions in crystallography over a long period of time. Ada Yonat for ribosome and for her work over a period of many years, her perseverance. It reminds me somewhat of my own experience.

Let us suppose that there is a celebration of Jerry Karle's birthday and you have to give a brief presentation about his merits. What would you say?

I would do it. What would I say? He has made significant contributions without any doubt. We collaborated for a period of about 14 years, we worked together and we complemented each other, and we were a very good team. Something like that came when we had a celebration of our 75th birthday. The Medical Foundation of Buffalo, as my institute then was known, decided that the theme would be direct methods in the 21st century. Jerry is a year younger than I am so they picked a date midway between our birthdays, when I was close to 76 and he was close to 75. We had this meeting in Buffalo and Jerry came and Isabella as well, and many people came worldwide. To a visiting crystallographer, to an outsider, it may not be known that Jerry and I are not on good terms.

The Swedish Academy probably knew about this. One of the things, which is done at the ceremony in Stockholm, is that one person for each category of the Nobel Prize is invited to give a three-minute talk. If it is a shared prize, only one gets to give the talk. Some time before our trip to Stockholm, I got a call from the President of the Nobel Foundation and he gave me this information and he said that Jerry and I would have to make the decision who would give that talk. I believe he knew that the relationship was strained. I called Jerry up immediately and to my surprise, he let me decide. And I did decide; I decided that I would give

Herbert and Edith Hauptman in 1995 (photograph by I. Hargittai).

the talk. Mostly I decided this because I knew that my wife would never forgive me if I decided otherwise; it didn't matter much to me. But I leaned over backwards to make my talk apply equally well to his background and to mine, and I showed it to him before, and he agreed that it was fine. So it didn't matter who actually gave the talk because it was impartial. Basically, what I was saying was that we were both greatly indebted to a lot of people and primarily our parents.

Why did it matter so much to your wife who gave this talk?

She feels that I've been upstaged so much by Jerry and Isabella and I've received so little credit compared to what he has received. He is more accepted as a member of the crystallographic community than I am, simply because of our background. He is a physical chemist, I'm a mathematician. The Academy of Sciences is a perfect example of what I'm talking about. I did not get elected to the Academy until 1988, three years after the Nobel Prize; by then Jerry and Isabella had long been members.

Do you have any plans for retirement?

It's virtually impossible for me to retire. We've just got a big grant from NIH. We had a program project for 9 years. It's a major grant, about a million dollars a year. It was just renewed for another 5 years. It means that I'm obliged to continue working for another 5 years. It's the biggest grant we have and a major part of our total budget. However, I don't see yet how I will be able to continue after 90; that's a pretty old age to be working, so it may be that by that time I'll have to retire. On the other hand I'm still fascinated by this never ending problem. I don't think much about retirement.

References

1. Cochran, W.; Woolfson, M. M. "Have Hauptman & Karle Solved the Phase Problem?" *Acta Crystallogr.* **1954**, *7*, 450–451.
2. Vand, V.; Pepinsky, R. "The Statistical Approach of Hauptman & Karle to the Phase Problem." *Acta Crystallogr.* **1954**, 7, 451–452.
3. Hauptman, H. *Chem. Intell.* **1995**, *1*(2), 26–30.

Jack D. Dunitz, 1999 (photograph by I. Hargittai).

22

JACK D. DUNITZ

Jack D. Dunitz (b. 1923 in Glasgow, U.K.) is Professor Emeritus of Chemical Crystallography of the Swiss Federal Institute of Technology, Zurich, Switzerland (ETH). He received his B.Sc. and Ph.D. degrees from Glasgow University in 1944 and 1947, respectively. For a decade after his doctorate he held various positions at Oxford University, California Institute of Technology, and The Royal Institution in London. He has been with the ETH since 1957. He retired in 1990. His research interests have included crystal structure analysis as a tool for studying chemical problems, and his recent work has focussed on problems of polymorphism and solid-state chemical reactions. A sampler of his many distinctions: he is a Fellow of the Royal Society (London, 1974), Foreign Associate of the U.S. National Academy of Sciences (1988), and a Member of the Academia Europaea (1989). He has received the Bijvoet Medal (University of Utrecht) and the Gregori Aminoff Prize (Royal Swedish Academy of Sciences). He was George Fisher Baker Visiting Lecturer at Cornell University. We recorded our conversation on September 6, 1999 at the ETH.

You began in Glasgow. How did you get to the ETH?

It is a long story. After my B.Sc. degree, I was a Ph.D. student with John Monteath Robertson, who was one of the leading figures in X-ray crystallography at the time. Although he had no direct role in protein crystallography, he more or less invented the method of isomorphous replacement. Of course, the name of Johannes Martin Bijvoet must also be mentioned in this connection. There was an idea in the air and different people used it in different ways. Bijvoet and Robertson developed the method, and Max Perutz applied it to protein crystallography. Curiously enough, when I was at Glasgow, we never used the heavy atom method, nor the Patterson's function. That's why I wanted to go to Dorothy Hodgkin, to learn how to use these methods. What we did learn at Glasgow was how to look at an X-ray diffraction diagram and see the molecular transform in it. That was also very useful.

Incidentally, at the 1999 IUCr meeting in Glasgow there was a special session honoring J. M. Robertson and his school, which spread all over the world. One of my fellow students was Sandy Mathieson, who went to Australia and became one of the early protagonists of structure analysis of natural products by X-ray analysis. Another was John White, who went to a junior faculty position at Princeton. There, he started B12 crystallography at about the same time as Dorothy Hodgkin but never got enough credit for this. It was not Dorothy's fault. They published their first papers together. But it seems ironic that Princeton did not give him tenure for pioneering work that eventually developed into the achievement for which Dorothy Hodgkin got the Nobel Prize. Other crystallographers who started as Robertson students were Jim Trotter, who went to Vancouver, George Sim who went to Illinois, then to Sussex, and then back to Glasgow. There was Michael Rossmann, one of the most outstanding protein crystallographers in the world, now at Purdue. Then Ian Paul at Illinois, George Ferguson in Canada, and others.

Jack Dunitz and Dorothy Hodgkin in Los Angeles in 1989 (courtesy of Jack Dunitz).

For a decade after my Ph.D., I had a gypsy-like existence, I was the wandering Jew, der Fliegende Holländer, somebody who has no fixed place to live and work. It was a marvelous decade, and it all happened by chance, except for the first step. At Glasgow, I had read Dorothy Hodgkin's papers on cholesterol iodide. It was one of the first three-dimensional analyses of a complex molecule — the penicillin work was not yet published. I wrote that I wanted to work with her and in late 1946, thanks to the generosity of the Carnegie Trust for the Universities of Scotland, I went to Oxford as a postdoc. Apart from Dorothy, I was influenced there by Gerhard Schmidt, who, although a few years older than I, was still working for his D.Phil., having lost valuable years because he had been interned in 1940 as an "enemy alien." Gerhard had learned organic chemistry from Robert Robinson and went on to found that marvelous school of structural organic chemistry at the Weizmann Institute.

At the end of 1947, beginning of 1948, Linus Pauling came to Oxford as Visiting Professor. *The Nature of the Chemical Bond* was the most influential book in my scientific education, and I had a tremendous hero-worship admiration for the man. I listened to his lectures spellbound, and, through Dorothy, I actually met the great man. That summer, partly through Verner Schomaker's initiative, I got an offer to come to Caltech as a Research Fellow, and there was no hesitation.

Linus Pauling and Jack Dunitz at the ETH in 1977 (courtesy of Jack Dunitz).

I stayed three years in Pasadena instead of the originally intended one. In those postwar years, Caltech was the world center of structural chemistry. Jerry Donohue, Kenneth Trueblood, Dick Marsh, Jim Ibers, Alex Rich, Martin Karplus, Edgar Heilbronner, Massimo Simonetta were there or about to arrive. Lipscomb had just left. These were people in transit, so to say. Robert Corey and Eddie Hughes were on the staff and so was Verner, who was a major influence on my development. He had a passion to get things right. He often saw that something taken for granted might not in fact be true, and he could worry about such questions for years until he got to the bottom of them. The main problem on which I worked with him at Caltech arose out of our earlier conversation in Oxford, where I had been working on the crystal structure of tetraphenylcyclobutane and had found long bonds in the cyclobutane ring, 1.57 Å instead of 1.54 Å. Just about that time Charles Coulson and Bill Moffitt had developed their "bent bond" theory, which explained why bonds in small strained rings should be shorter than normal, as in cyclopropane, 1.51 Å. So why were the bonds in tetraphenylcyclobutane long instead of short? We decided to determine the structure of cyclobutane itself. We found long bonds and proposed an explanation in terms of non-bond repulsion, which has stood the test of time.[1]

The sequence of my ten-year wanderings was Glasgow-Oxford-Pasadena-Oxford-Pasadena-Bethesda(Maryland)-London-Zurich. My second stint at Caltech started in September 1953, just after my marriage to Barbara Steuer. Then I spent a year and a half at NIH in Bethesda. This came about because I had shared an office at Caltech with Alex Rich. When he was offered a job at NIH to start a structural lab in the Institute of Mental Health, he invited me to come and help. So Barbara and I decided to try the East Coast where we had never spent any time before. During that period, I was getting various expressions of interest from American universities. We saw we could easily drift into settling permanently in America. We had no objection to living in America but didn't think we should drift into it. Rather, we should make a conscious decision, one way or the other, but to make a decision, we had to consider the alternative. So we decided to go back to the U.K. for a time, and Dorothy Hodgkin helped me find a position at the Royal Institution, where Lawrence Bragg had just been appointed as new director, following his retirement as Director of the Cavendish Laboratory in Cambridge. At that time, some people considered Bragg to be an old fashioned physicist. But he had a good eye for what was important, or rather for what was going to be important.

When he moved from Cambridge, he left two research areas behind: one was protein crystallography and the other was radio-astronomy.

But now I'm ahead of events. In 1951, after the first stay in Pasadena, I returned to Oxford, to Dorothy's lab. This was the time when my collaboration with Leslie Orgel began. It was the time of our work on ferrocene, which started because we didn't believe Woodward's structure and ended by proving its correctness and providing a molecular orbital picture for the molecule.[2] Later, we became interested in structural problems in inorganic chemistry, the structure of spinels, and so on. By that time, he had an official position at Cambridge, and I was already in London, and we used to meet for a day's work once a week or so, alternately in Cambridge and London. In our collaboration, Leslie was the theoretical part. He knew and developed theories and I knew the structural part. Leslie left theoretical chemistry once it became clear that the future of theoretical chemistry was going to lie in more and more accurate computations. After all, John Pople was one of his colleagues in Cambridge. But it was not only his dislike of computational chemistry that led him to leave the field. From the very beginning he was also attracted to "origin of life" chemistry, where he rapidly became a leading figure.

In mid-1952, Sydney Brenner arrived in Oxford to work with Hinshelwood on bacteriophage. He already had a degree in medicine from South Africa. Soon we were holding weekly mini-seminars, where we argued about all sorts of things, among them about DNA. Sydney talked, Leslie talked, I talked, often all at once. Partly through Sydney, who knew the current literature, we learned about Chargaff's rules and wondered about what they could mean. When a telephone call came from Cambridge one day in the spring of 1953 to say they had this marvellous structure and could I come to look at it, we all three went. We knew enough about DNA that when we saw it we knew at one glance that it was the right structure. It accounted for all the known facts about DNA, apart from having this beautiful self-complementarity to explain the mystery of heredity.

You were at Caltech from 1948 to 1951 and this was the time of the discovery of the alpha-helix. I would like to formulate a more general question about the role of structural chemistry in the emergence of molecular biology.

Linus Pauling was one of the founders of molecular biology and I reviewed his contribution in the obituary I wrote about him.[3] Of course, other people contributed to these beginnings as well. One thinks of J. D. Bernal and Bill Astbury. Astbury was probably the first to try to explain the properties of biological material in terms of molecular structure. I think of his work on fibrous proteins such as keratin, for example. But he never got the alpha-helix. He missed that crucial structural step. The Cambridge people also missed the mark. About six months before Pauling's alpha-helix, there was a paper by Bragg, Kendrew and Perutz who tried to list all possible conformations of a regularly repeating polypeptide chain. But they considered only those with an integral number of amino acids per turn of the screw. Pauling ignored this limitation but put in the condition that the peptide group had to be planar. He knew this from the structural studies at Caltech on small molecules containing the peptide group and he also believed in it from his resonance theory. He also insisted that all possible hydrogen bonds had to be formed. Pauling was one of the first to spell out the importance of hydrogen bonding in proteins. The formulation of the alpha-helix was the first and still is one of the greatest triumphs of speculative model building in molecular biology. It was the forerunner of the tremendous investment in computer-assisted molecular modeling in present day research.

Pauling used to tell me occasionally about his models and what one could learn from them. As I recall, he talked about spirals, and one day I said that the word "spiral" referred to a two-dimensional curve, like the logarithmic spiral. Since his polypeptide coils were three-dimensional, they were better described as "helices," I suggested. Pauling replied that the two words could be used interchangeably, but he thanked me for my suggestion because he much preferred "helix" and declared that he would always use it in future. Maybe he felt that by calling his structure a helix there would be less risk of confusion with the various other models that had been proposed. I must admit that I have no notes on the matter, and may be remembering it all wrong. But I still like to think that I helped to give the alpha-helix its name — my personal contribution to molecular biology. It may not be much of a contribution but it's better than none.

I should mention that at Caltech I came into contact not only with the chemists but with the biologists working around Max Delbrück. The very first evening I arrived at the Athenaeum, the Caltech faculty club, I sat down at a table and was joined by Gunther Stent and Carleton Gajdusek. They were so to say the first people I met there, and through them I got to know Max and became part of his circle in lunch time conversations and weekend trips to the California desert. There the talk was much about bacteriophage and about DNA. By that time there was sufficient evidence to suggest that DNA was the hereditary substance, but nobody had the faintest idea about how it worked. Other people thought it was proteins, and others thought it was proteins and DNA. It was talked about. That is why when I went back to Oxford, I had some background in molecular biology and was prepared to some extent for the fantastic developments that were soon to take place.

Structural chemistry seems to be present to a much lesser extent today in the forefront of molecular biology than it was decades ago, and the geometrical changes that we like to discuss do not seem to be important for biological behavior.

And not only for biological behavior — for chemistry, one might say. Of course, chemical reactions involve drastic changes in molecular structure, but the time has passed when experimental structural research was mainly concerned with bond lengths and bond angles. If you want bond lengths and bond angles in ground state molecules today, it's quicker and probably just as accurate to calculate them, either with molecular mechanics force fields or with quantum chemistry programs. With the latter, you can even obtain information about the structure and energy of transition states, at least in the gas phase. But for some time now the focus of interest has moved from bonding interactions within molecules to so-called non-bonded interactions between molecules. Supramolecular chemistry. Of course, the crystallographers have always been interested in intermolecular interactions, or at least they should have been, and I don't know whether chemistry followed crystallography here or the other way round.

With regards to biological behavior, non-bonded interactions affecting the conformation of proteins and nucleic acids are of the utmost importance, especially now that genome research is identifying amino acid sequences in proteins whose functions are completely unknown. To guess the function, you need the fold. But if you have only the sequence, what is the fold? Or rather, perhaps, what are the folds? — plural. There are proteins that may exist as monomers, dimers, trimers, or fibers. There are the amyloid structures in the brain, proteins that in the normal course of events are doing one thing but under different conditions make insoluble structures on the surface of neurons. When you change from one structure to another, you have conformational changes, so it is not completely true that the amino acid sequence completely determines the fold. It may allow for alternative folds. There are some interesting papers by David Eisenberg on what he calls domain swapping. This kind of change is going to be very important, but the associated changes in bond lengths and angles are not going to be of much interest. Perhaps I should modify what I just said. There will be such changes when, for example, a glutamic acid side chain, which has a certain pK_a in one environment, goes into another environment where it may have quite a different acid strength.

You have been much concerned with what you and Hans Bürgi called "structure correlation."

Structure correlation is based on the recognition that the various functional groups in chemistry are not fixed and rigid. When one examines the distribution of the structural parameters of a group observed in different crystal and chemical environments, one often finds correlations among these parameters. These correlations are characteristic of the group, they tell us something about the energy surface and in many cases they can be interpreted in terms of supposed mechanisms of chemical reactions. But I still haven't answered your question how I came to the ETH.

Yes, we had got as far as The Royal Institution in London. Then how did you finally end up at the ETH?

It happened as a result of having been at Caltech and at Oxford. The story, as I heard it afterwards, is that Professor Leopold Ruzicka, a powerful figure at the ETH and in Swiss chemistry, had been so impressed by Dorothy Hodgkin's analyses of penicillin and Vitamin B12 that he wanted to have an X-ray crystallographic research group in his own laboratory. He invited Dorothy herself to come to Zurich to lead such a group, and when this proved unrealizable he asked her to recommend someone as possible alternative. The story, again as I heard it, was that she gave my name, which Ruzicka had also heard from Edgar Heilbronner, with whom I had shared an office at Caltech. In December 1956, I was asked to come over for discussions here and was offered a position as ausserordendlicher Professor. In those days there were no negotiations. Ruzicka was in a hurry to find someone before his imminent retirement the following October. He more or less gave me an ultimatum, take it or leave it. I had two weeks to decide. Although I had promised to stay at The Royal Institution for five years, Bragg generously freed me from this obligation and advised me to go — it was too good an opportunity to pass, he said. So it came about that I arrived here on October 1st, 1957, the same day as Ruzicka retired, to become attached to this community of splendid individualists — my friends and distinguished colleagues in the Organic Chemistry Laboratory of the ETH. For more than forty years we have argued and discussed and learned together about chemistry and about everything else under the sun. I am also lucky to have been chosen over the years by marvellous students and postdocs, by incredibly skillful collaborators, who have been excellent teachers and from whom I have learned practically everything I know.

What would you single out as your most important contribution?

Structure correlation is probably one of the most important. People talk about Bürgi-Dunitz trajectories. There is, or used to be, a Dunitz conformation of the cyclodecane ring; that came from the time I worked on the conformations of medium rings. In that roundabout way, as things work in science, structure correlation arose very indirectly out of the other. The work on medium ring conformations was being done around the same time as people were trying to develop hydrocarbon force fields, and we worked together with them to some extent, especially with Shneior Lifson's group at the Weizmann Institute. They could tune their potential functions with our structures. This gave me the idea that from a systematic set of related structures we can say something about the shape of the potential energy function. So we started on an ambitious project to derive the potential function for out-of-plane deformations of the amide group. We began with the structures of small ring lactams and worked our way towards the larger rings. For the small rings the conformation was cis-planar and for the very large ones it was trans-planar, so somewhere in the middle it had to be between these and therefore non-planar. The medium ring size is very interesting for hydrocarbons and it turned out that this holds also for the lactams. When you come to 9-, 10-, 11-membered rings, the amide group is neither cis nor trans but distorted out-of-plane. When you start doing this, you are thinking already of looking for structural correlations among closely related molecules, so that was in the air.

In 1973, Eli Shefter was here on a sabbatical, funded by some pharmaceutical foundation, and felt he should do something connected with pharmacy. He and Hans-Beat Bürgi then determined the crystal structure of methadone and found an unusually short approach of the amino nitrogen to the carbonyl carbon, which was seen to be slightly displaced from the plane of its three bonded neighbors towards the approaching amino group. This reminded me of similar observations in published structures of some alkaloids. It was still before the time when one could search the Cambridge Structural Database, but we went to the literature, extracted the relevant data. They fitted in a nice sequence that could be interpreted as a series of points defining the reaction path for nucleophilic addition to a carbonyl group. This was new, exciting, and it seemed important. It was certainly something we could be pleased about.

Much earlier, I was pleased with the structure of ferrocene, even though we originally believed that the sandwich structure was wrong. I was pleased with the work I did with Leslie Orgel on the normal and inverted spinels, although that is still inorganic chemistry or mineralogy. The medium-ring project started as a direct result of my coming to the ETH. When I arrived I didn't really know what to do next. I read many papers about what was going on here, including Prelog's essay in *Perspectives in Organic Chemistry* about medium rings. I saw at once that this was an area where definitive structural information was lacking and that I should try to provide it.

At various times, I was absorbed by mathematical problems. They can worry one for years, and then, when you see the answer you can get a sense of illumination that can elevate you above the clouds for a few hours or even longer. Some such problems arose in connection with questions of the flexibility of rings; for example, why does the chair form of cyclohexane appear to be rigid while the twist-boat family is flexible? Then there was the equilateral pentagon problem, which arose out of a gas-phase electron diffraction study of arsenomethane by Waser and Schomaker in 1945. The molecule is a five-membered ring of As atoms; the As–As bond lengths are equal but the angles vary over several degrees. Why? One might try to find a chemical explanation but it would be a waste of effort. The result arises from geometric necessity. Of all equilateral polygons, the pentagon has the special property that if the figure is not planar then the angles cannot be equal. Although unknown to Euclid, this is an elementary geometric theorem. I knew it was true but I couldn't find a proof. When I did finally see how it could be done, I had that special feeling I mentioned. I didn't know the mathematical literature well, and it was quite possible that some Polish or Hungarian mathematician had solved the same problem years before. With Jürg Waser, I even wrote a paper on the problem. Then there was that little theorem about the tetrahedral arrangement in *The Chemical Intelligencer*.[4] When I read it I knew it couldn't be true.

Mostly in collaboration with Kenneth Trueblood and Verner Schomaker, I also worked for a long time on the motion of atoms and molecules in crystals. I believe my book contains the first easily graspable account of the famous 1967 Schomaker-Trueblood paper where they introduced the S tensor for non-centrosymmetric crystals. Cruickshank had shown in 1956 that you could account for the anisotropic displacement parameters of atoms in crystals in terms of molecular translation and rotation. This worked fine for a rigid molecule sitting on a crystal center of symmetry. In this case it is clear that the three rotation axes must intersect at the center. Then the question arose, where do the rotation axes intersect when the molecule is in a general position? Many pages were written about this question, and then in 1967 Schomaker and Trueblood showed that the search for an intersection point was illusory. The rotation axes need not intersect at all. In such a case one needs to account for the quadratic correlation between rotation and translation, and this is done by adding an extra tensor called the S tensor to the rotation and translation tensors. This was a pretty difficult paper to understand, and when I wrote my book on the Cornell lectures,[5] I spent several weeks trying to write it up in a way that was correct and also understandable. I wanted to understand it myself. Fortunately, Trueblood was here on a sabbatical at that time, so every day I brought a version for him to check. In addition to the rigid-body molecular motion, there is also the internal motion. When one puts it all together, the mathematics involves complicated matrix algebra, and there are always unexpected singularities cropping up. I used to like these problems, but they are too complicated for me now. As I get older I cannot cope with them any more.

I would like to ask you about your background. Vlado Prelog told me a story from your school years in Glasgow, when you were caught in a crossfire between Catholic and Protestant boys and when you told them you were Jewish, they asked you whether you were a Catholic Jew or a Protestant Jew. Then Prelog told me that you had asked not to use this story.

The way Prelog told the story was not quite accurate. I went to a good, several hundred year old grammar school in Glasgow, Hutchesons' Grammar School. I had no trouble whatsoever being a Jewish boy there, but, as I must have told Prelog, I could well have had trouble if I'd been Catholic. In Glasgow, there was a centuries old problem of Protestants against Catholics, and Catholics against Protestants, as in Northern Ireland but not nearly so vicious. But Scotland has been almost free of anti-Semitism. Earlier, at elementary school, when we lived on the borderline between a relatively comfortable middle class area and the slums, I knew that if I saw the ragged kids from the slums I had to get out of their way as fast as I could, not because I was Jewish but because I was decently dressed and had shoes on. They didn't know who I was — this was the period of the depression.

My father's family came to Scotland towards the end of the last century from a place called Grajewo, a small town North of Bialystok, close to the border of today's Poland, Lithuania, and Belarus. The family legend is that when my great uncle, my grandfather's brother, came to a small mining village in Scotland, he became acquainted with Keir Hardy, who later became the first Labour M.P. in Britain around the turn of the century. According to the legend, Hardy was influenced by political discussions with my great uncle, but when I took the trouble to read the official biography of Keir Hardy I found no mention of uncle Slonimsky. In czarist Russia, eldest sons were exempt from army service and brothers very often had different family names so that they could all claim to be eldest sons. As for the name Dunitz, I found about 65 people with this name in the Internet. But I have no idea where the name comes from. I doubt that it has any connection with the Nazi admiral Dönitz and although some people have suggested it is derived from the name of the Russian river Donets I doubt this too. My paternal grandparents, whom I never knew, went to Jerusalem around 1900 and are buried on the Mount of Olives. My father, left in care of an elder sister, was brought up in London and served in the British Army in World War I. Always somewhat impractical — a dreamer, one might say — he was not a good soldier but he was a good runner and used to win the half mile race for his regiment. Although he had no formal education, he had a tremendous appreciation for literature and music — Beethoven, Mozart, and Schubert. In literature, he was especially sensitive to new, unknown authors. I remember he told me about Solzhenitsin's *One Day in the Life of Ivan Denisovich* as soon as it appeared in English — before it was even reviewed. In my opinion it is Solzhenitsin's masterpiece. On my mother's side, my grandfather, Morris Gossman, had come to London from the Ukraine, from Zhitomir. He was a wanderer. From the 1890s until World War I, he travelled the world, looking for gold, in Australia, in South Africa, and in the Yukon. As far as we know, he never found any. My mother was the eldest of the seven children my grandfather produced during the intervals between his travels. Although greatly influenced by my father, she had her own definite taste in poetry, much of which she could recite from memory. It says something about progress in social health that although she was brought up in relative poverty in London's East End she lived to the age of 101, and it says something about her memory that she remained a formidable bridge player until well into her 90's.

You came to Switzerland more than 40 years ago.

It has been a great experience for me to be here. I was very fortunate to find here great colleagues with whom I could interact scientifically and personally. On the whole, I have the good fortune, and I call it fortune rather than luck, in having had a career in science in the second half of the 20th century. I know it has been a terrible century in many ways. But in the last 50 years, a person of modest talents could have a great time in science. He could be supported by the public to be paid to do things that he liked. That's been for me a marvelous piece of good fortune. I don't think there was any time before when that was so true and I don't think there's going to be any time in future where it will be so, because people are going to want value for their money in more predictable ways.

References

1. Dunitz, J. D.; Schomaker, V. *J. Chem. Phys.* **1952**, *20*, 1703.
2. Dunitz, J. D.; Orgel, L. E. *Nature* **1953**, *171*, 121.
3. Dunitz, J. D. *Biographical Memoirs, Royal Society London* **1996**, *42*, 315.
4. Dunitz, J. D. *Chem. Intell.* **1998**, *6*, 53.
5. Dunitz, J. D. *X-ray Analysis and the Structure of Organic Molecules*, Cornell University Press, Ithaca, New York, 1979; 2nd corrected edition, Verlag Helvetica Chimica Acta, Basel, 1995.

Hartmut Michel, 1998 (courtesy of Hartmut Michel).

23

Hartmut Michel

Hartmut Michel (b. 1948 in Ludwigsburg, Germany) is Department Head and Director at the Max Planck Institute for Biophysics in Frankfurt/Main and Adjunct Professor at the University of Frankfurt/Main. He received the Nobel Prize in Chemistry in 1988, jointly with Johann Deisenhofer and Robert Huber, "for the determination of the three-dimensional structure of a photosynthetic reaction centre." (There are entries on my meetings with Johann Deisenhofer and with Robert Huber in this Volume.) Following high school and military service, Hartmut Michel studied biochemistry at the University of Tübingen between 1969 and 1975, including a one-year research period at the Max Planck Institute for Biochemistry and the University of Munich. He did his doctoral work with Dieter Oesterhelt at the University of Würzburg between 1975 and 1977 on light-energy conversion in halobacteria. He continued working with Oesterhelt first at the University of Würzburg, then, when Oesterhelt had moved, at the Max Planck Institute for Biochemistry in Martinsried near Munich. In 1981, he established his own group for crystallization and structure analysis of membrane proteins, in Oesterhelt's department in Martinsried. He has been in Frankfurt since 1987.

Hartmut Michel is a Member of Leopoldina (the oldest German science academy, in Halle), the Academy of Sciences of Göttingen, and the Academia Europaea, among others. He is a Foreign Member of the National Academy of Sciences of the U.S.A., the Royal Dutch Academy of Sciences, and the Chinese Academy of Sciences. He has received numerous prizes and awards. The interview communicated below was conducted by e-mail in March 2001.

In my first question, I asked Professor Michel about the research that led to the 1988 Nobel Prize, about the circumstances of that work, about the contributions of the various participants in the work and about those of his co-recipients of the Nobel Prize and his former mentor, Dieter Oesterhelt.

The crystallization of the photosynthetic reaction centre was crucial for the structure determination. One has to keep in mind that the photosynthetic reaction centre is a complex of membrane proteins, and it was considered to be impossible to crystallize membrane proteins at that time. The reaction centre was therefore the first membrane protein or complex of membrane protein whose structure could be determined. I had started the crystallization of membrane proteins in 1978 when working as a post doc in Dieter Oesterhelt's lab in Würzburg. My attempts were caused by an accidental observation, which I had made with bacteriorhodopsin and which led me to try to crystallize membrane proteins and to develop strategies to achieve this. The results were two papers in the *Proceedings of the National Academy of Sciences of the U.S.A.* with D. Oesterhelt in 1980 (*77*, 338–342; *77*, 1283–1285). One described the formation and analysis of a new two-dimensional crystal form of bacteriorhodopsin and the other was about the first true three-dimensional crystallization of any membrane protein, namely bacteriorhodopsin.

Then, D. Oesterhelt made the decision to start a collaboration with Robert Huber [of Martinsried] in order to determine the structure because he had gotten an offer to become a director at the Max Planck Institute at Martinsried at that time. This replaced the collaboration with Richard Henderson with whom we had collaborated in the characterization of the two-dimensional bacteriorhodopsin crystals. We traveled once to Martinsried and the crystals were X-rayed in our presence. The resolution was insufficient, but Huber considered the method of crystallization as interesting. With the hope of getting good crystals, I skipped a planned postdoc time in the U.S., and accompanied D. Oesterhelt to Martinsried in August 1979. I was hoping to optimize the crystallization method there, because it should have been possible to check the quality of the crystals immediately by having access to R. Huber's X-ray equipment. It turned out that this was the wrong expectation. After two or three X-ray photographs, which still showed insufficient quality of the crystals I was no longer able to get my crystals X-rayed. Before Christmas 1979, Wolfram Bode, a senior coworker with Robert Huber, asked me to give him some mounted crystals because there might be some time on the generators between Christmas and the New Year. However, at Easter 1980, the crystals were still not X-rayed!

As a result, I applied for an EMBO [European Molecular Biology Organization] short-term fellowship to optimize the crystals and to X-ray them at the MRC [Medical Research Council] Laboratory of Molecular Biology [LMB] in Cambridge with the help of Richard Henderson. It was granted and I stayed for about four months in Cambridge during the summer of 1980. The crystals improved continuously, but the final breakthrough did not show up. As a consequence, Dieter Oesterhelt decided to buy his own X-ray generator. In order to facilitate its service, he made the arrangement with R. Huber that the new generator (about DM200,000) should be installed in R. Huber's department and should be equipped with two X-ray cameras (about DM20,000, each), to be provided by R. Huber. R. Huber should be responsible for maintenance, and one of the two cameras should be used by myself and the other by R. Huber and his department.

Until the X-ray generator could get operational, I tried to crystallize four other membrane proteins in parallel to bacteriorhodopsin, among them the photosynthetic reaction centre from *Rps. viridis.* This bacterium came to my attention because I had listened to a lecture given by E. Wehrli from the ETH Zurich at Burg Gemen near Münster, Germany, who had analyzed two-dimensional crystalline photosynthetic membranes from this bacterium by electron microscopy. I thought that this might be an interesting project and asked him for some membranes to start with. When the project looked promising, I started to grow the bacteria myself, developed a new method for isolation of the reaction centres, and tried to crystallize them with the methods I had optimized for bacteriorhodopsin before. In these experiments, I could use another detergent for crystallization that was not tolerated by bacteriorhodopsin but worked nearly immediately here.

After a short discussion, in July 1981, D. Oesterhelt agreed that this should be my project on which I could base a career. In September, I X-rayed the first RC crystal, which showed diffraction to 3.3 angstrom resolution, which could be considered to be good enough to determine the structure. In the beginning of 1982, I could start with the data collection using D. Oesterhelt's shared X-ray generator. I could take two rotation photographs per day (only X-ray films were available in those days), starting one in the morning, the other in the evening at around 22:00. Nevertheless, it took me more than three months to collect one data set. The other camera on D. Oesterhelt's generator was mainly used by R. Huber himself who collected data on alpha-antitrypsin. R. Huber repaired the generator and aligned the cameras personally, when necessary. During the days I worked together with one technician and one graduate student on the biochemical characterization of the reaction centres, and tried to isolate the genes, in order to get the necessary biochemical information (amino acid sequences, pigment composition) needed to interpret the (expected) electron density maps. This was due to D. Oesterhelt's wish that we should do all the genetic and biochemical work in his department. We therefore had to establish all the molecular biological techniques, and the gene coding for the H-subunit of the reaction centre was probably the first new gene cloned and isolated at Martinsried. As a further consequence, I lacked the time to work myself into the evaluation of the X-ray data and structure determination.

I then gave a talk in R. Huber's departmental seminar, introduced the project, and asked for help. Three people volunteered to help: Otto Epp, James Remington, and J. Deisenhofer, after some initial hesitation. At the end, Otto Epp took my X-ray photographs, scanned them, and measured the intensities and scaled the data. J. Deisenhofer joined in later with a new Japanese postdoc, Kunio Miki, and did the more sophisticated crystallographic computing. I searched experimentally for the heavy atom derivatives, looked through the board with close to 1000 heavy atom compounds, many of them synthesized by Robert Huber personally in the years before. I soaked the reaction centre crystals, took the X-ray photographs, and looked for changes in the diffraction patterns with occasional advice from W. Bode or R. Huber.

R. Huber considered the reaction centre to be a "dull photosynthetic protein which cannot do anything. If it would be a receptor, I would be personally interested," he said. D. Oesterhelt generously considered the project as one of the young people. J. Deisenhofer calculated the electron density maps, and we frequently sat together to try to interpret and to incorporate the new sequence information which we were gathering in D. Oesterhelt's department, into the model. R. Huber suddenly changed his mind when he interpreted the electron density map of phycocyanin (another dull photosynthetic protein as R. Huber had called it initially). This project had been started by W. Bode, with the electron density map calculated by Tilman Schirmer, a graduate student. R. Huber realized that the protein fold is nearly identical to that of the globins. He certainly also became aware of the general interest, which was brought toward the project by the scientific community.

Suddenly, when the interpretable electron density map had already been calculated by J. Deisenhofer, I found my X-ray photographs developed by R. Huber during the weekends, when he came to the lab before me. He told me that this is just a favour, "I realize how hard you work, and whether I only develop my own films or also yours, in addition, does not take more time." This was a true argument, but at the same time he started to treat me in a pretty ugly manner, and it looked like he considered me as a major opponent because most of the credit for the project was given to me at that time. However, he overlooked that J. Deisenhofer would be much more damaged than I when R. Huber would take over the project, because they were both X-ray crystallographers with R. Huber in the senior position. As a result, the relationship between R. Huber and J. Deisenhofer became also very bad. When it came to the authorship of the publications R. Huber insisted to be a co-author. His argument was that he had helped in the data collection (see above) and we should acknowledge his work in setting up and running the department, and in establishing the know-how by allowing him to be a co-author. I considered this to be an internal issue of his department. The result is known. The Nobel Prize was given to J. Deisenhofer, R. Huber, and myself. D. Oesterhelt could not have been included because he was not a co-author of the important publications.

Have you continued your research along the same lines as before?

We have continued with the same line of research. As from the beginning, we aim to understand the mechanism of action of membrane proteins not the structure. The knowledge of the structure is only the necessary prerequisite to understand the mechanism. At present we have determined the structures of six membrane proteins. The next most successful lab in the world has determined three membrane protein structures.

What are your plans?

My major scientific personal interest lies in understanding the mechanism of proton pumping by cytochrome *c* oxidase. We are interested in more membrane proteins, and turn from bioenergetics more towards receptors and transporters.

Did you anticipate the Nobel Prize?

There were phone calls to my secretary in August and September 1988, asking her for materials about me because they were expecting me to get the Nobel Prize in 1988.

How did the Nobel Prize change your life?

I have much less time; I sit on too many committees, talk too much to people from politics and society instead of scientists. I get the feeling that I react instead of act. I spend probably three out of five days on committees and with reviewing.

You had been appointed to a directorship before the Nobel Prize whereas Johann Deisenhofer had difficulties in securing a similar position in Germany. What was the difference?

It was obvious to many people in Germany that I had started the project, that the novelty was lying in the crystallization of the RC and not in the structure determination. A number of German X-ray crystallographers were just jealous. JD was under the shadow of R. Huber.

According to your biosketch, your family background did not have much indication that you would become a scientist. What turned your attention to science?

I have always been interested in nature and science. It was mainly by reading books and later by an excellent physics teacher at high school (gymnasium).

You mentioned the connection between Peter Mitchell's chemiosmotic theory and your work. Mitchell's chemistry Nobel Prize in 1978 generated some criticism among chemists that it was far from chemistry. Could you comment on this?

Peter Mitchell could have received the medicine/physiology prize as well as the chemistry prize. Perhaps there was no more convincing candidate for the chemistry prize available. What, of course, is wrong is the name "chemiosmotic" theory. Actually I would call it "electrochemical" theory. Then all the chemists would have been convinced.

There has been some overlap between your interest and Richard Henderson's at the LMB in Cambridge. Did your results have any impact on their work on bacteriorhodopsin? Would you care to say something about your interactions?

There was always a good and friendly relationship between Richard Henderson at Cambridge, Dieter Oesterhelt, and myself. The methods are competing. It was actually Richard Henderson, who influenced Dieter Oesterhelt to let me become independent. I do not work on bacteriorhodopsin, therefore there is no impact of my work on his work. Probably, without the success with the reaction centres, there would have been a Nobel Prize for the work on bacteriorhodopsin.

What was the crucial moment in your work on the crystallization?

There was no single crucial moment. There were many crucial steps: first the accidental observation, then seeing the first bacteriorhodopsin crystals,

Johann Deisenhofer and Hartmut Michel at the reception of the Nobel Foundation during the centennial Nobel celebrations in Stockholm, December 2001 (photograph by I. Hargittai).

then the first reaction centre crystals (around the same time my daughter was born), and most important, observing the first high resolution X-ray diffraction pattern of the reaction centre crystals. When showing the diffraction pattern at a conference in Erice, Sicily, in 1982, for the first time, I got a standing ovation.

Would you care to comment on your relationship with Johann Deisenhofer?

See the response to your first question. We had talked about the project even before we started our cooperation. We had developed sympathy for each other without which the project would not have run so smooth and fast.

Deisenhofer described the moment when the symmetry of the reaction center became visible as the high point of the work. Do you remember this moment?

Actually, I remember that moment only very vaguely.

How do you compare research possibilities today in Germany with those in Great Britain and the United States? Did you ever consider going to work in the U.S.?

The German system, especially at the Max Planck Institutes, has the big advantage of a high level of guaranteed basic funding so that you can really start risky (but important) projects. This is not possible when you have to apply for grants every three or five years and you have to show success after that time. This is the reason why the U.S. lags behind Europe (especially Germany) in membrane protein research. About 20 U.S. universities stated their interest in hiring me. I had firm offers from Purdue, Urbana-Champaign, Chicago, and Harvard, and negotiated with Stanford and Yale, before I decided to accept the offer of the Max Planck Society. Even Harvard could only supply empty labs, which would have to be filled by grant applications. With the Howard Hughes Institute now, the situation in the U.S. is improving, and even the NIH favours long-term, high-risk projects by setting up new programmes.

How do you compare research possibilities today in the institutes of the Max Planck Society with those at the universities in Germany?

The possibilities are much better at the Max-Planck-Institutes; there is more money for consumables and investment, and only a minor teaching load.

How far has former West German and former East German science blended during the past decade? Do you have associates from the former East? What is your experience?

There has always been some excellent science in the East. Nowadays science seems to be more appreciated in the East. I had excellent students from the former East. Their education is excellent. There has been some mixing, but from the sheer quantity one may overlook the scientists from the East. One problem is the abolition of the East German Academy. Many of the scientists there could not find equivalent positions afterwards.

You were born years after World War II had ended. Have you been interested in history? Have you been interested in the role academia played in the era of National Socialism? Are you familiar with Benno Müller Hill's book, Murderous Science*? Did you know that Josef Mengele was paid by the Deutsche Forschungsgemeinsschaft as a postdoc of a Professor von Verschuer while he was carrying out his experiments on humans in Auschwitz?*

I have always been deeply interested in history. I can answer all your questions with YES.

Do you have hobbies?

Outdoor activities, gardening, jogging, and travelling to interesting countries.

Johann Deisenhofer, 2000 (photograph by I. Hargittai).

24

Johann Deisenhofer

Johann Deisenhofer (b. 1943, Zusamaltheim, Bavaria, Germany) is Regental Professor and Professor in Biochemistry at the University of Texas Southwestern Medical Center at Dallas; he holds the Virginia and Edward Linthicum Distinguished Chair in Biomedical Science. He is also Investigator in the Howard Hughes Medical Institute. In 1988, Johann Deisenhofer, Robert Huber, and Hartmut Michel were jointly awarded the Nobel Prize in Chemistry "for the determination of the three-dimensional structure of the photosynthetic reaction center."

Johann Deisenhofer received his Diploma in Physics from the Technical University of Munich in 1971 and his Dr. rer. nat. degree in 1974 from the Technical University of Munich for research done at the Max-Planck-Institute for Biochemistry in Martinsried under Robert Huber. His thesis work was the structure refinement of the basic pancreatic trypsin inhibitor at 1.5 angstrom resolution. He worked as a Research Associate and then as Staff Scientist at the Max Planck Institute between 1974 and 1988. He habilitated at the Technical University Munich in 1987. He has been at the University of Texas Southwestern Medical Center at Dallas since 1988.

Dr. Deisenhofer is a Foreign Associate of the National Academy of Sciences (Washington, D.C., 1997) and a Member of the Academia Europaea (1989). He shared the Biological Physics Prize of the American Physical Society (1986) and the Otto-Bayer-Preis (Germany, 1988) with Hartmut Michel. He received The Knight Commander's Cross of the

Order of Merit of the Federal Republic of Germany in 1990 and the Bavarian Order of Merit in 1992.

We recorded our conversation in Dr. Deisenhofer's office in Dallas on July 11, 2000.*

What was the principal novelty in your X-ray structure determination of the photosynthetic reaction center?

There were two new aspects. One was the fact that this molecule is an integral membrane protein. Structures of such molecules had not been determined before. They sit in a membrane, which is a lipid bilayer about 50 angstroms thick. This lipid bilayer is a very different environment for a protein than water is. Many of the proteins we know are water-soluble so they present a polar surface to their environment. Membrane proteins have two types of surface: they present a hydrophobic surface in that part, which is inside the membrane, and a polar surface in the part that sticks out. That makes them very different to handle, to purify, to crystallize from water-soluble proteins. When we crystallize it, we have to crystallize the whole protein, both parts inside and outside of the membrane. We tried to create conditions in which both the hydrophobic surfaces and the hydrophilic surfaces of the protein were in the correct environment. That was done by coating the hydrophobic surface with so-called detergent micelles. These are molecules that have hydrophobic and hydrophilic ends. They are similar to lipids but they don't form planar bilayers. They tend to form spherical micelles. Under the right circumstances these micelles can form a belt around the hydrophobic surface of the protein, and thus replace the lipid. This complex of the protein and the detergent has many properties like a water-soluble protein and can be crystallized as water-soluble proteins. So this was a new feature.

Was this the largest molecule at that time whose structure was determined by X-ray crystallography?

Yes.

---

*This interview was originally published in *Chemical Heritage* **2001**, *19*(3), 38–41.

What was the other novelty?

The other new aspect of our work was that it provided the first detailed structure of the photosynthetic reaction center. Thus it helped to understand how light energy is converted into chemical energy in the primary process of photosynthesis.

Does this knowledge have practical implications?

We understand now how nature does it. Nature, of course, has specialized machines, chloroplasts or bacterial cells that make all the components; they assemble them, they store the energy, and they use the stored energy. To reproduce that in an artificial system is not straightforward. We can learn how nature is able to make systems that, for example, have the property of running with near infrared light, which has not yet been achieved for artificial systems. Photovoltaic cells need blue light or near ultraviolet light. Photosynthetic organisms can work with red and infrared light. This is

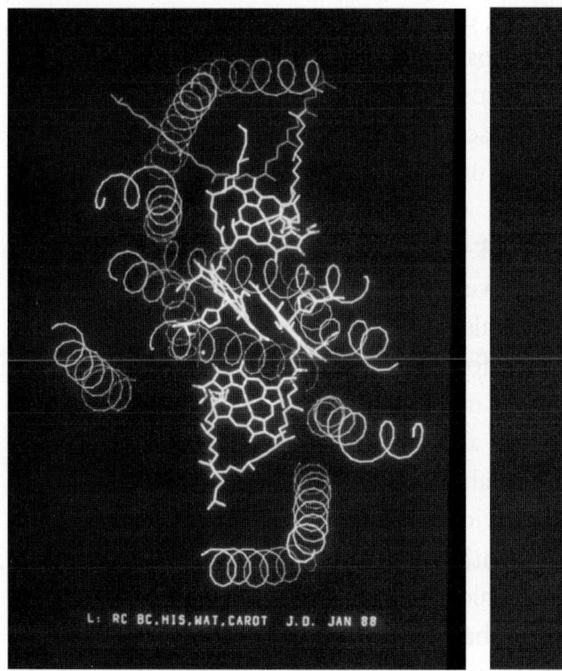

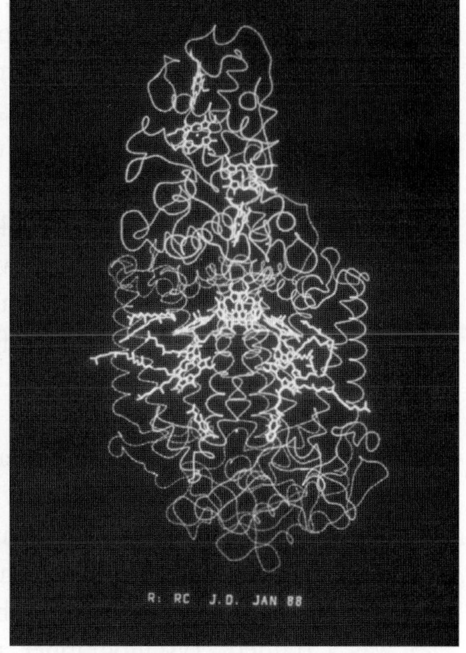

Overall view of the reaction center structure and view along the approximate symmetry axis. Protein chains are represented as curved lines (courtesy of Johann Deisenhofer).

one thing. The other thing is how nature makes sure that the optimum efficiency is reached. In photosynthetic systems every quantum of light leads to a charge separation across the membrane; there is practically no loss. This is something we can study and the principles from that can be applied to artificial systems. But this has not been done yet.

The Nobel Prize was shared by three of you. What was the division of roles and labor among you in your prize-winning research?

Hartmut Michel was the biochemist who had been studying membrane proteins for quite a while before starting the work on photosynthetic reaction centers. He was not a crystallographer but he had the firm determination to crystallize membrane proteins. He started with the well-known bacteriorhodopsin, which is a guinea pig for membrane protein work. Hartmut's boss at that time, Dieter Oesterhelt was one of the pioneers to recognize what bacteriorhodopsin is actually doing. Oesterhelt and Walter Stoeckenius, a researcher with whom Oesterhelt worked at that time in California, had discovered bacteriorhodopsin and then Oesterhelt proposed its function.

Bacteriorhodopsin is the protein component of the purple membrane; it resembles the visual pigment rhodopsin and acts as a light-energy converting system. It is part of a simple photosynthetic system in halobacteria that live in the Bay area near San Francisco and cause its characteristic orange-red color. Oesterhelt, by then in Würzburg, hired Hartmut to purify and crystallize bacteriorhodopsin. When Oesterhelt moved to Martinsried to become Director of the Max Planck Institute for Biochemistry, Hartmut went with him, and this is how we met.

Hartmut was able to grow crystals of bacteriorhodopsin but they were not suitable for X-ray crystallography for various reasons. He went to scientific meetings where people interested in membrane proteins came together and at one of these meetings he heard about photosynthetic reaction centers and about purification protocols available at that time, and the big advantages that the purple bacteria can provide. Under certain conditions their membranes are filled with photosynthetic reaction centers. Thus you can have a good source of protein, which is extremely important. He tried to work with them and soon he had success.

I was a trained crystallographer and my contribution was to solve the structure. Hartmut continued his crystallization experiments and he had a big impact in choosing the heavy-atom derivatives that we used and so on. He also dealt with data collection. My job was to process the data and to start building a model when we came close to the solution, and refine the model.

Huber, essentially, was my boss and, basically, that was his contribution.

So you could've gotten the Nobel Prize just the two of you, Deisenhofer and Michel, in principle.

In principle, yes.

But the three of you got it and three is the maximum number. Was anybody left out who might've been included?

In my view Oesterhelt should've been included if the rules had permitted this. He supported the project at a time when it was not clear whether it would work or not. He was the director of an Abteilung in the Max Planck Institute for Biochemistry and Robert Huber was also director of an Abteilung. Hartmut Michel worked in Oesterhelt's Abteilung and I worked in Robert Huber's Abteilung. It must have been a difficult choice for the Nobel Committee because there were many different ways in which that prize could've been given, to Oesterhelt and Huber, to Michel and myself, or some other combination.

Were you anticipating the Nobel Prize?

No. But people were predicting it in conversations. However, it was not clear to me which people would get it and, also, I did not expect it to come so soon.

Your results appeared in 1984.

It was a two-step publication. The main findings were published in two papers. One was a relatively small paper in the *Journal of Molecular Biology* [**1984**, *180*, 385–398] on the arrangement of the chromophores in the reaction center. It was the most interesting information for all the people who were trying to understand the reaction centers. Then, about a year later we published a paper on the complete structure of the protein subunits that was in *Nature* [**1985**, *318*, 618–624].

Who were the authors?

In both cases they were myself, then Otto Epp, Kunio Miki, Robert Huber, and Hartmut Michel, and the order was the same in both cases.

So Oesterhelt was not among the authors.

He was not. I believe, the reason that Oesterhelt was not on these papers is that originally, from Oesterhelt's point of view, he wanted this project to be for the young people. He promised Hartmut that he would not claim authorship. On the other hand, Huber decided that he needed to be a co-author.

You have mentioned somewhere that Rudolf Mössbauer's presence was decisive for you to go to Munich to study there at the Technical University.

That's right. I enrolled at the Technical University in 1965.

Did you have any interaction with Mössbauer?

Not directly. What attracted me was his plan to form a department system like in the American universities and get rid of the big professors.

But he didn't succeed.

He did not. What happened was that the number of professors increased but many of them still run their units as big professors. It must have been a bitter experience for Mössbauer when he realized that he could not change the German system.

You left Germany before the Nobel Prize. Why did you do that?

This is a complicated question. It can only be understood in the context of that time. I realized that with the work on the reaction center I had a very good chance to get an independent position somewhere. The Max Planck Society offered a job as a director to Hartmut Michel but, at that time, not to me. I had contacts with universities in Germany and with the European Molecular Biology Laboratory in Heidelberg. At that time I also made many trips to the United States, by invitation, to talk about this story. In the summer of 1986, two years after we had completed the structure I received a letter from this institution, the University of Texas Southwestern Medical Center. It was motivated, as I now know, by the plan of the Medical Center to build up a structural biology unit. They invited me to join it. At that time I'd never heard of the place. I asked a postdoc in Martinsried who came from Texas, and he spoke very highly about this medical school and I decided to visit and see it for myself. When I came I was overwhelmed by their recruiting techniques. They laid out big carpets and all that. This was a great contrast to how people are hired in Germany, where it's always you who is asking for something. Here it was the opposite, they were asking me to come. I returned home with this feeling and I came again in January 1987 and experienced the winter in Texas, it was 20 degrees centigrade and the sky was blue and I just had experienced several weeks of fog in Munich. I thought this was the place where one could live. Eventually I decided to come even though some possibilities in Germany had developed, but the conditions were not comparable.

Had they offered you a professorship in Germany before the Texas offer, would you have opted for Germany?

I would've, yes.

In your autobiographical note preceding your Nobel lecture you are saying that your experience with the photosynthetic reaction center changed your life and that it is hard to describe the excitement you felt during this work. Please, try to describe it.

The most exciting stage in our work is when you get the electron density distribution of the molecule but there is no model yet. Then you are building an atomic model into that electron density distribution. The electron density distribution is visualized as a set of contour lines that outline a globular object, which has a very confusing internal structure. The photosynthetic reaction center is a huge molecule of ten thousand non-hydrogen atoms.

I began to look through, trying to decide where to start and what features could be assigned with some certainty. Then I learned to recognize chlorophyll molecules; I hadn't worked with chlorophyll before. There were also hemes that were better known. I found there all the features that people who had been doing, for example, spectroscopy had described. All available physical chemical methods had been applied to analyzing the composition of the reaction centers. It was extremely exciting to localize these features and build models for them. When I stepped back to see the arrangement, the unexpected observation about it was that it was symmetric. There was a symmetry in the arrangement of the chlorophyll that nobody had anticipated. Nobody, to this day, completely understands the purpose of this symmetry. I think it can be understood only on the basis of evolution. I think that the photosynthetic reaction started out as a totally symmetric molecule. Then it turned out to be preferable to disturb its symmetry, sticking to an approximate symmetry but changing subtly the two halves of the molecule. Because of the difference in properties of the two halves, the conclusion had been, before the structure came out, that there cannot be symmetry; that it has to be an asymmetric molecule. Now when people looked at the structure, it looked totally symmetric to the naked eye. That realization was the high point I will never forget.

The technical device of model building of that time, the computer display, was much inferior to what we can have today. It allowed only a small part of the whole molecule to be looked at, at one time. So you were building details and little things and every now and then you summed up these details and looked at the bigger picture at a lower level of details. Those were the most exciting moments in the whole project for me.

Were you alone when you noticed the symmetry for the first time?

I was. I was sitting in a dark room and I was interacting with the machine.

What did you do at that moment?

I started smoking again. I'd quit smoking. But I could not live with this excitement without doing something like that. It was a bad decision because it was very difficult to stop smoking for a second time. It lasted, the smoking, that is, for half a year.

I also called Hartmut and showed him the whole thing. It was very nice to have a colleague like him who in many ways complemented my expertise. We could give each other many things in the course of this work. It was a relationship of complete trust. When such a story becomes known, there is always a temptation to claim the whole fame. This did not happen between us and I'm very glad that it didn't because it could've ruined everything.

Hartmut pioneered a lot of things during that time. For example, he was the first to begin DNA sequencing in Martinsried. He realized that if we were to have a complete model we needed to sequence the protein. The sequence of the proteins in other species of photosynthetic bacteria was known and it was estimated that our molecule was about 50 percent the same but it also meant that it was 50 percent different. He would come with pieces of the new sequence and we sat together and tried to find it in the electron density map. First, I'd built the model without knowing the sequence and, of course, I missed some amino acids and built some more that were not there. That was a very interesting and exciting time. In retrospect, it was the highest moment of excitement in science I ever reached; it never came again in the same intensity.

Do you continue research along the same lines nowadays?

I continue with biological macromolecules but in moving to a medical center I realized that the photosynthetic reaction center should not be my main interest. I also knew that studying more photosynthetic reaction centers was an entirely new issue. It required different approaches and sources of protein, purification, and it was a good point in time to switch. There were many scientists here that needed structural information on medically relevant proteins. Although nobody told me what to do, I sensed that the expectation was to focus on mammalian and human proteins and that's what I did.

One example for our current work is the structure analysis of human HMG-CoA reductase. This enzyme catalyzes an important step in our body's own production of cholesterol. Millions of people in the U.S.A. alone, who have high levels of cholesterol in their blood, take drugs that inhibit HMG-CoA reductase to lower the cholesterol level and thus reduce the risk of artherosclerosis. These inhibitors are known as "statins." We now know the tree-dimensional structure of the catalytic portion of human HMG-CoA reductase; we also know how substrates and inhibitors bind to the enzyme. This adds to our knowledge and will enable us to think about even better inhibitors.

As far as crystallographic computing methods are concerned, I always was and still am very interested in additions to our repertoire of methods, but the time has long gone since I tried to write computer programs myself. Young people are much better at this (and at many other things). Nowadays, our repertoire of software is so large that one can work as a crystallographer without knowing anything about programming languages. That is one of the big differences between now and the time when I entered the field.

What turned you originally to science?

My father came from a family, which ran a farm. Although he was the youngest of three brothers, eventually he continued the family farming after the war. He returned in mid-1945 after a brief spell as a POW. When the Americans released their POWs, they started with the farmers.

Did he serve in the Wehrmacht?

Yes. He was a sergeant. His interest was much broader than the farm and he transmitted his interest to me. I became an avid reader and read popular science books. The county library sent a bus with books weekly and I borrowed their books religiously.

How about religion?

I was brought up as a Catholic but I lost my faith when I was a teenager.

Do you freely talk about this in Texas?

Yes, among friends. It is not productive to go to a big party here and declare that religion is probably without basis. I usually go to the parties organized by the University; there are a lot of potential donors at these parties and it is important not to irritate them.

What do you enjoy doing most, nowadays?

Still, reading. I read a lot of scientific literature, then, in my free time, right now I'm reading a big book on Roman history, in German. Theodor Mommsen wrote this book in the mid-19th century and he got the Nobel Prize in Literature for it in 1902. His use of German is fantastic. If Thomas Mann had written history books, he might have written in this style.

What else should I have asked you about?

Maybe what made me decide to study physics? That started with a book by an astronomer, Fred Hoyle, Sir Fred Hoyle now. He worked in Cambridge,

U.K., and wrote excellent popular books. The first one I read explained astronomical observations you can make with your naked eyes. This book had the greatest impact of all the books I read on my later choice of my career. I chose physics as my broad subject and I thought I would specialize in astronomy later, but things turned out differently.

Robert Huber, 2001 (photograph by I. Hargittai).

25

Robert Huber

Robert Huber (b. 1937, Munich, Germany) is Director of the Section for Structure Research of the Max Planck Institute for Biochemistry in Martinsried (near Munich), Germany. He received the Nobel Prize in Chemistry in 1988 jointly with Johann Deisenhofer and Hartmut Michel "for the determination of the three-dimensional structure of a photosynthetic reaction centre." He received his Diploma in Chemistry, his Dr. rer. nat., and his Dr. habil. degrees from the Munich Technical University in 1960, 1963, and 1968, respectively. He has been a director at the Max Planck Institute for Biochemistry since 1972 and was appointed Professor at the Munich Technical University in 1976. He has received many distinctions and awards. He is a member of the "Orden pour le Mérite für Wissenschaften und Künste" of Germany, Foreign Associate of the National Academy of Sciences of the U.S.A., Foreign Member of the Royal Society (London); has received the Otto Warburg Medal of the German Biochemical Society (1977), the Keilin Medal of the Biochemical Society (London, 1987), the Sir Hans Krebs Medal of the Federation of European Biochemical Societies (1992), and the Linus Pauling Medal (1993/94), among many others. We recorded our conversations in his office in Martinsried on March 3 and 4, 2001.

I would like to ask you to single out a few examples from your research.

There are two cases that I felt most pleased about. One was the crystal structure analysis of an insect hemoglobin, in the late 1960s, when molecular evolution was not yet in evidence. It was a great surprise that an insect protein should have about the same structure as that of a mammalian protein.

I had, before that, as a very young person, in my Diploma Work, determined the molecular weight of an insect hormone, ecdyson, and found that Peter Karlson and Adolf Butenandt had reported an erroneous molecular weight for it. They had discovered this substance, the first known hormone in the insect world. At that time it was not yet known that the insect hormones would be related to human hormones; today this is commonplace. I worked on this problem in the late 1950s. Whereas Butenandt and Karlson had used classical methods, I used crystallographic techniques for the molecular weight determination.[1] This result had consequences for me as I got the feeling that I could accomplish something, and I decided to become a scientist.

When did you become interested in chemistry?

I was eight when the war was over and there was a gap in our schooling between 1944 and 1946. My life, as I remember it, started after the war. When I was a schoolboy I knew that I was interested in chemistry. It was not from my family background because my father worked in a bank and my mother was a housewife. But she bought me books and from them I learned to like chemistry. I also borrowed books from the public library. Of course, my interest in structural biology came later.

I went to a humanistic gymnasium. We had no chemistry; rather, we had Latin and Greek. One of our biology teachers offered a chemistry course, but he asked me not to attend it because after the first class he determined that I knew too much and corrected him frequently.

Did you come across Butenandt?

Yes, he was a big name, and he was the Director of the Institute for Biochemistry of the Max Planck Society. However, he was so high in the hierarchy that I communicated with Karlson who was his associate.

Were they upset when they learned about your molecular weight determination?

They were but also pleased that I had solved the problem after all.

I would like to ask you about the work that led to the Nobel Prize.

The photosynthetic reaction center was certainly the most important protein structure, which Hans Deisenhofer, my former doctoral student, and I and

Hartmut Michel determined. It was clearly a breakthrough in understanding the electron transfer in the process of photosynthesis. Our earlier works were the basis that all the tools and instruments and methods were available in our laboratory to enable us to take up such a large project, which at that time was the most complex system to date in crystallography.

In what sense was it a breakthrough?

There were so many data around about photosynthesis, especially about spectroscopy and electron transfer rates, but it was a black box. There were spectra and there were numbers, and with the three-dimensional structure this black box was suddenly illuminated. For the first time we saw and understood how light energy converted into electric current across the membrane.

You built up protein crystallography in this Institute after you had studied crystallography with your mentor, Professor Hoppe. But the cradle of protein crystallography was Cambridge. What was your main motivation?

I wanted to raise protein crystallography in Germany, in Bavaria, in Munich. This was my purpose from a very young age, and I did. This is why I stayed here. I am grateful that I had received the background for it from my teacher, Walter Hoppe, and the independence for it. I went for visits to other places but never for more than a few weeks.

Did Hoppe live long enough to see that you succeeded?

He died before the Nobel Prize but he lived long enough to see the solution of the structure of the photosynthetic reaction center. I took his widow with me to Stockholm.

The photosynthetic reaction center was the largest system of a crystallographic structure determination then. What is the largest system today?

Just a few months ago the structure of the ribosome (an RNA enzyme) was solved by parallel studies at several places, in Hamburg by the Max Planck group at DESY (Deutsches Elektronen Synchrotron) and at the Weizmann Institute in Israel, at Yale University, and in Cambridge, England. Three years before we had determined another complex structure, the proteasome,[2] which was the most complex one to date at that time. More and more complex systems are being studied and the attribute "most complex" is rather short-lived. This is the field into which structural biology is moving, the study of complex molecular machines. It is the multi-component complexes rather than the individual proteins that are being targeted, the stable species now, later perhaps the transient ones.

How much is structural biology involved in studying prion systems?

My laboratory is not involved in this currently but there is considerable work going on elsewhere. One idea is that different conformers of the same protein are involved of which one is a sick and amyloidogenic version, which catalyzes its own generation.

Without a nucleic acid.

That's quite established. What is not clear whether there are other proteins involved or not in this process, maybe just in catalytic quantities? Some years ago we worked with the so-called serpin proteins, serine protease inhibitors. We determined the first structure in the early 1980s. Later, it was found that this protein occurs in two conformations. One is a latent form, which is inactive, not acting as an inhibitor, and the other one is inhibitory. Here we have two forms of the same protein, with the same chemistry, and the difference in activity is related to a conformational difference. This is what I suppose to be for the prion as well, with the additional property of the sick version to be a catalyst and nucleus for amyloid deposition.

Can we consider this to be a case of polymorphism?

Usually you name polymorphism when you have different crystal packings formed by the same molecular structure. Here you have different molecular structures of the same protein.

I would like to get back to the determination of the structure of the photosynthetic reaction center. Was there a crucial moment in that study?

It was when Hartmut Michel joined in. He was an associate of Dieter Oesterhelt here in the same Institute. He had already worked for some time on the preparation of membrane proteins. One day he came and he had some crystals of the protein from the purple bacterium *Rhodopseudomonas viridis*, and we put these crystals into our X-ray camera. These first crystals were not too good but they were promising. He then improved his procedure and obtained larger crystals and after a while it was clear that this was a doable project. We then started the crystallography and we did it in a relatively short time. We were lucky in several aspects. One was that heavy-atom derivatization was successful quite quickly. In this, we, of course, followed Max Perutz's pioneering work, as does everybody else. His isomorphic replacement method in the determination of protein structures revolutionized and even created structural biology.

At this point I would like to mention an important contribution by my mentor. The method that is most frequently used in protein crystallography today was given the name Molecular Replacement Method later, but was introduced as Faltmolekülmethode by Hoppe.[3] The reason for its widespread application is that we are increasingly involved in the determination of structural variants and members of protein families. The method goes back to the time when I joined Hoppe's laboratory. He published it in 1957, truly classic papers. The basic idea is very simple. The Patterson function is the Fourier transform of the intensities. Patterson showed that it is the representation of all the vectors in the unit cell. If all the vectors would be resolved, the structure solution would be straightforward. But they are too crowded. Hoppe's idea was that if we know the basic skeleton of the molecular structure we could generate the intramolecular vectors from it and search for them in the Patterson function. This search gives you the orientation of the molecule. Then, knowing the symmetry of the intermolecular arrangement, you can generate the molecular vector sets, search for them, and so determine the crystal structure. Hoppe's clearly expressed basic idea was that the molecular Patterson function is composed of two sets; a translation-invariant intramolecular set and a translation-dependent intermolecular vector set. A consecutive search for these vector sets leads to a solution of the crystal structure. When I joined Hoppe in the late 1950s, there were no computers available, and we made these vector constructions by hand, on a piece of paper, and looked for superpositions with Patterson projections. When the first computers became available to us, I was keen on programing, I wrote the first program[4] to do the analysis, and applied it to the structure determination of ecdyson, which I mentioned at the beginning of our conversation, and found that it is a steroid and defined its exact structure and stereochemistry. It was the first crystallographic analysis of a very large organic molecule without the use of heavy atoms. In any case I wanted to acknowledge the very important contribution of my mentor, Hoppe, which is now the basis of many protein structure determinations. The name, which he gave the method, Faltmolekülmethode or molecular folding method, relates it to the mathematical operation of folding.

When was its application begun in protein crystallography?

It was a gradual process when it became clear that there were families of molecules that were similar. Hoppe's method was very useful and a short route to the determination of their crystal structures. I used it, for example, to determine the structure of the reaction center from *R. sphaeroides* using the diffraction data of Rees and Feher when I. Allen from their group visited Martinsried for a week.

Is the method known to have been originated by Hoppe?

It is now better known as Molecular Replacement Method. Michael Rossmann and David Blow working in Cambridge, England rediscovered the method

Group photograph of leading scientists in Martinsried in 1973. Back row, left to right: Huber, Eddan, Wünsch, Braunitzer, Hofschneider, Hoppe, Hannig, Kühn, Zillig. Front row, left to right: Dannenberg, Butenandt, Lynen, Ruhenstroth-Bauer (courtesy of Robert Huber).

in the early 1960s. They approached the problem from a different aspect, namely, internal symmetry. Think of the hemoglobin case, which is a tetramer with local twofold axes. Rossmann and Blow wanted to find out about the local symmetry from the diffraction intensities. Then they realized the additional potentials of their approach. They worked with protein molecules whereas Hoppe was a small-molecule person, and so was I at that time. The two fields did not communicate with each other.

Dieter Oesterhelt's name is missing from the Nobel roster. He was Hartmut Michel's mentor. Was he not a player in the photosynthetic reaction center project?

Not directly but he was important as the Doktor-Vater of Hartmut Michel and certainly brought Michel towards biochemistry and the crystallization attempts of membrane proteins.

Is there a parallel between Deisenhofer and Michel on the one hand and you and Oesterhelt on the other? Or is this an oversimplification?

I was involved in the project while Oesterhelt, with some purpose, kept out. He let Hartmut Michel do this entirely independently.

Was Deisenhofer not independent?

Deisenhofer was independent, too. This project was not simple crystallography, not at all. It was a most complex structure determination. Even the methods of measuring the intensities were not automated at the time. We had developed instruments, X-ray cameras and methods for that purpose also. We had a small workshop at that time, with a mechanic and an electronics person. One day the mechanic had a stroke. A week later the electronics person had a heart attack. They were very important in servicing the instruments. I was the only one in my department able to service the instruments. It was at a critical time for the photosynthetic reaction center work, around 1983, and I spent much time each day taking care of the instruments. It is a side issue, not even a scientific one, but it shows you that things may look different from the perspective of today's possibilities than they actually were. The work at that time required a background also concerning the availability of the samples for isomorphous replacement and methods to apply them. I have made many of these samples and I built up an enormous collection.

Could Deisenhofer have done the same?

Deisenhofer was more specialized in doing the computations.

According to Deisenhofer, the Nobel Prize could have gone to different combinations of people, for example it could have gone to Huber and Oesterhelt, and maybe 50 years ago it might have, it could have gone to Deisenhofer and Michel, or, as it indeed did, to the three of you.

I agree and I am just happy about the way it went by the Nobel Committee's decision, not mine. From the various responses I have received, it was generally acknowledged that solving this complex structure in such a short period of time required a background of a very productive large group.

Before the Nobel Prize, Deisenhofer could not find an independent position in Germany and a few months before the award was announced he had accepted an appointment in Texas. By then he was 45 years old. You were appointed director at the Max-Planck-Institut für Biochemie at the age of 35.

At that time I was the youngest director in the whole Max-Planck organization. Adolf Butenandt was President of the Max Planck Gesellschaft and I told him so, and I remember what he told me, "It will pass."

Was it anticlimatic to continue your research after the Nobel Prize?

It was not the case. Our research steadily went on; the group became a little larger; we learned how to make recombinant proteins and crystallize them, which opened many more protein targets for our investigations. My personal life did not change either. I was frequently invited before the prize and I received a few more invitations afterwards. I had to decline more invitations after the prize and I felt more free to decline them too.

Why did you feel so?

I already had the prize. I remember a Nobel laureate chemist telling me that since the prize, come October, he can sleep peacefully.

Nowadays it happens that people in molecular biology, without a broad background, go right to a frontier problem, solve it, and then disappear from science.

This may be possible when all the methodologies and instrumentation are available. I began my research at a time when methods in structural biology were poorly developed and this development just started. My interest in the development of techniques and methods has continued ever since. For instance, there was a lack of suitable heavy-atom compounds. Max Perutz used simple mercury and platinum compounds. I thought we should have many, many more and I decided to synthesize them, as they were not available. I also distributed them among colleagues all over the world. Heavy-atom clusters are particularly useful. The Ta_6Br_{12} cluster is an example, which then binds at specific sites of a protein. I started making these compounds when I was young and have continued making them. It became also clear that in protein crystallography we needed fast and automated intensity measurements. At the beginning, protein crystallographers used the small molecule crystallography equipment. These were single counter detectors, detecting one diffraction spot at a time. In a protein experiment, usually many diffraction spots are generated at the same time. Area detectors were developed; the simplest of which is the photographic film. In the early 1970s, we developed a program in order to evaluate automatically the diffraction patterns on these photographic films. This program was widely distributed in the world. At that time we still built Kendrew-type protein models, which were fitted into the electron density map. In the middle of the 1970s, interactive graphics displays became available, but there were no programs to use them. You had to display electron densities and to interactively fit the molecular models to them for our purpose. Alwyn Jones developed the first useful program to display electron densities and fit the molecular models. He is now professor in Uppsala and was a postdoctoral fellow at that time in Martinsried. We are now working on incorporating non-natural amino acids into protein structures. These non-natural amino acids may serve as heavy-atom derivatives or as spectroscopic labels. We are developing methods in order to prepare such proteins. Another most recent development is an instrument, which we constructed to improve crystal quality. We often find that protein crystals do not diffract adequately. You don't see any optical defect but in the X-ray beam they do not diffract. We have found that by a careful dehydration protocol we can improve the quality of these crystals. We worked on refinement methods as early as the early 1970s. The phases determined by isomorphic replacement with heavy-atom labels are often of bad quality. This is due to the errors of measurement and because of non-isomorphism and other factors. We learned how to improve the phases by making use of the known geometry of the building blocks of the structure, the amino acids adapting first a program written by Robert Diamond in Cambridge, England. The refinement is an iterative process leading to spectacularly clear electron densities and much more accurate models. The first crystal structure refined in this way was the basic pancreatic trypsin inhibitor in Hans Deisenhofer's thesis work. Almost all the published structures since about 1980 have been refined with this approach. My department has always had a great interest in methods, techniques, and instruments. Their availability was the basis of its productivity. People call it the powerhouse of protein structures, and the determination of the photosynthetic reaction center rested on the accumulated expertise there.

Hartmut Michel was Dieter Oesterhelt's associate and Johann Deisenhofer was yours. Oesterhelt was also a director at the Max Planck Institute so you were of equal ranking. Then came the Nobel Prize, which included Deisenhofer, Huber, and Michel. Oesterhelt was not a co-author of the papers reporting the structure determination of the photosynthetic reaction center, and more than three persons could not have shared the Nobel Prize in the first place. My question is, how much change did the Nobel Prize introduce into your careers? Was there a watershed effect?

Little changed. We had no personal relationship before, we met as colleagues; we are now on "Du" terms, which happened after the Nobel Prize. "Du" is the familiar address as opposed to "Sie" and it is still not very common among colleagues in Germany. Oesterhelt is very well respected for his work on *bacteriorhodopsin*, which he discovered together with the American biochemist, Stoeckenius. He has also remained involved in actual research as I have.

Is this typical for a big-name German professor?

It is not. This is more like Max Perutz. He is this big picture in front of me, both as a human being and as a scientist. I see him at least once a year. We are both in the *Order pour le Mérite* of Germany. There are about 70 members, half of them Germans and the other half from other countries. It has a French name; Friedrich the Great founded it as a military order; at that time the language of communication was French. King Friedrich Wilhelm IV added, under the influence of Alexander von Humboldt, the peace class for science and arts to the Order. It embraces all fields of culture and about one third of its members are scientists. Rudolf Mössbauer became a member couple of years ago at my suggestion. The number of members is restricted and the existing members elect a new member when a member passes away. When I became a member, the physicist Wolfgang Paul and the chemist Manfred Eigen were there already from among the Nobel laureates. The biophysicist Erwin Neher and the biologists Bert Sakman and Christiane Nüsslein-Volhard were recent additions.

You dedicated your Nobel lecture to your wife, which was a beautiful if rare gesture.

We have four children and, besides, she provided me with a perfect background for my work. Unfortunately, we no longer live together.

The Nazi concentration camp Dachau was in a suburb of Munich. Have you ever visited the site?

Yes.

Not very much has been uncovered about the attitude of German academia in the Nazi era.

It was only recently that a presidential commission of the Max Planck Society examined the behavior of the predecessor of the Max Planck Society, the Kaiser Wilhelm Gesellschaft, during the Nazi era. This commission says in a preliminary report that some intolerable experiments were made under the responsibility of some members of the Kaiser Wilhelm Gesellschaft.

Did these experiments take place in concentration camps?

The details are in the reports, which have just been published.

At some point Adolf Butenandt was the President of the Max Planck Society. Did you mention in our previous conversation that he prevented such an investigation?

I did not say that, but others have. In any case, it did not happen under his presidency. Now it is happening. I wish that other organizations would also carry out similar investigations of their past.

How about the Deutsche Forschungsgemeinschaft?

There has been a recent investigation.

Did you read Benno Müller-Hill's book Murderous Science[5]*?*

I read parts of it and some of his articles.

Do you agree with him?

I believe what he says; he is a serious scientist. However, he may be biased against the Max Planck Society, which he compared with the Russian Academies. At one point he made an analysis of costs and research productivity with the aim to show that the Max Planck institutes use their money less effectively than the universities when they get grants from outside sources in a competitive manner. I disagreed and had a dispute with him. In fact, much of my budget also comes from outside grants.

Do you think Müller-Hill was biased against the Max Planck Society in his book Murderous Science*?*

Maybe, and I am much happier that there is now this much broader based and objective investigation of the past of the Max Planck Society.

This is by the Society itself.

It is by a presidential commission of independent personalities.

Do you think that Müller-Hill was a catalyst in initiating such investigations?

That may very well be.

Nobel laureates are often approached to sign various petitions and declarations to lend them their weight. It was, however, unique when at the beginning of the 1990s, you actively participated in soliciting signatures from Nobel laureates in protest of the Yugoslavian war against the independence of Croatia.

I had a friend and colleague in Ljubljana, Slovenia, through whom I became aware of the situation in the former Yugoslavia. I also had contacts with the Rudger Boskovic Institute in Zagreb from which I had had a number of doctoral students in my group. Then I received a letter from my colleague there, Greta Pifat, who later served as Deputy Minister of Education in Croatia for a short time. I felt we had to do something against the Yugoslavian Army destroying Croatia. The idea of an appeal with Nobel signatures was born and the project became very successful. The first signature was by Linus Pauling, which was very important because it convinced people that this was very serious. Altogether we collected 112 signatures, the largest number of Nobel signatures ever brought together in a single project. I believe this appeal had an impact. At that time it was a big question whether the Western powers would recognize the independence of Croatia. I contacted Hans-Dieter Genscher the then Foreign Minister of Germany, and they overcame the resistance of the English and the French, they forced a joint declaration by the European Union in support of Croatia. It was a delicate matter because of history during World War II, and I could understand Genscher that he did not want to act too fast; but Germany clearly wanted the recognition of Croatia.

Four years from now, in 2005, you will have to retire. What are your plans?

First, I am looking forward to the remaining four years. I like to work with young people around me. After that I'll have an emeritus status at the Institute. It includes an office, half a secretary, and, if I want it, a few square meters of laboratory space. My future will depend on how I will feel at that time. I will carefully weigh other offers from European universities. I'll have plenty of time during the next four years to explore my possibilities.

References

1. On the ecdyson research, see, e.g., Karlson, P.; Hoffmeister, H.; Hoppe, W.; Huber, R. "Zur Chemie des Ecdysons." *Justus Liebig Annalen der Chemie* **1963**, *662*, 1–20; Huber, R.; Hoppe, W., "Zur Chemie des Ecdysons, VII: Die Kristall- und Molekülstrukturanalyse des Insektenverpuppungshormons Ecdyson mit der automatisierten Faltmolekülmethode." *Chemische Berichte* **1965**, *98*, 2403–2424.
2. Groll, M.; Ditzel, L.; Löwe, I.; Stock, D.; Bochlev, M.; Bartunik, H. D.; Huber, R. "Structure of the 20s Proteasome from Yeast at 2.4 Å Resolution." *Nature* **1997**, *286*, 463–471.
3. Hoppe, W. *Acta Cryst.* **1957**, *10*, 750; *Z. Elektrochem.* **1957**, *61*, 1076.
4. Huber, R. "Die Automatische Faltmolekülmethode." *Acta Cryst.* **1965**, *19*, 353–356.
5. Müller-Hill, B. *Murderous Science: Elimination by Scientific Selection of Jews, Gypsies, and Others in Germany, 1933–1945.* Cold Spring Harbor Laboratory Press, New York, 1998. German original: *Tödliche Wissenschaft.* Rowohlt Taschenmbuch Verlag, Reinbek, 1984.

Manfred Eigen, 1997 (photograph by I. Hargittai).

26

Manfred Eigen

Manfred Eigen (b. 1927 in Bochum, Germany) is Professor Emeritus at the Max Planck Institute of Biophysical Chemistry in Göttingen, Germany. He was co-recipient of the Nobel Prize in Chemistry for 1967 together with R. G. W. Norrish (1897–1978) and George Porter (b. 1920). Eigen received half of the prize and Norrish and Porter shared the other half, "for their studies of extremely fast chemical reactions, effected by disturbing the equilibrium by means of very short pulses of energy." [An interview with George Porter appeared in *Candid Science: Conversations with Famous Chemists*. Imperial College Press, London, 2000, pp. 476–487]. Manfred Eigen studied at the University of Göttingen from 1945, receiving his doctorate, under Arnold Eucken, in 1951. His dissertation was about the specific heat of heavy water and aqueous electrolyte solutions. His many distinctions include the Otto Hahn Prize for Chemistry and Physics (Germany, 1962) and the Linus Pauling Medal of the American Chemical Society (1967). He is a Member of the "Orden pour le Mérite für Wissenschaften und Künste" of Germany. He is also a member of many learned societies, including the Göttingen Academy of Sciences and the "Leopoldina" of Halle, and he is a Foreign Associate of the National Academy of Sciences (U.S.A.), Foreign Member of the Royal Society (London), The Russian Academy of Sciences, and the French Academy of Sciences. Professor Eigen is an accomplished pianist who has performed with famous orchestra and conductors. At the conclusion of our meeting, he presented me with a CD with his playing Mozart, accompanied by the New Orchestra of Boston, directed by David Epstein in 1991 (KV 414) and by the Chamber Orchestra, Basel, directed by Paul Sacher in 1981 (KV 415). We recorded our conversation in Professor

Eigen's office in Göttingen on December 10, 1997. The following narrative is based on parts of this conversation.

The Max Planck Institute of Biophysical Chemistry in Göttingen is well outside the old town. As I walked into its big, modern, and impersonal building, I noticed modern art everywhere on the walls, which made the environment much friendlier. The Institute came out of the old Institute of Physical Chemistry, which was in downtown Göttingen, in Bunsen Straβe. In the mid-1960s Professor Eigen was receiving attractive offers to go to the United States, but people tried to keep him in Germany and asked him what they could do for him? Eigen told them that biochemistry and biology were too weak in Göttingen, so they made a proposal that Eigen should found a new institute, which was then called Institute for Biophysical Chemistry. Eigen's idea was to be close to the University, which just had moved out its campus also to the countryside. First, Eigen went to the farmers and bought all the land for the Max Planck Society. He paid them DM12 per square meter. At the time of our conversation the price of land around was DM500 per square meter.

They united three groups in the new institute. One was the old physical chemistry, another was spectroscopy, and the third was neurophysiology. They chose the study of information as a unifying subject on the molecular, cellular, and the central nervous system level. Their divisions range from pure physics to biology and medicine. The Institute opened in 1970 and has been very successful. Fitzpeter Schäfer invented the dye laser there. The application of nuclear magnetic resonance tomography for examinations of the whole body was accomplished there and Jens Frahm, holds most of the patents for it. Erwin Neher and Bert Sakmann got the Nobel Prize for Physiology or Medicine in 1991 "for their discoveries concerning the single ion channels in cells." Then, Eigen's group continued their studies on fast reactions and worked out the evolutionary biotechnology. They have groups in biological areas, in genetics, yeast research, *Drosophila* research, in the study of microtubulenes, the eukaritic cells, and many other areas.

Eigen called his Nobel lecture "Immeasurably Fast Reactions" and the expression itself goes back to a note in the textbook of physical chemistry by Arnold Eucken. When Eigen was a student, Eucken's book was their bible, and it talked about "unmessbarschnelle Reaktionen." If one wants to start a reaction, first the reactants have to be gotten together. If the reaction is slow enough, the reactants are just put together in a test tube.

Arnold Eucken (courtesy of William B. Jensen and the Oesper Collection, University of Cincinnati).

However, reactions can be much faster. It was about the late 1940s and early 1950s, when Eigen was entering the field and at that time people were using flow methods. They flowed together the reactants in a mixing chamber where they mixed in a millisecond and this was about the limit. The method was first developed in the 1920s.

Starting a reaction by mixing the reaction partners had limitations due to the mixing time. Also, there are many reactions that are much faster than having half times of milliseconds. Eigen thought that there had to be ways to study faster reactions. His doctoral work though was in a different area. Eucken was interested in developing a theory of water structure. He based it on data of specific heat and density and structure, and whatever other information he could get hold of. In order to test his theory, he needed data on heavy water, D_2O. Eigen's project was to build an adiabatic calorimeter with which he could measure specific heat very precisely, up to the fifth decimal, to see the very subtle difference between H_2O and D_2O.

Eucken wanted him to go to as high temperatures as possible. They had a sealed Jenaer glass vessel to protect their sample from impurities. Its volume was half a liter in order to have a big heat capacity. They did not know what temperature the glass vessel would stand. Eucken thought that even 200°C would be possible, but Eigen wanted to go only as high as 175°C in the first experiment. However, there was a big explosion at 169°C, and the whole apparatus landed on the ceiling.

Heavy water at that time was very much at premium so Eucken was very unhappy and thought that the explosion blew up Eigen's project. Although Eigen promised him to build a new calorimeter in three weeks, the heavy water would have been impossible to replace. Fortunately, however,

Eigen did not risk their heavy water in the very first experiment. When Eucken learned about this, from that point on he let Eigen do whatever he pleased. Eigen received his doctorate at the age of 23.

Eigen likes to quote an amusing episode from his work concerning accuracy. Eucken told him to take a very large piece of graph paper to plot everything very carefully. However, Eigen told his professor his precision was much better than what he could do on paper. He and a mathematician colleague had developed a numerical method, which could do everything superior to plotting the data on graph paper. When Eigen drew the line across the data points, Eucken took his ruler and checked whether the curve was properly linear or not. He found it quite good except of a millimeter difference at the end. Eigen, however, claimed that if was Eucken's ruler rather than his curve that was off by 1 millimeter. The professor went to the machine shop, had his ruler checked, and it was his ruler that was not precise enough.

Eigen was measuring the specific heat at the saturation pressure, and what he wanted to have in his table was either the specific heat at constant volume or at constant pressure, C_v or C_p. Knowing the pressure, the density, and the compressibility, they could calculate these quantities. Compressibility they could measure most precisely by measuring sound velocity in heavy water. There were physicists in the physical chemistry institute who were experts of measuring sound velocity. They worked for the British Navy investigating sound absorption in seawater. They had a puzzle that seawater had a very high sound absorption, especially in two ranges, in the megacycle range and in the hundred-megacycle range. Distilled water had a much lower sound absorption, by orders of magnitude. Sodium chloride solution had even lower sound absorption than distilled water. Eigen by then had had experience in sound velocity measurements in heavy water, and wanted to help his colleagues in solving their puzzle.

When the theoretical physicists invited Eigen to give a seminar, he talked about ionic hydration because after the heavy water experiments he looked into the specific heats of ionic solutions. In the same colloquium, two of his colleagues, Konrad Tamm and Günther Kurtze, talked about sound absorption in seawater. There was a big discussion of the cause of the two maxima. A theoretical physicist, by name of Becker, brought up the question whether it might be the consequence of hydration as Eigen had just talked about ionic hydration. They knew that it could not be the sodium chloride, so the next candidate was magnesium sulfate. Seawater is bitter because of its magnesium sulfate content.

Tamm and Kurtze asked Eigen to teach them about ionic hydration. Eigen devised an experiment with which it was possible to test Becker's suggestion within one day. They knew already that the sodium chloride solution did not show the sound absorption although its ions were also hydrated. That meant that the sodium ion did not do it and the chloride ion did not do it either. They then tested magnesium chloride solution and sodium sulfate solution, but neither solution showed sound absorption. This gave Eigen the idea that it must be the interaction between the magnesium and sulfate ions. Supposing that this interaction goes in several steps, they could imagine that the sulfate ion enters first the outer coordination sphere of the magnesium ion, then the inner sphere. This means a faster process and a slower process. They worked out a theory of coupled reactions and arrived at a qualitative description of the effect. At that moment it occurred to Eigen that this was a fast reaction. A megacycle corresponds to about 10^{-7} seconds and a hundred megacycles to about 10^{-9} seconds. So they were measuring a fast reaction, an inorganic reaction of entering a coordination sphere of a divalent metal ion. They proceeded immediately with a lot of other reactions. Eigen had especially good data for ammonia in water solution, yielding NH_4^+ and OH^- ions. The volume difference was well known for this reaction. He used Debye's theory of diffusion controlled reaction and predicted the maximum, and they got it within a few percent. Thus they had a method to study fast reactions. For more complicated systems the sound absorption was not the most convenient method because at low frequencies large amounts of the sample are needed. However, the main thing was that Eigen recognized the principle, that is, not to mix the reactants. Rather, they had to bring the reactants to equilibrium or to a steady state, perturb it, and measure how it returned to the steady state.

By then Arnold Eucken had died and Friedrich Bonhoeffer, who was building up the Institute for Physical Chemistry, supported Eigen's research. In the meantime Eigen decided not to use the sound experiment anymore. Rather, he started using electric fields which provided a much greater velocity, that of light, and decided also to do temperature jump experiments. Bonhoeffer gave him all the freedom to build up these experiments. Leo De Maeyer joined him in the following years and they studied a lot of inorganic systems and measured all the substitution coordination reactions. Word about their success was spreading fast and visitors from all over the world came, including Gerald Schwarzenbach, Janik Bjerrum, and others. Eigen and his colleagues opened a new field. This was in the 1950s. In the 1960s, the organic chemists came and they did lots of organic reactions.

Finally, the biochemists came and very soon they were in the midst of enzyme kinetics.

In 1954, there was a famous meeting in Birmingham, England, of the Faraday Society, "The Study of Fast Reactions." It was just in time for Eigen to present his new methods. He was 27 years old. Eigen was familiar with the research of Porter and Norrish, who were the stars of the field, and before the meeting he visited them in Cambridge.

The title of Eigen's lecture in Birmingham was "The Study of Fast Reactions with Half-Time as Short as 10^{-9} Seconds." Before his lecture, there were other presentations. The first paper reported measurements in the seconds range. They also called it fast reactions. Then came Roughton of Hartridge and Roughton. He said since the previous paper used the term fast reaction and theirs was in the millisecond range, he called them "very fast reactions." The next speaker was Norrish, so he used the expression "extremely fast reactions." Then came Eigen and his British colleagues suggested calling his reactions "damned fast reactions," or even "damned fast indeed reactions."

The Nobel Prize was in 1967 and Eigen called his lecture "Immeasurably Fast Reactions" in memory of Eucken who had used this expression. By the time of the Nobel Prize, Eigen and his colleagues had already done quite a bit of biochemistry. J. Monod came out with the alosteric model and D. E. Koshland came out with an alternative model. They both tried to interpret the fact that certain enzymes exhibited cooperative behavior in binding their substrates. Eigen's group did experiments and measured glyceraldehyde-phosphate dehydrogenase and, for that enzyme, they proved the Monod mechanism to be correct. Later, they also found the Koshland mechanism in other cases.

Eigen has also been interested in the chemical basis of the origin of life, which seems to be a very different question from his previous studies, in reality it is not. Two of his papers were named "Citation Classics" by *Current Contents.* One paper was "Proton-transfer, acid-base catalysis, and enzymatic hydrolysis. Part I: elementary processes." *Angew. Chem.* **1963**, *75*, 489; *Int. Ed. Engl.* **1964**, *3*, 1–19. It was one of the most cited papers in the field. The German and the English versions have been cited more than 285 and 965 times, respectively, by 1990. Then Eigen worked on evolution and published another paper, "Self-organization of matter and the evolution of biological macromolecules." *Naturwissenschaften* **1971**, *58*, 465–523. This paper was the most cited paper in that journal in over 490 publications by 1990.

Eigen and his colleagues went deep into biochemistry in the 1960s. They could measure enzyme mechanisms. They could measure the single steps in multi-step reactions and they called their method relaxation spectrometry. As a result of these studies they realized that the enzyme reactions are optimal. They are fast, as fast as possible, but this is only one aspect. In fact, there are two counteracting principles to consider. One is that the enzyme has to be very specific for its substrate. This is a very specific catalysis and only this way can a complex reaction scheme be controlled.

Specificity in chemistry is provided by binding. The stronger the binding, the higher the specificity. However, if the binding is strong, the dissociation will be slow. On the other hand, the aim is to turn over the substrate into product as fast as possible, and this is the other principle. Many of these processes require splitting a water molecule, so a hydrolytic process is a good example. Eigen and his associates calculated that they could easily get down to a millisecond, corresponding to a turnover number of 10^3 per second. Since recombination can be diffusion controlled, and many of the enzyme-substrate interactions are diffusion controlled indeed, which is about 10^9 per second, there is a six order of magnitude ratio for the stability constant of the enzyme-substrate complex.

There is an optimal adaptation of the enzymes for their job. They are not necessarily the fastest possible, but they do an optimal job. Whenever

Manfred Eigen explaining at the blackboard (photograph by I. Hargittai).

one attempts any change in the enzyme, mutations or other changes, it is found to be impossible to improve them anymore. They can be made faster but there will be some loss in specificity. Eigen thus concluded that the biological macromolecules are optimized systems. He then asked the question about the origin of optimization? He was not a theologian and found the answer in Darwinian principles. However, Darwin worked with living beings, and he did not speculate about the origin of life or the conditions inside the cell.

Eigen eventually found that it was replication that governed the optimization for molecules. He came out with this theory in *Naturwissenschaften*, and showed that Darwin was valid even for molecules if they were reproducing molecules. In this case it was possible to describe the process with a mathematical theory. He showed the necessity of a certain error threshold; if the mutation rate is too high, information is lost, and if the mutation rate is too low, the progress rate is insufficient, and so on. This is the point where Eigen's studies of fast reactions and the molecular evolutionary theory are connected.

According to Eigen, evolution did not take as many billion years as many people believe. They did some dating, using RNA data, and found that the genetic code existed already four billion years ago. Since the earth is 4.5 billion years old and since it had to cool down first, there was not much time for all that chemistry to come about. Then it took about three billion years to get from a single cell organism to man. They had worked out a theory of evolution but they needed experiments also to show that their model was right and was relevant too. They had to show that those molecules, like nucleic acids and proteins, do evolve in this way, that they use error thresholds. They found that viruses were the best examples of doing this. In the course of their experiments they also realized that they could be used as a technological device. They developed an evolutionary biotechnology. They have an evolutionary experiment, which takes 70 minutes, leading to a new molecule, optimally adapted to its environment.

Thus it is not quite correct to say that Eigen went from very fast reactions to very slow ones as some of the processes in the evolution are rather fast. He and his associates have founded a company in Hamburg. Its name is Evotec from Evolutionary Technology, and at the time of our conversation it employed 80 people. The company optimizes proteins, produces new substances, screens others, creates fast techniques of single molecule detection, and develops anti-virus strategies and diagnostic tools.

Professor Eigen has been in Göttingen since 1945, but he also spent shorter and longer periods of time in the United States. At one time he was the Andrew D. White Professor at Large at Cornell University at the Chemistry Department. Peter Debye got him there. Eigen spent a semester at Harvard University, another at Stanford University, and had many offers from other places. Lately he goes regularly to the Scripp Research Institute in La Jolla.

Eigen met Debye for the first time in 1954 at Yale University at a celebration of Debye's 70th birthday. Eigen remembers as Lars Onsager was talking to him, somebody, it turned out to be Debye, tapped on his shoulder and said, "Don't worry, you'll never understand him, but be sure, he's right." In the early days, Eigen had done some work on electrolyte theory but Debye told him, "Don't do that. Never try higher approximations for the electrolytes."

Once Eigen was invited to a student discussion in Aachen, Germany, together with Edward Teller on the hydrogen bomb. It was after the 1968 student revolution. The students were all against Teller, and they were so rude to him that Eigen felt it necessary to defend him, although he was supposed to be Teller's opponent. Eigen thinks that after the political changes of 1989, people will have to reevaluate Teller's role because of his impact in bringing down the Soviet Union.

In concluding our conversation, Professor Eigen told me a story about Heisenberg, which he thinks is very characteristic. When Heisenberg heard of Eigen's theory in 1971, he invited Eigen to Munich. Since Heisenberg was already very sick, he asked Eigen and another 30 people to his house. They had a lively discussion and Eigen had the feeling that he may have gotten his message through. At the end Heisenberg was seeing him off to the door and Heisenberg said that he was very impressed with what Eigen had said to him, but then added, "Do you really think that life was not created by the Lord?"

John C. Polanyi, 1995 (photograph by I. Hargittai).

27

JOHN C. POLANYI

John C. Polanyi (b. 1929 in Berlin) is University Professor at the Department of Chemistry, University of Toronto, Toronto, Canada. Dudley R. Herschbach (b. 1932), Yuan T. Lee (b. 1936), and John Polanyi received jointly the chemistry 1986 Nobel Prize "for their contributions concerning the dynamics of chemical elementary processes." The location of the conversation with John Polanyi, on Tuesday, August 1, 1995, was his quiet office but his schedule was hectic. The interview is augmented here by the brief speech he gave at the Stockholm City Hall on the occasion of the Nobel Prize award ceremonies in December, 1986.*

Let's jump in the middle. What's the relationship between reaction dynamics and molecular structure?

I wish I knew. The beginnings of reaction dynamics, in which I was involved, hardly required penetrating insights into molecular structure, since the rules were being established for reactions in which an atom attacked a diatomic molecule. The type of question being asked was, does the atom approach the diatom along the axis of the two atoms or at right angles to that axis? What sort of forces does the atom feel, as it approaches? Is it immediately trying to climb over an energy barrier? When does it start to feel some

*This interview was originally published in *The Chemical Intelligencer* **1996**, *2*(4), 6–13

attraction, as the new chemical bond begins to form? In answering these questions one must understand the structure of the *transition state*; it simply isn't enough to look at the structure of the molecule under attack. In fact, whatever the reaction, we directed our attention to the attacking atom and to the two atoms that constituted the bond under attack. The first thing we did was to write

$$A + BC \rightarrow AB + C.$$

People laughed, because that was how Dudley Herschbach, Yuan Lee, myself, and many others in the dynamics community routinely began our lectures.

Does the reaction dynamics approach now extend to complicated molecules?

Yes, it is beginning to, but the simple framework of A + BC remains useful. Our finding was that most of the interesting things were related to the bond that was being formed and to the bond that was being broken. If you start to hang complex structures on the attacking species or on the species under attack, the focus of your attention remains on the atoms that comprise the bond being formed and those that comprise the bond being broken. As reaction dynamics invades organic chemistry, where more intricate rearrangements are observed, this will cease to be so true.

At the end of your Nobel lecture, as if projecting into the future, you mentioned two directions. One was transition-state spectroscopy *and the other was* surface-aligned photochemistry. *That was nine years ago. Was this a prediction?*

It was in a sense, because both these fields were very much in their infancy. The question was, how far would they go? As it turns out, both of them are maturing today in a very satisfying way.

I was involved with both early on. In the case of transition-state spectroscopy, we had the idea that we could see the very short-lived intermediate between reagents and products by means of something like line broadening. The spectral line broadening would be due to the strong repulsion for a millionth of a millionth of a second (10^{-12} seconds) between the pair of particles that were in the process of separating as reaction products. So we looked at spectral lines in chemiluminescence and found very great broadening, that is to say, what are called "wings" on the lines emitted in a chemical reaction that was forming electronically excited atoms.

This was a beginning to a field that has matured in surprising and marvelous ways. One way is due to Neumark and his coworkers at the University of California, Berkeley, who obtain the transition states of chemical reactions not by bringing the reagents together but by first forming a negative ion incorporating them and then removing an electron. It's a beautiful method. It plunges you right into the heartland of a chemical reaction. Another approach is well advanced in the laboratory of Zewail and coworkers at Caltech and involves the use of femtosecond lasers to actually clock the reaction as it passes through the transition state. Again a fabulous development. Neither of those things were anticipated.

The way new fields are born is as a consequence, first of all, of a surmise that there is something new that can now be done. In the case of transition-state spectroscopy, the surmise was that one could study the interaction of light with the very short-lived collocation of atoms that we are talking about here. These are subpicosecond collocations which constitute the successive intermediate configurations between reagents and products. As it turned out, it could be done in much better ways than were initially dreamed of.

So much for transition-state spectroscopy, a field in which I am still enthusiastically involved.

At the same time I am heavily involved with what we call "surface aligned photochemistry." There are quite a number of machines in my laboratory devoted, as in the past, to getting information about the motions of newly born reaction products. Previously, however, we were getting some evidence about the molecular dance in cases in which the dancers could come onto the stage in any configuration they wished; that's what happens in gas-phase chemical reactions. But if you start with your reagents in the adsorbed state on a single crystal, the forces that adsorb them also order them. Then you trigger the reaction by sending a pulse of light through the ordered species. That's usually an ultraviolet laser pulse. The reaction now starts to unfold, that is to say, the dancers start to engage in their dance. The novelty is that they have started their movements this time from defined positions, marked by chalk on the stage. If you look at their unfolding motions following this ordered start, it should now be easier to infer what they did in the transition state, which is the most mysterious and interesting stage in a chemical reaction.

Do calculations parallel the experiments?

Our work has always included experiment and theory proceeding in parallel. We're just writing a paper, which is part experimental, part theoretical, describing how photofragments leave the surface after a photolysis event. In the same paper we report the calculations (the theory) of the initial geometry of the molecular species, which we then check experimentally with polarized infrared spectroscopy. We then simulate the action of the light theoretically by considering the effect of changing a bond into an antibond. Our theoretical molecule flies apart. Next the classical and quantal equations of motion are used to predict the outcome. Finally, and crucially, we compare the computed outcome in terms of energy distributions and angular distributions with those we measure in the lab. This to-and-fro between theory and experiment has been typical of our *modus operandi*.

What molecules are involved in such a study?

The simplest ones are hydrogen halides. We just had a discussion in this room, before you came in, about hydrogen iodide adsorbed on lithium fluoride and on sodium fluoride. The crystalline surface of lithium fluoride is an alternating pattern of big F^- ions and little Li^+ ions. Even a perfect crystal will have, therefore, a rather rough surface. When we change our surface to *sodium* fluoride, we are changing the unit cell size. In each case we adsorbed the same molecule, which is HI at the moment. The HI molecules try to get comfortable on the surface by bringing the iodine over Li^+, to maximize attraction. They also try to bring their hydrogen end over F^-, and that's an operation that involves them in making compromises, since the fit is imperfect as with a grown-up riding a child's bicycle. Then, as one increases the coverage, the adsorbed HI molecules become aware of each other's presence and the alignment is affected not just by the adsorbate-substrate forces but by adsorbate-adsorbate forces.

One of the variables we use in order to change the pattern of adsorbate is the nature of the substrate, but it's a pain in the neck to change the substrate. Another variable, as I was saying, is the coverage. It's very common in surface science for molecules at low coverage to lie down and then at higher coverage, as they start to crowd each other, to stand up. If these molecules are the reagents for some photoreaction, then, with changing

John Polanyi in the lab at the University of Toronto, about 1982 (courtesy of John Polanyi).

coverage, they change their relative orientation. Why does this matter? Because we have a basis for expecting the outcome of the chemical reaction to change with orientation and hence it must change between low coverage and high.

In a sense the crystal is *catalyzing* the photoreaction. More interesting still, this is *geometrical* catalysis, brought about by aiming things differently. It is a form of catalyis that should be of particular interest to people in the dynamics community, like myself. We have been discussing the effect of different arrangements and alignments of molecules on chemical reactions for decades. Now we are developing a means of manipulating our reagent species by adsorbing them on different surfaces and with different coverages on a given surface so as to produce different alignments and arrangements at will.

Apparently, you are a public figure in Canada. You have a busy waiting room with magazines like in a dentist's office. What's your experience in communicating with people outside the scientific community?

Perhaps you'll concede that my magazines are better than most dentists'?

You are right that I've been engaged in various political debates, such as those sponsored by Pugwash, for over 35 years. My experience is that one can readily get access to senior politicians. The real question is, can one get them to listen? I would claim that if you make a sufficiently cogent argument they will have to listen. I think we scientists have some training in organizing our thoughts and in trying to persuade difficult audiences.

In fact, our colleagues constitute just such an audience. When you go to a scientific meeting with a new idea, you don't expect people to applaud. What they do is to tear it apart. That's their function. In science we arrive at the truth through an adversarial dialogue. So we are used to having to make a case, and shouldn't be frightened to do so.

In the past, it is true, we were frightened to speak out on larger issues. We felt that we'd be trafficking in our reputation as scientists in order to get a hearing and that as a result we could bring science into disrepute, since people could say that we had abused our credentials.

As with most criticisms, there is something to this one. We do have to be careful about this. We have to explain what our expertise is.

But were we to take the opposite view that science is a sort of priesthood and that to keep it in high esteem we must keep it pure by ensuring that no scientist or group of scientists meddles in things outside their own discipline, then we would be involved in a different sort of irresponsibility.

The fact is that science is having a colossal effect on the world scene, and as a result we cannot responsibly opt out of the debate on world affairs. Earlier today, I had somebody in here who wanted to talk about the fiftieth anniversary of the dropping of the first atomic bomb in Japan. I was 16 in 1945 when that bomb was dropped. Though it came at the end of a huge and terrible war, it was a transforming moment in other respects, too. My own thinking was deeply affected by it, as it should have been. It transformed, for example, the relations between nations. But it was only one of a whole series of technological changes, to which we scientists have contributed, that have changed the world. It would be irresponsible, therefore, for us not to be involved in the debate that follows.

Often, all that we have been able to contribute has been technological solutions. But even these can have tremendous impact. I have, for example, been involved in a lot of arms control discussions with Russian scientists. They may perhaps seem peripheral now, but at the time two colossal adversaries were threatening each other and the world. The danger of the arms race was very real. Nonetheless, the political community, our leaders, were saying that we can't do anything about it for a whole lot of technical reasons. It was necessary for the technical community to say quite specifically, "yes, we can." If we fail to stop testing of nuclear weapons, or if we fail to reduce the number of nuclear weapons, it's not because it's impossible to verify these things. We explained in some detail how we could do it. If then we *didn't* do it, it was because we didn't want to. By clearing the way on the technological level, one has an undoubted influence on the way history unfolds.

Let's speak about your teachers. Was your father your teacher?

Formally, he was my teacher for one year. I entered Manchester University in 1946 when I was 17. He lectured to me in the first year. That was the last year he lectured in science. Then he transferred to philosophy. He also taught me a great deal in conversations despite my many absences away from home, first in boarding school and then for three years as an evacuee in Canada.

Most of what he taught me about physical chemistry I learned at one remove from him. I was a student for six years in the Department that he had shaped in Manchester. My professor, Meredith Evans, was one of his favorite students and my Ph.D. supervisor, Ernest Warhurst, was another student of his. What I learned from his students gave me a sense of scientific values — where the field was going, what were the important questions to tackle, and, to a degree, how to tackle them. Without those things I would have been lost. But it happens that I didn't get them directly from him, but from people who owed a lot to him.

When you speak about transition-state spectroscopy, it seems to me to have a close relationship to Michael Polanyi.

It does, of course, but I don't think that's the closest I got to his interests. He would have thought it far-fetched that one might get light to interact with this subpicosecond entity which is neither reagents nor products. Though it was not first done with lasers, it was the existence of lasers — of which of course, he never dreamed — that got people thinking about "seeing" the transition state.

I find myself now at the age of 66 engaged with great excitement in some novel experiments in which we are trying to look at transition states for sodium-atom reactions. It is this project that brings me eerily close to my father's interests of 1929 and subsequent years.

When I was being conceived (I was born in 1929), my father was establishing himself as the most perceptive interpreter of sodium-atom reactions, which he understood as being in a sense the simplest of all reactions. They are so simple that even a physicist can understand them. The sodium, which is easily ionised, comes up to a molecule with high electron affinity, and an electron jumps across. Then the positive sodium ion is drawn to the negative molecule. Because the electron hops a large distance, my father coined the term "harpooning" for this. It is also called this because the positively charged sodium hauls in its negative catch. This is a uniquely simple reaction. It is different from most reactions, which are fascinating because they are *not* sequential events. Harpooning reactions can however be described as sequential. Step 1, reagent approaches; step 2, the harpoon jumps across; step 3, the alkali fisherman pulls in the catch. The end.

Today, in my lab, we are finding that it is possible to access the harpooning event, not by taking the reagents and bringing them together, but by forming a loose complex which is in the configuration of the transition state, that is to say, by starting in the middle of the reaction. That is what we are currently doing. And that is indeed a lineal descendent of my father's interests.

I am, however, only one of many who have seen the extraordinary possibilities offered by harpooning reactions. For example, Dudley Herschbach began his life as a dynamicist by studying that type of reaction. One should also add that my father himself was part of a continuous progression. What drew him to sodium reactions was that Fritz Haber had been studying an unexplained chemiluminescence from them. This was in Berlin and my father was in Haber's Institute as a young researcher. The history, as is usual in science, constitutes an unbroken chain.

Was he the determining influence in the direction you took in science?

He personally wasn't. But where I trained for six years was. If the question is whether he was the determining influence in my going into science, then, yes, but I should qualify that answer. At the time when I learned most from my father, in my late teenage years, his interests were even livelier in non-scientific fields than in scientific ones. He had another son, George, who went into the humanities, equally under his influence. I could just as easily have gone into economics or philosophy or theology and have ascribed it to my father's stimulus. He was, of course, delighted to see me go into science, just as he would have been delighted to see me go in many other directions.

Perhaps I am being disingenuous. I can only say that if he steered me towards science, I didn't notice.

How did he make the transition from physical chemistry to philosophy? Were you a witness to this?

We seem destined to discuss transition states. Yes, I witnessed this one directly. I got back to England right at the beginning of my fifteenth year, and until I was well into my twenties I saw a good deal of my father. That was the time, beginning in 1944, when he was making the transition. The fact that he made that transition isn't so surprising. There are a lot of scientists who have started to ruminate about how discoveries are made, how people learn anything, and the role of logic in this as compared with faith. And all this was of interest to him too.

What is striking, in my view, is the originality and impact that he had in his new field of epistemology, the theory of learning. He would have said confidently that what he did in that area was much more important than what he did in science.

I have a sense of wonder at all he did in science, and yet I believe he may easily have been right that his contribution to epistemology will turn out to be more lasting. The sales of his books and the interest in his ideas continue to be great. Eventually his name will, of course, be forgotten, but his philosophical ideas will live on as a significant contribution to the development of philosophical thought.

What is remarkable, then, is the quality of the contribution he made in his decades as a philosopher. Actually, his first book on a nonscientific theme was being conceived in the 1930s when he attacked the Russian economic system and at the same time confronted the leading British social scientists of his day, Sydney and Beatrice Webb, who'd published a learned volume explaining how the Soviet five-year-plan constituted a superb innovation and was bringing prosperity to the USSR. My father took this thesis apart in a series of essays, which became a book in 1940, that went far beyond economics and inquired why it was that British liberals, the so-called Fabians, were so careless of the freedoms that they enjoyed; the book was called *The Contempt of Freedom*. It was an influential book and

John and Michael Polanyi (courtesy of the Hungarian National Museum, Budapest).

a prescient one. It is forgotten today. His best known book is, instead, *Personal Knowledge.*

As with new scientific theories, my father's thinking was initially rejected by the professionals. He was not embraced by the philosophers of his day, who felt that he was an ignorant outsider. This lasted for a large part of his time in philosophy. The people who paid attention to his work were closer to theology. This was in part because the philosophy of the time was "linguistic analysis." That brand of philosophy, centered on the study of the structure of language, passed. I don't know whether my father contributed at all to its passing. It is an interesting question. Whatever the case, there followed a school of philosophy far more friendly to his ideas.

Do you share his interest in philosophy?

Just as a human being; not as a philosopher. I have no qualifications or ambitions in that area. It was a very brave and extraordinary undertaking on his part to do something so ambitious in the realm of philosophy. Unless you read widely and deeply, you are vulnerable to attack. But he did read widely and could withstand a skeptical and at times hostile audience.

It wasn't only Michael Polanyi who made the name of Polanyi famous. It was a very extraordinary family in Hungary. Are your children aware of their family background?

They are well aware. The spirit of curiosity and creativity is very evident in them. Our daughter is a journalist and TV producer (in science, to her surprise), and our son has done a lot of things starting with science and leading into the heartland of politics and social science.

Earlier you mentioned that your father shaped his department in Manchester. Have you shaped your department at the University of Toronto?

No I haven't. The structure of science in the academic world is totally different here. In England, as on the continent, the Head of the Department shaped the interests for the whole department. That's not the case here. Nobody who has worked with me since I've been here, starting in 1956, has been appointed to a position at the University of Toronto. There is no "School." It's not like that. But I do have students in other places working on things that had their origins in interests they encountered while working in this laboratory. I'm happy about that.

Whatever the structure of the academic community, one expects one's influence to be fleeting. What people hope for is that this fleeting influence will, on balance, be positive. Perhaps it is as well that they never find out.

Nobel Banquet, Stockholm, December 10, 1986 by John C. Polanyi

Your Majesties, Your Royal Highnesses, Ladies and Gentlemen, late in his career as an actor Richard Burton was asked by an interviewer what it was in his work on the stage that had given him the keenest pleasure. Burton thought for a while, and then replied: "the applause."

That is not so ignoble a confession as it sounds. The applause is a celebration not only of the actors but also of the audience. It constitutes a shared moment of delight.

In some countries the actors are permitted to participate in the applause. This is what, by mutual agreement, the three of us — Dudley Herschbach, Yuan Lee, and myself — wish to do tonight.

Alfred Nobel in inaugurating his prizes, and thereafter Your Majesties, the Nobel Foundation and the people of Sweden in giving them this elegant and open-hearted expression, have shown the world how to celebrate.

What you have undertaken to celebrate is, of course, the truly remarkable aspect of this occasion. I know of no other place where princes assemble to pay their respects to molecules. Yours is a rare enthusiasm, expressed with such a degree of conviction that the world has come to share it.

When, as we must often do, we fear science, we really fear ourselves. Human dignity is better served by embracing knowledge.

We three have known each other for decades. Now, because of you, we regard one another with a new sense of wonder. Because of you our wives hesitate for just an instant before summoning us to do the dishes. Thanks to you the wider community of reaction dynamicists, who share our interests and have contributed in a vital fashion to the development of this field, declare themselves proud.

We applaud you, therefore, for *your* discovery, which has made a memorable contribution to civilization — I refer, your Majesties, and our Swedish hosts, to the institution of this unique prize, for which we, in the company of many others, thank you.

Dudley R. Herschbach, 1995 (photograph by I. Hargittai).

28

DUDLEY R. HERSCHBACH

Dudley R. Herschbach (b. 1932, in San Jose, California) is Professor of Chemistry at Harvard University. He was awarded the chemistry Nobel Prize in 1986, jointly with Yuan T. Lee of the University of California at Berkeley and John C. Polanyi of the University of Toronto, "for their contributions concerning the dynamics of chemical elementary processes." There is an interview in this Volume with John C. Polanyi.

Dudley Herschbach received a B.S. degree in mathematics in 1954 and an M.S. degree in chemistry in 1955 from Stanford University. His Master's thesis, under Harold Johnston's direction, was titled "Theoretical Pre-exponential Factors for Bimolecular Reactions." Herschbach went to graduate school at Harvard University and received an A.M. degree in physics in 1956 and a Ph.D. in chemical physics in 1958. His Ph.D. thesis, under the direction of E. Bright Wilson, Jr., was titled "Internal Rotation and Microwave Spectroscopy." Between 1959 and 1963, Herschbach taught and did research at the University of California at Berkeley and in 1963, he returned to Harvard University where he has been Frank B. Baird, Jr., Professor of Science since 1976. He is a Member of the American Academy of Arts and Sciences, the National Academy of Sciences of the U.S.A., has received the Pure Chemistry Prize of the American Chemical Society, the Linus Pauling Medal, the Michael Polanyi Medal, the Irving Langmuir Prize in Chemical Physics, and numerous other distinctions.

Because of scheduling difficulties, we met and recorded a conversation at Logan Airport of Boston on July 23, 1995, and the narrative below is based on parts of this conversation.*

When Dudley Herschbach left E. Bright Wilson and found himself on his own, he wanted to do something unusual, and that became reaction dynamics. Already as an undergraduate student at Stanford he had learned from Harold Johnston about the challenge of chemical kinetics. They had to postulate all the molecular steps in the 1950s because there was no way to directly observe the reactive intermediates. It was much of a guessing game. They did what they could, varied the concentrations, looked at isotope effects, but they could never be sure that they included all the steps. Most chemical reactions are a whole network of elementary steps and an elementary step is a happening in molecular terms but there was no way to show that. So when Herschbach heard, in a physics course, about molecular beams, he said to himself that this was the way to study molecular reactions. He could cross two beams in vacuum, and though most of the molecules would miss one another, those that did actually interact would do so in a collision and he could detect the free flying products. Then he could work, like a physicist, backward from the information on velocity and angle, and later, with spectroscopy, he could even do rotational motion and vibrational motion of the products. With all that information he could then learn about the forces that govern the elementary act of making and breaking a bond in a single collision.

Herschbach originally went to Harvard because he wanted to learn about molecular structure and molecular dynamics of stable molecules as a preparation for kinetics. One particular reason why he liked to work in internal rotation is because it is somewhat like a chemical reaction in which the methyl group is turning from one potential well into another. Having brought up in microwave spectroscopy, he got used to calculate everything first. The microwave spectrum is a jungle, and one first has to know where to look. So he calculated everything he could about molecular flows, pumping speeds, and all the rest. He and his students found that the alkali reactions should not be difficult to study this way, and, luckily, the first experiments worked. A lot of advances in other areas, such as lasers and supersonic

*This narrative was originally published in *The Chemical Intelligencer* **1996**, *2*(4), 12

beams, went into their success. Fundamentally it was indeed information about what happens in a truly single elementary event. They knew what reagents came together, they knew what products came out, and they did not have to guess anymore. In a few cases, they were even able to measure every step of the elementary process that was postulated from more traditional kinetics and check it out. In some cases the old ideas were wrong and in other cases they were right.

From the very beginning, they kept in mind the history of molecular structure and chemical bonding investigations, as exemplified in Linus Pauling's work. In the early days there was the transition from the era of thermodynamics to the era of bonding and spectroscopy, and Pauling had a major role in this. He used first diffraction methods, X-ray crystallography and gas-phase electron diffraction, and he was able, through combining them with theoretical ideas, to build up an insightful set of notions of what governs molecular structures. The culmination of that was his prediction of the α-helix, from what he knew of the structures of small molecules, combined with symmetry considerations. Herschbach had this wonderful example in mind when he determined that it would be most important to understand the dynamics of chemical reactions, what governs making and breaking bonds, distributing energy in the products between translation, rotation, and vibration, and how that is governed by the electronic structure of the reactants.

In the very early days, when they were studying alkali reactions, these reactions seemed to be completely unrepresentative, just an eccentric family of reactions. As is well known, the alkalis are eager to give away their valence electron, and there is no surprise when they react with the very electrophile halogen molecules. What Herschbach and his students learned was, however, that there was an enormous variety of reaction dynamics even within that family of reactions. They could understand it in terms of one simple idea, the molecular orbital of the target molecule that receives the alkali electron. If this orbital had a node between the originally bonded atoms, such as antibonding sigma, then when it received the transferred electron there was a strong repulsion between the originally bonded atoms, and the target molecules started flying apart almost as if they were photodissociated. One of the fragments would have a negative charge because the electron was attached to it and the alkali was now a cation. They would combine to make a salt product. Because of the electron transfer, this was the ideal case of probing the connection of reaction dynamics to electronic structure. It only depended on the nature of the orbital that received the electron and that defined what the forces were like. They were able to relate whether the new product went forward, backward, or stayed around, like a lawn sprinkler. They could relate all this to the electronic structure.

The magnitude of intensities that they were detecting in these experiments is an interesting question. By today's standards, it was a huge signal, but then it was not considered to be so huge, nonetheless it sufficed to be detected with an electrometer. They were producing currents in the order of 10 million ions per second whereas later they were able to detect one ion per second. In terms of what chemists were used to thinking, even the 10 million ions per second were extremely low intensities. For the alkalis, they had hot wire detectors, which only detected alkalis. Already Otto Stern had used that, and others have as well. What Herschbach added was a more incisive understanding of molecular collisions, and how to design experiments where they could learn something from the scattering patterns, how to extract that information about the forces, and how to relate it to the electronic structure.

After developing their so-called supermachine, which could detect, with tremendous sensitivity, even nonalkali species, they studied reactions like $H + Cl_2$. They found a scattering pattern, with scattering angles and intensities of HCl, which, except for the scale, was almost identical with KI from the reaction of $K + CH_3I$. Dynamically speaking, it was the same reaction. They showed that the difference in scale was related to masses and exothermicities, but otherwise they were the same reactions. Herschbach find this an important example in showing the limitations in the information content of our ways describing chemical reactions by our usual chemical equations. The usual chemical equations tell us about the atoms that are involved and about the compositions of the molecules. However, they tell us nothing about the reactions. In this sense, we are still using 19th century notation in chemistry. There is need for a notation that would allow us to see that $H + Cl_2$ is the same reaction as $K + CH_3I$. At first glance a chemist would not have anticipated that.

In fact, in the case of the reaction $K + CH_3I$, Herschbach and his group arrived at confirming the picture that Michael Polanyi had already had in the 1920s. The target molecule is CH_3I, which is electrophilic, the alkali atom comes near, transfers an electron and it goes into the antibonding sigma orbital (the terminology is different from the 1920s), and that is the same orbital that is excited in photodissociation of CH_3I. There is very strong repulsive force between the carbon and the iodine in methyliodide. So it starts coming apart. The electron affinity of iodine is far greater than that of the methyl group, hence the electron is associated with the iodine and, as it falls apart, the alkali cation comes in and combines with the iodide ion. The characteristics that are measured, the velocity of the product and its angular distribution, are determined by the strong repulsive force that is released when the electron enters the antibonding sigma orbital.

In the case of the reaction $H + Cl_2$, there is no electron transfer but the frontier orbital (in Fukui's language) that receives the third electron, is an orbital that has a node between the two chlorine atoms in the transition state. So, again, there is strong repulsion between them, and it is the same reaction even though the full electron was not transferred.

There is then the *principle of orbital asymmetry.* In the reactions of ICl, the two ends of ICl differ substantially in electronegativity. It was found that hydrogen atoms, methyl radicals, oxygen atoms, or other halogens, when they react with ICl, prefer the iodide and not the chloride, even though the bond to chloride is much stronger. The principle of orbital asymmetry explains this because, looking at the molecular orbitals of ICl, one finds that the orbitals involved in the reaction are antibonding and are mostly iodine orbitals. The bonding orbitals are mostly chlorine orbitals. Whether the reagent wants to take an electron out or put one in, it deals with the iodine. This explains that reactions of ICl with many different reagents and a lot of other reactions in analogous situations.

Michael Polanyi was an early influence on Dudley Herschbach. He cherishes the memory of all his five meetings with Polanyi. The first time they met was in 1962 when Michael Polanyi came to Berkeley to give some lectures. Polanyi visited Herschbach's laboratory and Polanyi was telling him stories about his son John. Polanyi was surprised that John became a scientist because, he said, John in his teenage years used to bitterly criticize his father, saying that he was writing papers, all the time, that were not connected with the real world.

At the time of Michael Polanyi's visit to Berkeley in 1962, Polanyi had already switched to philosophy. Herschbach has read some of his books, among them his major book, *Personal Knowledge.* Herschbach thinks that Polanyi's books help making people aware of what scientists really do. Scientists get excited about their ideas and they want to see them work. Yet they have the discipline, and they must have the discipline because the scientific community as a whole insists on it, to test their ideas. These ideas do not always pass the test and the scientists have to give them up or modify their ideas.

In contrast to John Polanyi, who came from an exceptional family of intellectual giants, Herschbach came from a family where he was the first scientist, possibly even the first university graduate. It hurt but he was not handicapped by it. His parents were perfectly willing to see him doing anything that he wanted to do. Early on, what got him started in science was an article in the *National Geographic* magazine about the stars. His grandmother gave it to him. Then he saw a little book in school about the planetary system, and he talked with his teachers, and all this was very encouraging. The librarian of their little town of 3000 knew that he was interested in science so she saved science books for him. He grew up in Campbell, California, not far from San Jose and not far from Coopertino which is very famous now as the center of Silicon Valley. At the time he was a child, Coopertino was a crossroad. Two two-lane roads crossed, and there was no building within a mile of that intersection except a general store with a solitary gas pump in front.

Herschbach had a wonderful high school teacher of chemistry with a Master's degree in chemistry from Berkeley. He came back from the war just as Herschbach was starting high school in 1946. The teacher had very high standards and he did not give partial credits for any work in his course. He told his pupils that in the artillery corps he had learned that if you had the right method of your calculations but you wound up shelling your own troops, it did not count. Starting high school after World War II, and having so many teachers who had experienced war, was probably a big plus. These teachers had something to share with their students and their wartime experience convinced them of the values of civilization and the importance of education. Several doctors and scientists came out of Herschbach's class although theirs was a farming community and they had to bus the children in a 30-mile radius to get enough of them in this high school. When Herschbach returned for a visit a few years ago, he found five high schools in the same district. The area has just exploded.

When Herschbach arrived at Stanford University after high school, he was very much taken by its intellectual atmosphere. He had a roommate who had gone to one of the best academic high schools in San Francisco and Herschbach could see a completely different attitude. His roommate was not nearly as excited as Herschbach was, who was just dazzled by all the things he learned academically because he was quite ignorant. Today, Herschbach sees an advantage in being innocent when one goes to college.

As for his interests beyond his research, Herschbach finds it almost heart-breaking that the general public is either apathetic or antagonistic to chemistry. It must have something to do with the fact that people like to tell about the terrible time they had had with chemistry in high school. However, there is no reason why chemistry should be a miserable experience in high school. Especially with computers today, chemistry should be made much more palatable and much more appealing, and Herschbach is trying to help, with his broad lecturing activities, changing it into a more attractive experience.

Henry Taube, 1996 (photograph by I. Hargittai).

29

Henry Taube

Henry Taube (b. 1915 in Canada) is Professor Emeritus of Chemistry of the Department of Chemistry, Stanford University. He received the 1983 Nobel Prize in Chemistry "for his work on the mechanisms of electron transfer reactions, especially in metal complexes. His other distinctions include the National Medal of Science (1977), the Robert A. Welch Award in Chemistry (1983), and the Priestley Medal of the American Chemical Society (1985). Our conversation was recorded in Dr. Taube's office at Stanford University on February 28, 1996.*

In the published Nobel lectures, most laureates describe their background and life, whereas you give only a very few facts.

I'll try to fill in some of the details. My parents were born in Russia. According to one account, their forebears left Germany from near Danzig during the Empress Catherine's reign. Immigration of Germans was encouraged in the hope that it would lead to improvement of farm practices among the Muzhiks. The family lived near the town of Rovno in Northwestern Ukraine.

My parents were peasants and had no education apart from being taught to read High German by their Lutheran minister. This was in preparation for confirmation. The home language was Low German, my own first language. Life for them under the Czar was miserable. They escaped from

*This interview was originally published in *The Chemical Intelligencer* **1997**, *3*(4), 6–13

Russia in 1911 or so, as did many friends and relatives. Most settled in the U.S., but my parents chose Canada. The first stop was Winnipeg, Manitoba, where my father worked as an unskilled laborer. After two or three years they moved to Neudorf in Saskatchewan, where many of German extraction had already settled. During the four years there, my father worked as a farmhand and my mother cleaned houses, washing floors for others who themselves were far from well-to-do.

For the first two years after moving to Neudorf, my family lived in a rented sod hut, where I was born in 1915. By the time I was four years old, my father had accumulated the means to rent a farm, and we settled down in a two-room shack near the town of Grenfell about thirty miles south. This was my home until at 13 I left for Luther College, which is located in the provincial capital, Regina. I had the benefit of a one-room school for my early education. There were about two dozen pupils distributed sometimes over eight grades. The classes were called up in turn to face the teacher's desk, and the recitations of the lessons could be heard throughout the room. I listened to those of the classes ahead of me and advanced rapidly, skipping two grades. Whatever advantage this may have been was partly lost because the teacher was not licensed to teach high school so after I completed the eighth grade I was idle for a year, after which a qualified teacher was hired.

None of my classmates continued for a higher education after leaving Bradley School, and most stayed at home to help out on the farm. My father was unusual in trying to provide a better life for his children than he had had. We were four brothers, two born in Russia. August, the older of the two and about 15 years older than I was, left home soon after we moved from Neudorf to Grenfell. He settled in Milwaukee, Wisconsin, where we had relatives, after crossing the border to the U.S. illegally, a rather common practice for new immigrants. August had only a brief formal education in night school but became quite well read and prospered in the U.S. While we were still in Neudorf, Edward, the second oldest, so impressed Mr. Dancy, the school teacher there, that the Dancys, who had no children of their own, persuaded my parents to let Edward be part of their family, but without formal adoption. I only saw Edward during the summer holidays, but his visits, in which he shared some of his learning, greatly impressed me. He was, as far as I know, the first in the history of our family to get a higher education, and he progressed to a Ph.D. in German literature. It was my father's hope that I would become a Lutheran minister, not so much because he was deeply religious as because in his experience, being a minister was as much as one could hope for in the way of a good life. My father had planned that Albert, less than two years older than I was, also would go to the university, but in the stock market crash in 1929, my father lost the little he had managed to save. This happened at the start of my second year at Luther College. After this, my father was no longer able to pay for my stay there. Mr. Paul Liefeld, the chemistry teacher, persuaded the school to appoint me as a helper in the laboratory. This covered tuition and keep and, with some help from Edward, who by now had a university position — help which was continued for two years at the University of Saskatchewan at Saskatoon — I was able to continue my education.

Did you come across Gerhard Herzberg there?

He was one of my teachers in my four years at Saskatoon. Herzberg was the clearest lecturer I had had up to that time, and perhaps the clearest ever. I took two courses from him, one in atomic spectroscopy and one in nuclear physics, and did well in them. Professor John W. T. Spinks, who was adviser for my research toward the master's degree, had studied with Herzberg in Germany and was responsible for his coming to Saskatoon.

By then you must have been in the sciences.

I had started out as a candidate for the ministry, but when I was about 15, I found that I couldn't continue. Even though the Ohio synod of the Lutheran Church was one of the more liberal branches, they then still preached a literal interpretation of the *Book of Genesis.* Even as a child, I read everything that came my way, and at some point I came across Darwin's teachings and they made sense to me in spite of my trying hard to keep my faith. I remember vividly the incident that changed the direction of my life. I had gone into the library of Luther College, browsing as I often did, and picked up a book written, as I remember, by a Lutheran minister, dealing with evolution. I even remember the blue cover of the book and particularly the illustration on the left-hand margin of a left-hand page, drawings of perhaps three dozen mice running down the page. Each had a tail. An experiment had been done to cut off the tails of successive generations of mice. Despite this, the mice still grew tails, and hence the idea of evolution was shown to be nonsense. I also remember a feeling of sadness as I put the book back on the shelf and decided I had to turn to something else. When I told my father, bless him, he made no effort to dissuade me and simply expressed the hope that I would prosper at whatever I next chose to do.

I am among those who are rather slow in finding a career. I had been introduced to chemistry, and, as measured by grades, I was good at it and it was easy. But it didn't occupy my attention the way English literature did. My interest in literature was fostered by the physics teacher, Mr. Behrens. He would come into the lecture room beaming and smiling but, to the detriment of my science career, would talk mostly about poetry, not physics. A particular favorite of his was Emily Dickinson, whose poems he would often recite. It was a thrilling experience for me to see how much enjoyment he got out of the use of words put together well. It made me feel that literature is a wonderful thing to devote one's life to, but, as things worked out, when I left Luther College for the University of Saskatchewan, Saskatoon, I registered in chemistry.

Why?

The choice was made for a trivial reason. I had come to the university planning to take courses in a range of subjects, this again a result of reading, but specializing in English literature with the hope of eventually becoming a writer. On coming to register, I was confronted by an enormous crowd of students. I was shy and rather small and had more than the usual difficulties in registering. I went from one line to another and never seemed to get it right and then I spotted one of my former classmates from Luther College standing on the sidelines looking on. I walked over to him and asked him, have you registered? He said he had. I asked him, what did you register in? He replied, chemistry. I said, show me.

I did do well in the chemistry courses in Saskatchewan but — and I hesitate to say this because the Department, particularly the Chairman, was good to me — I didn't really get strongly motivated until beyond the Master's degree. For work toward the Ph.D. degree, I went to the University of California in Berkeley, where chemists such as G. N. Lewis, W. F. Giauque, J. H. Hildebrand, W. M. Latimer, and my Ph.D. mentor W. C. Bray were on the faculty and they were approachable and inspiring. In those days, what is called the "old boy network" still operated, and G. N. Lewis played a role in recruiting graduate students at Berkeley. He was Dean of the College of Chemistry, which meant that he had equal standing with the Dean of Arts and Sciences. He was a scientist of the highest distinction and a powerful personality. It was a condition of his accepting a position at Berkeley that the status of what was a department be elevated to that of a college. Lewis got to know the future chairman at Saskatchewan, Thorburger Thorvaldsen, when they were colleagues at MIT and, because of this connection, acted favorably on any recommendation that Thorvaldsen made for admission to graduate study in the College of Chemistry at Berkeley. Thorvaldsen recommended me to Berkeley without my knowledge.

I became deeply interested in chemistry soon after I came to Berkeley. Just the general atmosphere of the college was conducive to this; chemistry was in the air. When members of the faculty would meet, let's say in the halls, the discussion would be about some seminar topic that hadn't been quite resolved or about a puzzling observation that someone had made, sometimes in the freshman laboratory. What got to me early was not only their enthusiasm for the subject but their frankness in admitting the limits of their understanding and their willingness to learn from each other. There was little pretense, and they didn't feel that they had to impress others. At any rate, the fire was lit there quite early in my stay.

How about your present family?

I would like to say something about the Great Depression first. My graduate work was completed during this period. Nowadays we hear that opportunities for chemistry Ph.D.s are not as good as they were five or six years ago. This is a far cry from what it was like during the depression. There were no jobs, and because you knew there were no opportunities, you learned not to worry about it. I just felt that nothing could be done about it anyway and enjoyed my graduate work at Berkeley. Miraculously, I was saved again by the "old boy network." I'd applied to every department I knew about in Canada, and I heard from Saskatchewan and one other that they didn't have any openings. Lewis, without consulting me, had made representations to Cornell, where the great Peter Debye was Head at the time, and I was offered a job.

I married during my last year in Berkeley, in 1941, and my first wife and I had a child, Linda. We were divorced about three years later. I remarried in 1952, and Mary brought one daughter, Marianna, into the family and we have two sons, Karl and Heinrich. My given name was actually Heinrich. I'm so grateful that none of my children followed my profession because it's a no-win situation: if they're better than you, it can be a bit of a problem, and if they are worse than you, it's terrible.

Professor Lindquist, introducing your Nobel lecture, mentioned the continuity of electron transfer studies, from Arrhenius to Werner and from Werner to Taube. There is, though, a 70-year gap between Werner and Taube.

Henry Taube and his wife at the reception of the Royal Swedish Academy of Sciences during the centennial Nobel celebrations in Stockholm, December 2001 (photograph by I. Hargittai).

I really like to call my interest the study of oxidation-reduction (redox) reactions, and insofar as such reactions between metal ions are concerned, there was little prior work when I began my own. Werner had encountered redox reactions in the course of his research and, in fact, made good use of them in the preparative procedures he developed. I'm not sure that he thought much about the fundamental difference between them and the reactions he was primarily interested in, which were substitution reactions of metal complexes (coordination complexes). Many of the basic ideas underlying redox reactions were developed in the study of nonmetal chemistry. For example, my mentor at Berkeley became an expert in the redox chemistry of halogenates, after working with Professor Luther in Germany in the first part of this century.

In most such reactions, there is transfer, say, of an atom, in this case an oxygen atom, from one molecule to another. They can formally be regarded as involving electron transfer, but the electron transfer act is not obvious, and I've sometimes described them as proceeding by "covert" electron transfer. They are of a different kind from many of the reactions of metal-containing complexes, in which a single electron is transferred from one metal complex to another, without bonds being disrupted at either metal center. Such reactions can be said to proceed by "overt" electron transfer, and the term "electron transfer" as it was used when the subject opened up in the 1950s really applies to them. They are easier to understand than those in which there is bond making and bond breaking during the change in oxidation state.

I spent many years on the study of redox reactions in general before I started to work on the special class of electron transfer reactions. An experiment I did in 1954 — I use the personal pronoun because, although I had a coworker, I did most of the laboratory work myself — attracted a great deal of attention to the subject. Ironically, it did not involve electron transfer in the strict sense of the term, but it did introduce a new dimension to the subject of redox reactions of metal ions. It was rather certain at the time that some such reactions of metal-ion complexes do go by simple or overt electron transfer. It was speculated that in other cases, electron transfer could take place by an atom bridging two metal centers in the act of electron transfer, the bridging atom then being transferred from one metal to the other. My contribution was the unequivocal experimental demonstration that it really can occur.

Though the experiment was very simple, it needed a great deal of preparation before the underlying strategy could be devised. That preparation came to me when early in my stay at the University of Chicago I gave a course in what is called coordination chemistry, chemistry of the kind Werner did, where the interest is in metal ions and the atoms or groups directly bound to them (ligands), often six in number. I became interested in the issue of why some metal complexes in water as a solvent exchange ligands rapidly, while for others these substitution reactions are very slow. I also saw that this difference is very important, because it can be exploited in doing meaningful experiments. One of my contributions to the field was to show that the difference in the speed of substitution need not have a direct connection to thermodynamic stability. In thermodynamics you are concerned with initial and final states. In thinking about rates, you are concerned with the difference in energy between the initial state and the activated complex, a configuration passed through when the system reorganizes from the initial to the final state. The electronic structure of the metal ion undergoing substitution can affect this difference much more than it affects thermodynamic stability. As a simple illustration: if there is a low-lying vacant orbital on the metal ion, then in the act of substitution the incoming group can be bound to the metal center. This orbital is not directly involved in the bonding in the initial or final states. These ideas and others were formulated and published in an article in *Chemical Reviews* that appeared in 1952. It did change the thinking of some, first because of the sharp distinction I made between kinetic and thermodynamic stability, which of course I had been taught at Berkeley, and also because, on taking account of the differences in electronic structures of metal ions, the differences in rates of substitution began to make some sense.

The important point is that it was my understanding of the differences in ease of substitution as depending on electronic structure that enabled me to select the reagents for the experiments on the "bridging" mechanism for electron transfer that I began in 1954.

So the gap between thermodynamics and kinetics is quite conspicuous here.

Particularly here. As an important side issue, I want to mention that in thinking about the simple electron transfer reactions, theoreticians such as N. S. Hush and R. A. Marcus did deal with the contribution of driving force to the rate of reaction. The underlying principles are general and have been extended to other reactions.

In your Nobel lecture of 1983, you use this Fe^{3+}/Fe^{2+} reaction. This is also the first equation in Professor Marcus's Nobel lecture of 1992.

I believe that Marcus became interested in the field in trying to understand electron transfer in electrochemical reactions and then extended his insights to electron transfer in general. It should be kept in mind that Noel Hush, whom I've already mentioned, at the same time and independently of Rudy [Marcus], also developed the theory and that they agreed on the underlying principles. My own first experimental contributions were made at about the same time, and the theoretical advances were essential to the rapid experimental advances, which began at this time when others entered the field.

The Fe^{3+}/Fe^{2+} reaction system also influenced my own work. I was at Berkeley when the first attempt to measure the rate of electron exchange in this system was made. This attempt was made by Seaborg and coworkers, making use of a radioactive isotope of iron, produced artificially. The reaction proved to be too rapid for the method then used. Dodson, in about 1950, 10 or so years after the first attempt, succeeded. The history is interesting in its own right. Most of those who might have been interested in doing further work on the reaction were called away to do war research. Many of them worked on the chemistry of the emerging elements, and they were forced to think about ions in solution. Much of the chemistry of plutonium took place in water solution, and the chemists had to learn about the different oxidation states and their interconversions. Redox potentials, rates of reaction, solubilities, complexation, for plutonium and also for other elements, were all very important to the war effort. In a way, a large number of physical chemists were converted in the course of this work to physical inorganic chemists. Several of them, when they returned to academia at the end of the war, chose to work on the $Fe^{2+/3+}$ electron exchange. They were attracted by the excitement of measuring the rate of a process in which the products are the same chemically as the reactants; that is, no transformation of matter has taken place.

Did you participate in the war effort?

I did in a way. I was at Cornell at the time and had the job of teaching general chemistry to armed forces personnel. A program had been started to prepare candidates for officer training school (O.T.S.) by subjecting them to a year of study at the university level. They were very reluctant students because they were, in effect, selected for the program, and few of them wanted to be part of it. Lecturing to them was an altogether disagreeable experience for me. The program did not last long, and I understand that few of the students in it made it to O.T.S. An exception was a small group, members of the Air Force, who were headed for medical school, and they took the subject seriously. One of them returned to university after the war and became a very good physical chemist — my only known success from a great deal of effort.

The remainder of my time was nominally devoted to a project supported by the Navy, but in fact I managed also to continue my own research interests.

Did you teach general chemistry again?

I did at the beginning of my Stanford career, but I don't think I did a good job, even though I was rather popular. In looking back, I think I made a mistake in emphasis, a mistake that others of my time made. There is so much to marvel at in the transformations of matter, but, instead of emphasizing this aspect of the subject, I spent a great deal of time at the beginning of the course trying to teach arithmetic. While it is true that many of my students were very deficient in the subject, I'm not sure that I corrected this deficiency by trying to make it up in a chemistry course.

Being an experimentalist, when did you stop doing experiments with your own hands?

Even after I gave up doing experiments in my own laboratory, when I was about 45, I continued as a consultant during the summers in the Los Alamos National Laboratory. Some of my work on the interaction of solvent water with metal ions was done at Los Alamos, using nuclear magnetic resonance techniques. I believe ours was something of a pioneering effort in that field.

In any case, I was in my most creative period while I was still working with my hands. I still follow the progress of my coworkers in the lab, on a daily basis, and try not to get in the way.

Could you say something about structure in solutions?

Structure in solution is essential to understanding reactions, but it is much harder to determine than in solids. This is particularly true for metal ions which readily undergo substitution of the groups bound to them. To illustrate: in anhydrous ferric chloride ($FeCl_3$) in the solid state, each ferric ion is surrounded by six chloride ions. On dissolving the solid in water, there is immediate (on the time scale of less than a second) replacement of chloride by water molecules, and, in dilute solution, the dominant form of ferric ion is the hexaaqua species. On increasing the concentration of chloride ion, species appear in which one chloride is bound, followed by species with two chloride ions, etc. These substitution reactions are also on the time scale of seconds or less. In such a system, one has to reckon with a number of species in rapid equilibrium. Determining their relative concentrations (i.e., the compositions of the solutions) as a function of concentration is itself a difficult task, let alone establishing molecular structure, that is, determining shapes and measuring bond distances. Even to date, the compositions of solutions of ferric chloride have been defined only over a narrow concentration range, and detailed information about structures is still lacking. This situation is in marked contrast to that prevailing for hexaamminecobalt trichloride [$Co(NH_3)_6Cl_3$], a compound first prepared in Werner's time, and perhaps in his laboratory. In the solid, six ammonias are bound to the metal ion, and the chloride ions are excluded from direct contact with it. When the compound is dissolved in water, the integrity of the hexaammine is retained, and even though the complex is highly unstable in the thermodynamic sense, it breaks down to form cobaltic hydroxide and ammonia on a time scale of the order of weeks or months. In acidic solution, this complex is even less stable thermodynamically — at equilibrium, the products would be Co^{2+}, NH_4^+, and O_2, i.e., substitution followed by oxidation of water by Co^{3+}(aq) — but this reaction is on the time scale of years, even decades. It is an excellent illustration of the point that thermodynamic instability and reaction rate are not necessarily closely coupled.

In the experiment of 1954 that I mentioned earlier, I used a derivative of hexaamminecobaltic ion in which an ammonia on cobalt is replaced by a chloride ion. This species also retains its integrity when a solid containing it is dissolved in water. The situation for it is much like that which the organic chemists usually enjoy. Carbon tetrachloride in water is thermodynamically unstable. Given enough time, perhaps centuries, it will decompose to carbon dioxide and hydrochloric acid. Nobody worries about this reaction, because it is very, very slow. There is an obvious advantage in working with molecules of this type in studying reactions, because you know what you have in solution. Much of my research since 1950 has involved molecules of this kind. Even though substitution reactions may be slow, electron transfer reactions can be very rapid. The reaction of chloropentaamminecobaltic ion with the counter reagent I used in 1954, Cr^{2+} in water, is complete on mixing. The experiment was first done using test tubes, following the reaction by color changes which take place on mixing the reagents. I have always been attracted to the chemistry of the so-called transition elements — mostly high-melting metals, iron, cobalt, nickel, chromium, etc., — because their compounds are highly colored, and one can learn a great deal by simple mixing experiments.

Please tell us about your current work.

I have a small number of coworkers now, and though my interest in electron transfer reactions continues, I am now more heavily involved in preparative chemistry. A number of years ago, I saw a research opportunity in the chemistry of osmium. Although there has been a lot of research activity in the organometallic chemistry of the element over the past few decades, the traditional coordination chemistry of the element — let's say the kind that can be practiced in water as a solvent — has been largely neglected. But I want to emphasize that I came to the opportunity by a logical route. Beginning in the 1960s, my coworkers and I began to develop the traditional coordination chemistry of ruthenium, the element above osmium in the periodic table. Many unusual effects were discovered in the course of this work, and much of the work that led to the Nobel Prize was done by taking advantage of the new chemistry. A few key experiments with osmium showed that the chemistry of osmium would prove to be even more interesting, and this turned out to be the case. A former coworker, Dean

Harman, now at the University of Virginia, in working with osmium, uncovered a new field of organometallic chemistry which provides overlap of that field with traditional coordination chemistry. The compositions in standard organometallic chemistry can be quite complicated, a metal joined to carbon, and the co-ligands often diverse in nature, some of them exotic. As an early discovery, Harman, by a very simple reaction, prepared what may be the most stable complex in which benzene occupies a single ligand site on the metal, in this case osmium. It is certainly the simplest in composition, the five co-ligands being ammonia. It is unusual too in that the chemistry of the species can be studied in water as a solvent.

Harman is making a distinguished career in developing organometallics of osmium amines further. In my own laboratories we have continued along more inorganic lines. By reducing the number of ammonia molecules bound to the metal, many new reaction possibilities have been opened up. This is the direction of our current work.

A very recent and surprising development in chemistry is the discovery of a new state of combination for the element hydrogen (modern usage is dihydrogen). The discovery was made by G. Kubas and coworkers at the Los Alamos National Laboratory in the early 1980s. The essence of the discovery is that dihydrogen can form stable compounds in which it is bound side-on to a metal ion, occupying a single ligand site. In the first compound to be characterized, the central metal atom is molybdenum, and the co-ligands are three carbon monoxide molecules and two phosphines. The separation of the hydrogen atoms in the complex is only 10% greater than it is in free dihydrogen. The subject has grown enormously since the original discovery. As his last contribution to my program, Harman discovered the simplest dihydrogen complex known, in which dihydrogen is attached to Os^{2+}, the co-ligands being five ammonia molecules. We have since extended the chemistry to species in which only four ammonias are bound, so that a position is left open for a variable ligand. By changing the nature of the latter, we have learned a great deal about the nature of dihydrogen as a ligand. This has been a very exciting development for my coworkers and myself.

Any interests outside chemistry?

Gardening. I spend about two days a week tending a large garden. Mary and I continue to do all the work ourselves. I was raised on a farm and became used to hard manual labor, and I find that I need it for my physical well-being. Though I am not as vigorous as I was at 65, I feel as good as I did then, perhaps even better. At 65, there was more stress in my life. At that time, I was reaching retirement and was considering leaving chemistry to take up horticulture. I think I made the right choice in staying active in a subject I have so much enjoyed. I assumed emeritus status at Stanford University at 69 because I was given the opportunity to act as scientific adviser to Catalytica Inc., a local chemical company, and welcomed the opportunity to do something applied. I've had a very satisfying relationship with the company and enjoy my involvement there, particularly because the company itself is a rather new enterprise.

Any other outside interests?

I have from time to time tried to write fiction but find that I lack the imagination to do anything worthwhile. I also enjoy listening to music, especially vocal music. I think the human voice is the most marvelous musical instrument there is. I have a very large collection of classical vocal records, close to 10,000. My oldest record dates from 1897. I started collecting in about 1948 because of nostalgia for some records I heard as a child. My father had rewarded my brother Albert and me for our diligent help in harvesting by buying a phonograph with records. They all happened to be classical music, including some vocal. I especially remember some by the great Irish tenor John McCormack, whose records I have collected avidly.

When I stepped into your office this morning, you were rather upset about a referee report on your manuscript. Any comment?

I wish the referees understood my papers better. The referee report I got today is so absolutely off the wall that I don't quite know how to handle it. Publication is suggested but with drastic revision, and I feel that I can't revise it along the lines suggested. I even have trouble understanding some of the comments.

My impression is that you keenly follow the recent literature.

It is so vast that I can't possibly keep up. But I do like to hear about new developments. It's a very exciting time in chemistry, particularly with the advances that are being made in understanding the complex molecules involved in the chemistry of life and the application of this knowledge to the art of healing — now becoming the science of healing — it's absolutely incredible. Because of this and other advances, surprises even in traditional areas, I am reluctant to leave the subject.

Rudolph A. Marcus, 1996 (photograph by I. Hargittai).

30

Rudolph A. Marcus

Rudolph A. Marcus (b. 1923 in Canada) is Arthur Amos Noyes Professor of Chemistry at the California Institute of Technology. He received the 1992 Nobel Prize in Chemistry "for his contributions to the theory of electron transfer reactions in chemical systems." Dr. Marcus is a Member of the National Academy of Sciences of the U.S.A. (1970) and a Foreign Member of the Royal Society (London, 1987). His other distinctions include the Wolf Prize in Chemistry (Israel, 1985) and the National Medal of Science (1989). Our conversation was recorded in Dr. Marcus's office at the California Institute of Technology on February 19, 1996.*

If we can jump right in at the middle, in addition to electron transfer reactions, what other reactions have you studied?

I have also studied unimolecular reactions. There are also atom transfer reactions and proton transfer reactions.

Is electron transfer the most common?

It's fantastically common. It cuts across many fields. It's in biology, in photosynthesis, in respiration; it's also in inorganic chemistry, in electrochemistry, in the solid state; it's in the hopping problems, charge transfer spectra. It just goes into many, many areas.

---

*This interview was originally published in *The Chemical Intelligencer* **1997**, *3*(4), 14–18

How did you choose this field?

By accident. It was 1955. I happened to be reading through an issue of the *Journal of Physical Chemistry*, which was a symposium issue on isotopic exchange reactions. The symposium took place in 1952.

Previously I had published a theory of unimolecular reactions, which is still the standard theory today. So I saw this symposium issue, and I knew nothing about electron transfer. I noticed a theoretical article by Bill Libby, who later received the Nobel Prize for radiocarbon dating, and he had a proposal to make, trying to explain some of the experimental results of the field. In his proposal he had the words "Franck-Condon principle," applying it to the rate of the chemical reaction. I was immediately intrigued, and, as I looked closer, I noticed that he had some electrostatic calculations to determine the barrier to electron transfer. There was something I didn't feel was quite right, so I followed it up. I could follow it up because in the two previous years I'd written a couple of articles on the electrostatic properties of polyectrolytes, so I knew electrostatics. There was something wrong in Libby's article although the suggestion to apply the Franck-Condon principle was right — all the nuclei very heavy and the electrons very light, so when the electron jumps there is just not enough time for the nuclei to move. That was great; it was the details, implementing it, that somehow didn't seem right. During the next month I figured out what was missing, what was wrong, and I was able to calculate the electrostatics for this type of problem: for the electron to transfer, satisfying the Franck-Condon principle and still satisfying energy conservation, we have to have a fluctuation of the medium, of the solvent, to something in between that for the reactant and that for the product. This fluctuation has to happen around each reactant. What I did was to bring that into account. It was the part that was missing in Libby's article. Otherwise, you violate the law of conservation of energy. I found a way of doing that, and that's how I got involved.

Did you ever tell Libby about it?

Yes, and Libby told me, quite a few years later, that as soon as my article came to his attention, he immediately took it to Condon because Libby was really not a theoretician. He was mainly an experimentalist, and a great idea man. So he took it to Condon, and Condon said it was OK. Then, many years later, around 1980 or so, Libby asked me about writing a joint paper, but I was involved in other projects then.

Back to your article, didn't they send your manuscript to Libby to review?

I doubt that they did. There was so much else besides the Franck-Condon principle there. The review of my manuscript was quite interesting. It was a three-page review. The reviewer asked all sorts of questions, which were good questions. But throughout the review he said things like, and this was his opening sentence, "This article was obviously written by a physicist who doesn't know any chemistry." This was the *Journal of Chemical Physics.* I never really did find out definitely who wrote the review.

What was very good about the review was that it asked all sorts of questions. In the process of my answering those questions, it became a much more readable article.

Do you remember this story when you referee papers?

Yes, and I often show my comments to my wife, who is an excellent writer, and she often rephrases things. She always insists that initially I bring out something positive, so I have developed that style under her influence. She's been a tremendous help.

Where did you do your most important work?

The work that the Nobel Committee cited appeared from 1956 to 1965. I spent almost the entire period at Brooklyn Polytechnic, which I left in 1964.

Can you tell us anything about Herman Mark there?

Great guy, a wonderful individual. He was not only a great scientist but a great human being too. There was also this story about him. When his family was getting out of Austria after the German occupation in 1938, they had converted the family wealth into platinum and made hangers of platinum. His family drove out, with clothes on the hangers!

He built a wonderful center for polymer science at Brooklyn Polytechnic and made the place stimulating in many ways. There were other outstanding people at Brooklyn Polytechnic, like Ewald in crystallography — he was Head of the Physics Department; Fankuchen, who had worked with Bernal, and Ben Post and David Harker were other crystallographers there.

Could you say something about electron exchange with no change in the chemical bond?

What I'd worked on were electron transfer reactions in which, during the electron transfer, no bond is broken; the bond is just stretched or compressed. These are also important changes and can be a major barrier to the reaction in some cases. In some other cases though, there is no change indeed, and it is just the solvation outside. There are cases where bond stretching dominates and others where solvation changes dominate. Since then, of course, Jean-Michel Savéant, an outstanding electrochemist in Paris, has extended this to some of his reactions which involve electrons transferred to or from electrodes, along with breaking of the bond.

You have stressed biology in addition to chemistry in your work.

It's because, for example, in photosynthesis, the electron transfer is the key part. A lot of people have done experiments on the reactions and structures, and now having all that information, the question is, how does it function in terms of electron transfer. Where does it fit in with the ideas of simple electron transfer in solution? When you look at the structure and you look at the size, look at the polarity and nonpolarity of various parts, a lot of facts fit together. So many people feel today that they more or less understand what's going on even though there remain unanswered questions.

It has been suggested that Henry Taube's work is in a logical sequence starting with Arrhenius and continuing with Werner. Can we place your work in this perspective?

Certainly the focus of my work is reaction rates and modern reaction rates work started with Arrhenius. There are certain key figures who made important contributions on the way. For example, Norrish, Porter, and Eigen received the Nobel Prize for fast reactions. Getting into the field of fast reactions was a step on the way. Hinshelwood and Semenov received it for chain reactions. That's a step on the way. There were people back in the 1920s who also made important contributions, such as Rice, Ramsperger, and Kassel. There are many names, such as Eyring, Evans, Polanyi, and Wigner, and all of them can be regarded as part of the tree, and I would be some place on that tree.

Can you see the continuation? How is it branching off?

I have a feeling that the way things have gone, the branching is toward addressing questions that we had years ago but the techniques weren't available. For example, it has now become possible to study what's going on in a very fast time, experimentally. Who would have dreamt before that you could see one day all those free radicals? A few have been known from outer space. The same about fast reactions in solutions. When I was a graduate student and we could measure reactions that took a few minutes, we were delighted. Now people are measuring things that take a thousandth of a second and even reactions that take less than a millionth of a millionth of a second. Part of the direction is getting into more detailed things, and part of it is getting into more complex systems. In biology, for example, proton pumping is being investigated, the use of and conversion to ATP. One of the things for the future is the increased complexity, putting things together, learning about the reaction mechanism in complex structures. Just recently, there was a symposium here on protein folding, and there are many ideas about how to treat this very complicated problem. You need to know some chemistry from the point of view of understanding some of the things, but theory does come in, as does measuring things very quickly. There has been a lot of new technology.

The other direction is the more detailed knowledge about simple systems, for example, the movement of energy in molecules. Now people will be able to get that information. It is very exciting.

During your graduate studies you did some actual measurements of reaction rates. So you came from an experimental background.

When I was a graduate student in Canada, at that time there was no theoretical chemist in Canada. We had little or no training in theoretical chemistry. It was at McGill University. Canada was very strong in experiment and reaction kinetics. So those of us who might have had any interest in theory never even thought about it. There was no role model around. After getting my doctorate, which was in the experimental study of reactions in solutions, I went to the National Research Council in Ottawa, to study reaction rates in gases. After spending a couple of years there, another postdoc and I formed a two-man seminar to select some theoretical papers, and we were presenting them to each other. That was Walter Trost, who later became Vice President of the University of Calgary and died a year or two ago. He lived long enough to send me a nice letter when I received the Nobel Prize. He was a very bright fellow and very enthusiastic. Then I started applying some of what I learned in our seminar to some of what I was doing in Ottawa. Nothing innovative, just applying existing theory. Then I thought that maybe I could get a postdoc in theoretical chemistry.

Can you tell us something about your family background?

My mother was born in Manchester, England, and my father was born in 1895 in New York City. My grandparents come from Lithuania, which was then part of Russia. They were Jewish. They left in the early 1890s. A brother of my paternal grandfather went to Sweden and eventually got a doctorate in theology at the University of Uppsala. I learned more about him when I went to Stockholm in 1992 and found out, for example, that this great-uncle had written some 40 books in Sweden. In the Swedish *Who's Who* of 1939, he was listed under a new name, Steen — he had converted — but in brackets he gave Markus, with a *k*. He was one of my idols when I was growing up and two of my father's brothers, both medical doctors, were my other idols. Since my father was not a professional, I found these other idols.

Interestingly, I never corresponded with my uncle in Sweden, but an aunt of mine did until World War II, but then, not being sure which way Sweden was going to go, she didn't want to correspond because she was afraid that it might get him into trouble.

How about your children?

We have three children. The oldest is Alan, who was a producer-director and is now a university lecturer in film; the next is Kenneth, who also got his doctorate and teaches history; and Raymond is a graduate student at Indiana University, writing his thesis about the Nixon period. We have one grandchild, Sara, Alan's daughter. All of our sons are excellent writers and that's from their mother, but I'm the one who goes skiing with them.

How did you come to Caltech?

From Brooklyn Polytechnic, I went to the University of Illinois at Urbana, and once, at a National Academy of Sciences meeting, (I was elected in 1970, and this was in 1977), I saw Harry Gray and we were talking. I had met Ahmed Zewail, who was also on the faculty here, and I had enjoyed our interaction. So they and probably some other people thought that it might be a good idea to have another theoretician around. I moved in 1978.

Were you surprised when you got the Nobel Prize?

Yes and no. For quite some time people had been writing me notes — they are not supposed to do that — but they were saying that they nominated me and hoped that I'd get it. I received quite a few of those notes from places all over the world. When you hear something like that, you can't help thinking about it. On the other hand, I also recognized how many

Rudolph Marcus and his group at the University of Illinois (courtesy of Rudolph Marcus).

good people are around, outstanding scientists. There is certainly an element of luck involved.

How many students do you have?

I have four students and two postdocs.

Do you have financial support?

Yes, to support coworkers and to buy computers and travel.

What would it mean to you if restrictions were introduced for foreign students?

I would have fewer students.

Why wouldn't you then take more Americans?

You take the best of those who apply. A lot of people in theory are from abroad. I can only guess the reason. Probably it's easier to teach math and physics with little equipment than it is to teach experimental chemistry. On the average, the people applying from abroad are extremely well trained in math and physics, and probably extremely well qualified. Foreign students here are very good. If they weren't good, they wouldn't be admitted.

Ilya Prigogine, 1998 (photograph and courtesy of Vladimir Mastryukov).

31

Ilya Prigogine

Ilya Prigogine (b. 1917 in Moscow, Russia) is Director of the International Solvay Institutes of Chemistry and Physics, Brussels, Belgium, and of the I. Prigogine Center for Statistical Mechanics and Complex Systems, The University of Texas at Austin. He is a Belgian citizen. He received the Nobel Prize in Chemistry in 1977 "for his contributions to non-equilibrium thermodynamics, particularly the theory of dissipative structures."

Ilya Prigogine has published more than 600 papers. He is a member of over 60 national academic and professional organizations. To date, he has 45 honorary degrees. He was made Viscount in 1989 in Belgium. Five international institutes bear his name.

His main monographs include *Introduction to Thermodynamics of Irreversible Processes* (Charles C. Thomas Publisher, 1954), *The Molecular Theory of Solutions* (with A. Bellemans and V. Mathot, North-Holland Publishing Co., 1957), *Nonequilibrium Statistical Mechanics* (John Wiley & Sons, 1962), *Order Out of Chaos* (Bantam, 1984), *From Being to Becoming* (W. H. Freeman & Co., 1980), and *The End of Certainty* (The Free Press, 1996).

We recorded a phone conversation on April 13, 1998.*

*This interview was originally published in *The Chemical Intelligencer* **1999**, 5(4), 13–17

A recent dictionary of scientists lists you as a Nobel laureate in physics.

I have a degree in chemistry and another in physics. My first scientific publications were mostly on mixtures. In 1957, I published a book on the molecular theory of solutions. Also, my early work in thermodynamics was closer to chemistry than to physics. This has changed in the later part of my life when I became mainly involved with the microscopic roots of the arrow of time associated to entropy. This part is closer to physics. Anyway, the distinction between chemistry and physics is somewhat arbitrary. My later work is dominated by my belief that the direction of time is of fundamental importance in nature. This was the main conclusion of my studies of non-equilibrium thermodynamics.

Please tell us more about the importance of the direction of time.

The 19th century left us with a conflicting heritage. Classical mechanics and even quantum mechanics and relativity are time-symmetrical theories. The past and the future play the same role in them. On the other hand, thermodynamics introduces entropy, and entropy is associated to the arrow of time. So we have two descriptions of nature. Simplifying somewhat, we may say that the first emphasizes "being" and the second "becoming." This leads to many questions. What is the role of entropy and of distance to equilibrium in nature? And a second question is, how does the time-symmetry breaking of entropy relate to the laws of physics?

Thermodynamics had been studied both in far-from-equilibrium and in near-equilibrium situations. A near-equilibrium world is a stable world. Fluctuations regress. The system returns to equilibrium. The situation changes dramatically far from equilibrium. Here fluctuations may be amplified. As a result, new space-time structures arise at "bifurcation" points. We considered the possibility of oscillating reactions as early as in 1954, many years before they were studied systematically. We introduced concepts such as self-organization and dissipative structures, which became very popular. In short, irreversible processes associated to the flow of time have an important constructive role. Therefore, the question that arises is how to incorporate the direction of time into the fundamental laws of physics, be they classical or quantum.

What is the connection between these studies and the research about the beginning of life?

There is an important connection. Life developed under non-equilibrium conditions. Consider first an example. Far from equilibrium, you have chemical oscillations in which millions of millions of molecules change their color simultaneously. This type of coherence is possible only if there are long-range correlations. They occur only far from equilibrium. Similarly, biomolecules, with their complex structures, would be impossible to build in equilibrium conditions. They would have a negligible "probability." This is no more so in far-from-equilibrium conditions. However, the detailed mechanism by which biomolecules appeared is still a controversial problem. But surely, biomolecules are non-equilibrium structures maintained from one generation to the next by self-replication.

There are considerable efforts to find self-replicating simple systems, apart from the large biological molecules.

Indeed, even some very simple models lead far from equilibrium to Turing type structures in which one type of molecule grows, divides, grows, divides, and so on. They have been studied by Harry Swinney here at Austin in the Physics Department and had been predicted by one of my students, John Pearson, some time ago. These are very impressive results. However, there are still conflicting views on the mechanisms which produced biomolecules.

After I described non-equilibrium structures, I turned to the question, "What are the roots of these structures? What is the structure of nature that permits this variety of structures, permits the emergence of this enormous variety of behavior in far-from-equilibrium conditions?" This leads to the question about the relation between dynamics, classical or quantum, and thermodynamics. The relation between "being" and "becoming." I have recently published a "semi-popular" book on this subject (Prigogine, I. *The End of Certainty*; The Free Press, New York, 1997). Note that classical physics was mainly concerned with very simple systems, called integrable systems. Examples are the frictionless pendulum or the two body motion. In these systems, interactions can be eliminated by a suitable transformation of variables. That is the reason why H. Poincaré called them "integrable." But these systems form a small minority. Most systems are not of this type. Already the three-body problem is not integrable. If we start with three bodies interacting through gravitational forces and in a bound state (energy negative), after some time one body is expelled and the two others come closer. But nobody knows which body will be expelled and when this will happen. We encounter already here time-symmetry breaking and emergence of probability concepts. More generally, we come to the question: "What is the mechanism of breaking time symmetry?" Is this through *our* approximations as assumed in the traditional view or do we have to extend the basic laws of dynamics when we deal with situations where we expect the arrow of time to occur.

Simple examples are the so-called deterministic maps leading to chaos. The simplest example is the so-called Bernoulli map. A number between 0 and 1 is multiplied by 2 every second and reinjected into the interval 0–1. We have then "sensitivity" to initial conditions. Two trajectories as close as you want diverge exponentially. There are trajectories but to obtain them for arbitrary times you'd have to know the initial conditions with an infinite precision, whereas you obviously know the initial conditions only with a finite precision. For this reason, we have to introduce a description in terms of probability, which leads also to time-symmetry breaking. Probability is no more the sign of ignorance, it is the outcome of dynamics. Deterministic chaos is really the simplest example. There are many other types of systems where irreversibility emerges. Dr. Tomio Petrosky and I have shown that non-integrable systems in the thermodynamic limit (that is, for large systems) become integrable on the level of probabilities. They unify dynamics and thermodynamics. The problem of irreversibility has many other aspects. Chemists are always interested in excited states. Here also we encounter irreversibility associated to spontaneous decay. Quantum mechanics has let us understand the stable states as eigenfunctions of the Hamiltonian operator but has not given us a description of unstable states with finite lifetimes. That's also one of the problems I'm interested in.

How important was the discovery of the Belousov–Zhabotinsky reactions?

I think it was one of the most important discoveries of the century. It was as important as the discovery of quarks or the introduction of black holes. The significance of the Belousov–Zhabotinsky reactions is in demonstrating a completely new type of coherence. It shows that in nonequilibrium, coherence may extend over macroscopic distances in agreement with our theoretical results I mentioned. Again, at equilibrium coherence extends over molecular distances while in the Belousov–Zhabotinsky reactions this coherence extends over macroscopic distances, of the order of centimeters. This is a striking example of non-equilibrium structures.

There were great difficulties in getting the first paper published.

This was because their results seemed to be in contradiction with thermodynamics. It would have been a contradiction indeed, had oscillations been observed close to equilibrium, but it wasn't. It is far from equilibrium where these long-range correlations appear. The observation of this process is one of the great discoveries of the century.

There seems to be a gap between chemistry and physics in describing symmetry. In chemistry a perfect crystal is an excellent example of symmetry whereas for a particle physicist it serves as a convenient demonstration of broken symmetry.

There are many different ways to consider symmetry breaking. It is true that the crystal breaks the symmetry of the dynamic equations describing the interactions. But this is only an example. For me the most interesting thing is that far from equilibrium you automatically break symmetries. For example, in thermal diffusion, you create a situation where there are different concentrations in the "hot" part than in the "cold" part. Therefore, space symmetry is broken. Another example is the changing role of time. In the Belousov–Zhabotinsky reaction, for example, two instances of time are no longer playing the same role. At one point you have blue molecules, then yellow molecules, then blue molecules again, and so on. This is a time-symmetry breaking, but there are many other possibilities for symmetry breaking. The non-equilibrium structures have opened an entirely new chapter in symmetry breaking, and this may be not so familiar to some physicists.

Let us again emphasize that we observe irreversibility in many situations; in thermodynamical systems, which are large systems, gases or liquids, and so on. We meet also irreversibility in an excited hydrogen atom, because it is interacting with an electromagnetic field and which has an infinite number of degrees of freedom. This leads to the decay of the excited state. Unstable elementary particles present also time-symmetry breaking. If the system would be time-symmetric, the particles would not transform into each other. If you excite a harmonic oscillator, it will remain in the excited state as time is going on. Symmetry breaking comes from decay, from the transition from the excited state to the ground state with emission of photons. In all these cases the time-symmetry breaking is due to resonances. In Poincaré's words, resonances lead to non-integrability. Non-integrability is the basic criterion for irreversibility. They present resonances. For these systems, the basic description is in terms of probability distributions. Trajectories or wave functions become "stochastic realizations." The probability distribution may present more than one single maximum. The experiment may then realize one or another of these maxima. At the cosmological level, there may be universes which would have more positrons than electrons or more antimatter than matter. Our universe happens to have more electrons than positrons, more matter than antimatter. All these bifurcations occur in non-equilibrium conditions.

People like to speculate about the origin of the preference for left-handed amino acids in living matter, etc. Would you care to comment on this?

I have not studied this problem, but one of my former students, Professor Dilip Kondepudi, has performed many interesting experiments. One of the most spectacular experiments was with sodium chlorate, $NaClO_3$. In the liquid phase, the substance is achiral but in the crystalline phase this substance is chiral. When you cool it down, it forms 50% of the "left" and 50% of the "right" crystals. Suppose you stir the liquid lightly. It will form either about 100% of left or about 100% of right crystals. It seems that if the first crystal is left, it infects the others, and the same happens if the first crystal is right. Again, this is a typical case of a non-equilibrium system in which a few molecules determine the fate of millions of millions of molecules. You can perform this infection with a foreign chiral molecule as well. This is a very good example of a bifurcation in which the direction of events is determined by a small fluctuation at the molecular level. This is again typical for far-from-equilibrium situations.

Some people had proposed that the preference of left-handed amino acids may be related to electroweak interactions which stabilize very slightly the left-handed amino acids. Indeed, bifurcations are very sensitive to very small differences of energy. If the transition is going very slowly over the years, then even such very small differences in energy will introduce a slight effect in favor of one of the two amino acids.

There are numerous examples of such bifurcations in biology. One of my former students, Professor Jean Louis Deneubourg, did the following experiment. He took an ant nest and placed a pile of food in its vicinity and connected the nest and the food with two similar bridges. At the beginning, there were about the same numbers of ants on the two bridges. After a while, however, all the ants were on one of the two bridges and the other bridge was empty. Chemical signals encourage ants to take the bridge that the other ants had taken.

I would like to emphasize that bifurcations don't appear, as a rule, alone, in isolation. A bifurcation is followed by another bifurcation, by another bifurcation, and so on. Physics becomes increasingly associated with a narrative element. I like to tell my students that nature is more interesting than the stories of Sheherazade in the *Arabian Nights.* She was telling a story and, after an interruption, another story. In nature we have the cosmological story, inside of which we have the story of matter, inside of which we have the story of life, and so on. The nature emerging now before our eyes is much more unstable and much richer than we would have ever thought.

Can the general public perceive and appreciate this new knowledge?

This is a difficult question. The first part of my work (non-equilibrium physics) had a very positive response. My recent extension of classical or quantum physics is still controversial as many physicists can just not believe that this is possible. For them, classical or quantum physics is "final." I don't believe that there are final theories. Every theory is based on idealizations which may or may not be applicable. But I have not learned about any precise criticism about our work. Anyway, my books have been rather successful. My latest book, *The End of Certainty*, has appeared in 15 languages, so my message must have come across. Also, these ideas have diffused into economic and social sciences as well as into epistemology. Human society is a non-equilibrium interactive system. Of course, we have to be very careful not to fall into the trap of metaphors. However, there are useful analogies. When you go from the Paleolithic age to the Neolithic age, you have an increase of the flow of energy, a different utilization of natural resources and, as a consequence, the emergence of different civilizations. This is, again, a "bifurcation." We are at the point when we are going through a bifurcation again, with the information and communication explosion in our society. We can be certain that humanity will eventually reach a new, different structure but nobody knows yet what it'll be like.

What turned you originally to chemistry?

There had been chemists in my family. My brother had studied chemistry before he had gotten interested in African birds and became a well-known ornithologist. He passed away a few years ago. My father was a chemical engineer. I started by studying archaeology and history and played a lot of music. Then I became interested in philosophy. I read Heidegger, Whitehead, and Bergson. There everywhere I encountered the problem of time.

However, I entered the university in the years before World War II, and everybody was then saying that it was not the right moment to become a philosopher. Since there had been some chemists in my family, I went into chemistry and took up physics at the same time. I soon found out, to my great surprise, that time meant something very different in philosophy and in science. In philosophy, it was considered to be a very difficult question, the basic question of ethics and human responsibility. Indeed, time is our basic existential dimension.

However, when I tried to raise the question of time in talking to my physics teachers, they were astonished and explained to me that the problem of time had been solved by Newton and completed by Einstein, and there was nothing to look for anymore. For me this was a contradiction. My whole life has been devoted to trying to understand and resolve this contradiction. This required a change of our concept on nature. This is my intellectual history.

I suppose that everybody starting with this contradiction between these views on time would have gone in the same direction. What are the alternative possibilities? Einstein said that we are automata but we don't know that we are automata and time is an illusion. Other people, like Descartes or Kant, stated that we are in a dualistic universe. Even in current literature, like in Hawking's *A Brief History of Time*, you find this duality. On the one hand, it is stated that the universe can be understood geometrically, while, on the other hand, there is the anthropic principle, which introduces evolution. My main point is to go beyond this duality and to reach a unified view by extending the theory of dynamical systems.

There seems to be an increasing polarization of the general population between a small group of people who know everything and the rest who are becoming increasingly ignorant in this highly technological society.

This is a very important question. Information used to be scarce. Now we are submitted to a lot of information. It reminds me of a study I did many years ago. When you are on a lightly traveled highway, you do more or less what you want. But when you are on a crowded highway, you are driven by the others and you drive the others. The first I called an "individual" regime and the second, a "collective" regime. Between the two, there is a bifurcation. We have so much information that we are increasingly being driven by information and are driving information. Creativity is in danger. Originality and, in general, thinking become more difficult than in my younger days.

Some 25 years ago, you contributed to a book honoring the 80th birthday of Michael Polanyi. What was your connection to him?

I admired him very much. He was interested in my early work in thermodynamics and invited me to Manchester when he was still Professor of Physical Chemistry. It was some time between 1945 and 1948. It was an exceptional period in Manchester. In addition to Polanyi, there was also Evans and Turing and others. I just happened to be there when he was planning to go to Chicago and he didn't receive the American visa because of his connections with the Communist Party in Hungary when he was very young. The denial of his visa astonished him and made him very angry.

Polanyi made the switch from science to philosophy in 1948. For you, science and philosophy have been together from the very beginning.

I had a better training in philosophy than in science, and it was an accident that I went into science. Many people maintain that the history of Western thought has been an unhappy history due to the conflict between science, which is alienating, and metaphysics, which is antiscientific. There is a lot of antiscience feeling in postmodernism. C. P. Snow discussed the problems of *The Two Cultures.* He says that scientists read poetry but poets and philosophers don't read science. I don't think it is this simple. The scientists' ideas about the universe are difficult for the philosophers to accept.

There is a gap between philosophical thinking and scientific thinking. I hope that my work has contributed to narrowing this gap. We no longer believe that time creates a separation between man and nature. On the contrary, time is related to novelty, to creativity, and you find this everywhere in nature and to a special extent in man.

I take it you are an optimist by nature.

How could I not be an optimist? I'm 81 years old. When I was young, there was Stalin and Hitler and Mussolini. Also, the main overall change over this century, despite the tragedies like the wars and the Holocaust, is a decrease of inequality. There is a decrease of Eurocentrism. We now respect other civilizations. Science and technology, which used to be also the instruments of European imperialism, have become the instrument of diffusion of culture. There have been important advances. Overall, this century is a turning point in human history. But where we go from here, that's a different question and I'm not a futurist.

Anatol M. Zhabotinsky, 1995 (photograph by I. Hargittai).

32

Anatol M. Zhabotinsky

Anatol M. Zhabotinsky (b. 1938) is best known as the Zhabotinsky of the Belousov-Zhabotinsky, or BZ, oscillating reactions. The BZ reactions involve the oxidation of various organic acids and ketones by bromate in the presence of cerium or ferroin ions. Waves of oxidation are easily observed as a color change from red (ferroin) to blue (ferriin). Originally from Moscow, Dr. Zhabotinsky currently does research at the Department of Chemistry of Brandeis University in Waltham, Massachusetts, and that is where we recorded our conversation on July 24, 1995. Dr. Zhabotinsky kindly prepared the material of Boxes 1 and 2.*

Please tell us about your background and what led you to the Belousov-Zhabotinsky reactions.

Although I've done my best work in chemical oscillations, I'm not a chemist by education. I graduated in 1961 from the Department of Biophysics at the Faculty of Physics of Moscow State University. My interest in oscillations came from my father, who was a physicist and started his work with the famous theoretical physicist Leontovich and then worked with Prokhorov, a Nobel Prize winner. My father is now 78 and lives in the suburbs of Moscow but still goes regularly to his Institute. I am 57.

---

*This interview was originally published in *The Chemical Intelligencer* **1996**, *2*(3), 18–24

When I was a boy, I liked biology. However, when it was time to enter university, I had no incentives to study biology. It was the Lysenko period. I wanted to choose a field where I could do some decent work, and the closer it was to mathematics, the cleaner it was. Chemistry was also in a rather poor shape at that time. All fields that were not connected with the military were badly damaged and many were completely destroyed. So, when I entered Moscow State University in 1955, I chose physics.

But in 1958, when I had to choose my field of concentration, the situation had changed dramatically. At that time, there was a "thaw," and all the facets of normal life started to revive step by step, including science. For me, personally, a decisive event was the creation of the Department of Biophysics in the Faculty of Physics.

This was quite an unusual event. Several eminent and influential physicists, like Igor Tamm, were very actively helping to reestablish modern biology in place of the medieval preaching of Lysenko. They supported the idea of creating a biophysics department in the Faculty of Physics because they were of the opinion that it would be easier to teach physicists biology than vice versa. But the initiative came from a group of students in my class. This was completely out of the ordinary in a country where any initiative was permitted only from the top, usually from the very top. On the other hand, several of these students were children of influential members of the Soviet Academy of Sciences, and one was the son of Malenkov, who was the prime minister at that time. The famous mathematician Petrovskii was the Rector of the University, and he was also strongly supportive.

I joined the Biophysics Department, and it was real fun. We had several professors who made us understand the meaning of the real scientific discovery. One of the main attractions was the summer visits to the field laboratory of Timofeev-Resovsky. He gave us brilliant lectures on the basics of genetics and theoretical biology. He was a world-class scientist who had participated in the scientific revolution of the 1920s and 1930s, and knew almost all of its heroes. He was a man of outstanding integrity and also a very colorful person and a terrific storyteller.

After graduation, at the beginning of 1961, I was "distributed" — this is the precise translation of the official Soviet term for the mandatory procedure of sending graduates to designated jobs — to a medical institute which was dealing with cancer treatment. As a biophysicist, I was sent to the radiology department. There I found extreme incompetence all around, and I tried to leave as soon as possible. In the meantime, I spent a lot of time in the library trying to devise my own project.

I decided that I could study biochemical resonance, that is, the enhancement of natural metabolic oscillations by external periodic forcing. I found a couple of publications on damping oscillations in the dark reactions of photosynthesis and in glycolysis which could be a starting point for such a project. During this time, I kept close connections with the Biophysics Department, particularly with my former biochemistry professor Simon Shnol, who was very supportive. When my project proposal was ready, I approached him and asked for his advice. He told me that the project was very interesting but it would be impossible to perform an experimental work with enzymes because the necessary equipment and supplies were not available. Instead of biochemistry, he suggested that I try some oscillating chemical reactions. He told me in so many words that there was a chemist, Boris Pavlovich Belousov who had reported an interesting reaction that oscillates. Many people had seen this reaction, but nobody understood it. Shnol said, "This is a reasonable thing for you to try. I guess it will take a couple of months for you to figure out what is going on. Then you will write a paper. This will be useful for you and interesting for all of us." I agreed.

In the meantime, I managed to change jobs. My second job was again in a cancer research institute, but it was a completely different story. The chief of our department was Leon Shabad, a world-renowned expert in cancer hygiene. He loved strict order and was able to maintain it, and at the same time he had close and good personal relations with all the

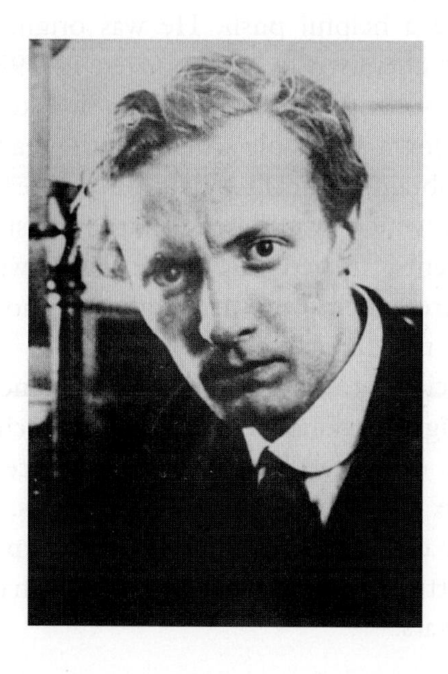

Boris P. Belousov in the 1920s (courtesy of Simon Shnol).

members of his department, including the low-ranking ones, which was totally foreign to the Soviet medical community. He assigned me the task of organizing the analysis of cyclic hydrocarbons in car exhausts by fluorescence and gave me full independence to pursue this project. I purchased equipment, assembled a spectrofluorometer, and increased the sensitivity of the measurements by three orders of magnitude in comparison with the method that was used before. Shabad was happy with my performance, and I had my own little laboratory and a lot of free time for my own research.

About two months after I changed jobs, Shnol gave me Belousov's recipe on a small piece of paper together with a tiny amount of reagents. Belousov's recipe was a really old-fashioned one. Weigh this much of this salt and this much of that substance, dissolve them in moderately diluted sulfuric acid, and add water to the final volume. I did it and observed the oscillations in the solution color. The solution was very hot, and this accelerated the reaction. The period of oscillations was rather short.

I devised a setup to record the oscillations by monitoring the optical density and redox potential of the solution and was able to detect the oscillations over a wide range of concentrations of the initial reagents. But when the time came to find out the chemical mechanism of the oscillations, I was uncertain about how to start, because I was rather ignorant in inorganic and organic chemistry, having taken only a general chemistry course in my university years. At that moment Lev Blumenfeld, the Chairman of my Biophysics Department, gave me a helpful push. He was originally a quantum chemist — before quantum chemistry was eliminated in 1950 — and very bright and very quick in solving problems. When I asked his advice, he almost immediately wrote down a possible scheme of the reaction. I started to check his scheme and obtained the first data on the reaction mechanism. Eventually, his scheme, which was based on the most well known reactions of the major reagents, turned out to be 100% wrong. In spite of that, it was really of great help, and it taught me how to proceed when confronting a really new problem.

Meanwhile, the Soviet Army decided to improve its performance by drafting all the young men with higher education diplomas in technical and natural sciences who were not exempted from military service. So, I became a graduate student at the Institute of Biological Physics, with Shnol as my adviser, in order to be exempted from conscription. In fact, I continued to work in Shabad's department, and practically nothing changed in my routine for the next two years.

When I wrote my first manuscript, I encountered an ethical problem. I was unaware of Belousov's publications, and I had not received his recipe directly from him. So I decided to put his name in the subtitle of the paper, which read "A study of kinetics of the Belousov reaction," and gave the manuscript to Shnol. I wanted him to be a co-author because it was his idea to do this study and he was my official adviser. He declined, saying that he had done nothing, but he was pleased to see the manuscript completed. I also asked him whether I could send the manuscript to Belousov before submitting it for publication. He told me that we can send it by the same route that the recipe had come.

About two weeks after having sent the manuscript, I received it back with a nice short letter from Belousov saying that he was glad that somebody continued the work that he was too busy to carry on. He also attached a manuscript of his own in which I found a reference to the only published communication by Belousov on the oscillating reactions. I immediately went to a library and found his communication in a booklet entitled "Short Communications on Radiation Medicine," an obscure and very unlikely place for such a paper. The booklet was published by a medical research institute where Belousov was the head of the analytical chemistry laboratory. I added a reference to Belousov's paper to my manuscript and sent it off to the Russian journal *Biofizika*. The paper appeared two years later.

Was there any difficulty in getting your paper published?

No. It was only slow but that was normal at that time, if you didn't have any connections to somebody on the editorial board.

But I have heard that Belousov had had great difficulties in getting his original discovery published, and this is why his communication ended up in this obscure unrefereed publication.

As far as I know, Belousov had sent his manuscript to two chemical journals in succession, both Russian of course, and in both cases it was rejected. I sent my manuscript to a biophysical and not a chemical journal. At that time chemists believed that the oscillating reactions contradicted the second law of thermodynamics (Box 1), but biophysicists were unaware of that.

Box 1
Are Homogeneous Chemical Oscillations Possible?

The first observation of chemical oscillations was made at the end of the 17th century by Robert Boyle. In the 19th century, several oscillating chemical reactions were discovered, mostly heterogeneous ones. At that time, oscillating chemical reactions did not attract much attention. Chemical kinetics did not yet exist as a branch of science and there was no commonly accepted view of the "proper" way for a chemical reaction to proceed.

At the beginning of this century, attempts were made to understand the mechanisms of chemical oscillations. However, the mechanisms of heterogeneous reactions are too complicated since they involve phase transitions and transport processes. Homogeneous reactions are much simpler, and it was natural to begin theoretical study with them.

In 1910, Hirniak proposed that cyclic reactions can oscillate. He used the simplest example, the cyclic interconversion of three isomers (Fig. 1). If clockwise reactions are relatively rapid and counterclockwise ones are relatively slow, it is possible to observe damped oscillations in the system. Really it seems evident that, if one puts all the molecules in form X_1, then most of them will be converted to X_2, then to X_3, then back again to X_1 and so on.

However, it was shown very soon, that detailed balance puts strong restrictions on the rate constants in this system: the product of the clockwise constants must be equal to the product of the counterclockwise constants:

$$k_{12}k_{23}k_{31} = k_{13}k_{32}k_{21}. \quad (1)$$

This condition immediately forbids any oscillations in the system. Later, it was shown that it is impossible to have any oscillations in the vicinity of the thermodynamic equilibrium state. This thermodynamical analysis made a very strong impression on the majority of chemists, who interpreted it as being valid for all homogeneous closed chemical systems.

At approximately the same time, Lotka proposed his famous models of oscillating chemical reactions based on irreversible autocatalytic processes. The first model included one autocatalytic step and gave damped oscillations. The second model became a paradigm in oscillating chemistry. It consists of two consecutive autocatalytic steps, resulting in undamped oscillations. The Lotka models attracted great attention from theoretical biologists, because

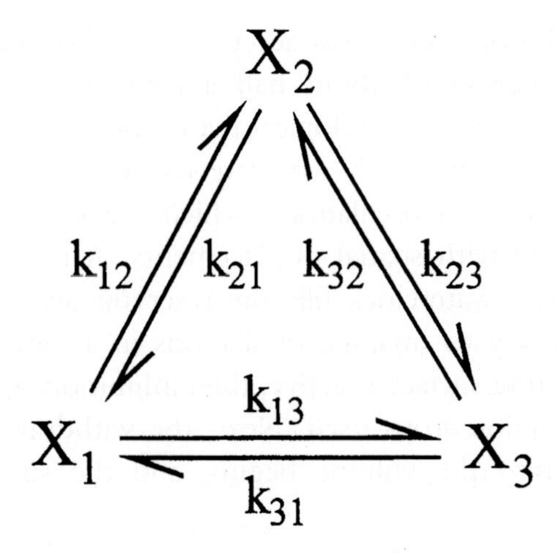

Fig. 1 Scheme of the cyclic reversible isomerization.

they connected two important features of living systems: multiplication and oscillations in population densities. Chemists, as a rule, did not accept the Lotka models, arguing that the corresponding overall mechanisms were oversimplified. It is, in fact, quite easy to write a complete chemical scheme that can be reduced to a Lotka model. However, this was done only several dozens years later.

In 1921, Bray published the first description of an oscillating reaction in the liquid phase, the catalytic decomposition of hydrogen peroxide under the influence of iodate ion. Amazingly, the initial response of the chemical community, instead of undertaking a normal study of the reaction, was to try to prove that the cause of the oscillations was some unknown heterogeneous impurity.

Thus, from the 1920s until the mid-1960s, it was the unfortunate custom to ascribe all concentration oscillations observed in chemical and biochemical systems to some invisible but important heterogeneous processes or simply to technical errors. The majority of chemists believed that oscillations were impossible in homogeneous chemical systems. They believed this in spite of the clear statement by Bonhoeffer in 1948 that there is no fundamental difference between homogeneous and heterogeneous systems. Bonhoeffer pointed out that heterogeneous systems are more convenient for oscillations, but no more than that.

What was the origin of this incorrect idea of the impossibility of chemical oscillations in homogeneous systems? As one can see, there is no logical basis for this conclusion, yet it was accepted by an overwhelming majority of chemists for a period of about half a century.

One can surmise that a psychological reason played a main role. It seems likely that people subconsciously associate oscillations with movements in space. Indeed, mechanical oscillations, which are the most familiar ones, are always connected with spatial displacements. A pendulum swings here and there, water in a waterclock fills the reservoir and then leaves it, and so on. Thus, it is easy to imagine oscillations in a heterogeneous system. For instance, a reacting surface is active, then inhibition by reaction products occurs, and the reaction stops itself. Next, the withdrawal of the inhibitor from the surface into the volume begins, and the surface becomes free and active again.

In a homogeneous system, there are no movements in space, only variations of concentrations in time. "Common sense" argues that oscillations need movement in space. Thus, if some plausible reasoning implies that the homogeneous phenomenon cannot exist, "common sense" is ready to agree with it, even without checking the correctness of the reasoning.

Prepared by Anatol Zhabotinsky

===

When did you meet Belousov?

I never met him.

Didn't you try to meet him?

I wanted to meet him, especially later. In the meantime we communicated by mail and by telephone. Then at one point I told him that I would really like to meet him any time and any place that would suit him. He was very kind and told me that he would like me to come to his home but at that time it was under reconstruction. Later there were similar exchanges, and eventually I understood that there would never be such a meeting.

Did you know anybody who had met Belousov in person?

Yes. Shnol did meet him once. When he asked Belousov about his reluctance to meet with me, Belousov told him that he had had several good friends and he lost all of them, and he didn't want to make new friends.

How did he lose his friends?

He was a military chemist and he was high-ranking, something like major-general. At the end of the 1930s, about 90% of the top-ranking officers were executed in the Soviet Union, and many of his colleagues must have been among them.

When was he born?

In the 1890s, I don't remember the precise year. In 1980s, the best literary magazine in Moscow, *Novii Mir* (*New World*), published an article on his life. What I remember is that he had been involved in some revolutionary activities before 1917 and was forced to emigrate. Then he entered the Zurich Polytech. However, he didn't have enough money to get his diploma. After the revolution he returned back and became, as I said, a military chemist. For a long time his lack of a formal degree didn't cause any problems, but after World War II regulations changed and formal certificates became important again. His status as a laboratory head was confirmed by high authorities, but his scientific career may have suffered because he didn't have a formal degree.

When did he die?

In 1973.

Let's return to your own career. Your first paper was published in the Soviet periodical Biofizika. *What was the next one?*

My second paper was also published in 1964. It appeared in the *Proceedings of the Academy of Sciences of the USSR*, and later it became a Citation Classic.

What was the essence of your work?

First I reproduced Belousov's results. Then I modified the system and made it much more convenient to study. I showed that it was self oscillating and determined several key steps in the reaction mechanism. After that, I wrote my first paper. However, I did not have a complete understanding of the chemical mechanism of the oscillations. Then I found the last key part of the mechanism, namely, what species was the inhibitor. It gave me a qualitative understanding of the entire chemistry involved in the mechanism (Box 2), and I was able to find a large set of similar oscillating reactions with various reactants and catalysts. At that moment I felt that I really had it done. So I wrote the second manuscript, and I wanted it to appear fast. For me, there was only one way to do this — to have it published in the *Proceedings of the Academy of Sciences*. For this purpose, I needed a full member of the Academy of Sciences to present my manuscript. I decided to approach Frumkin, a renowned physical chemist and director of the Institute of Electrochemistry, who himself had published a couple of papers on electrochemical oscillations. This was at the beginning of 1964. It took me two months to make an appointment with Frumkin, but once he had given his approval, it took only three more months for the paper to appear. Frumkin was known to read everything before he would sign anything. Perhaps this is why he asked me not to bring any more manuscripts to him, and he showed me the high stack of manuscripts on his desk, submitted by members of his institute. He was very nice indeed.

Box 2
Mechanism of the BZ Reaction

Figure 2 shows the basic mechanism of the cerium-catalyzed BZ oscillating reaction. The reaction consists of two main parts: the autocatalytic oxidation

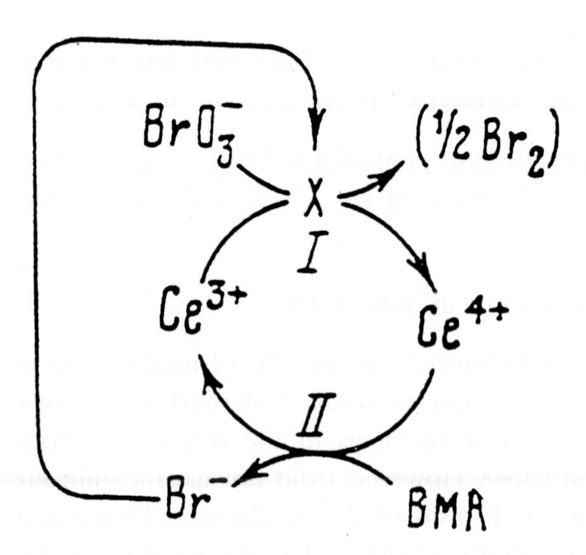

Fig. 2 The first mechanistic diagram of the BZ reaction. X is $HBrO_2$ or BrO_2 and BMA is bromo-malonic acid.

of cerous ions by bromate, and the reduction of ceric ions by malonic acid. Bromine derivatives of malonic acid are produced during the overall reaction. The ceric ion reduction is accompanied by production of bromide ion from the bromine derivatives. Bromide ion is a strong inhibitor of autocatalytic oxidation owing to its rapid reaction with the autocatalyst, which is presumably bromous acid or some oxobromine free radical.

An oscillatory cycle can be qualitatively described in the following way. Suppose that a high enough ceric ion concentration is present in the system. Then, bromide ion will be produced rapidly, and its concentration will also be high. As a result, autocatalytic oxidation is completely inhibited, and the ceric ion concentration decreases owing to the reduction of ceric ion by malonic acid. The bromide ion concentration decreases along with that of ceric ion. When the ceric ion concentration reaches its lower threshold, the bromide ion concentration drops abruptly. The rapid autocatalytic oxidation starts and raises the ceric ion concentration. When this concentration reaches its upper threshold, the bromide concentration increases sharply, completely inhibiting the autocatalytic oxidation. The cycle then repeats. Pulse injections of Br^-, Ag^+, and Ce^{4+} result in phase shifts in oscillations of the Ce^{4+} concentration (Fig. 3), which confirm the mechanism.

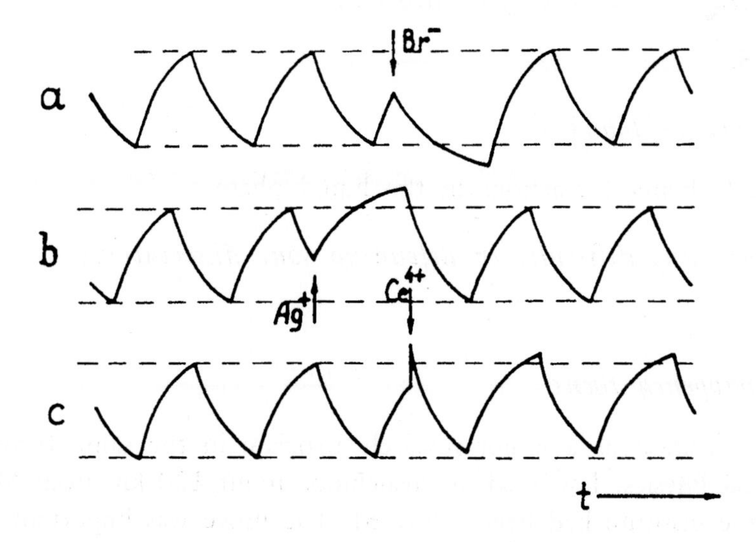

Fig. 3 Phase shifts in oscillations of ceric ion concentration caused by pulse injection of Br^-, Ag^+, Ce^{4+}, shown by arrows.

The first mechanistic scheme of the BZ reaction was published in 1972 by Field, Körös, and Noyes:

$$H^+ + Br^- + HOBr \leftrightarrow Br_2 + H_2O \quad (R1)$$

$$H^+ + Br^- + HBrO_2 \leftrightarrow 2HOBr \quad (R2)$$

$$H^+ + Br^- + HBrO_3 \leftrightarrow HBrO_2 + HOBr \quad (R3)$$

$$2HBrO_2 \leftrightarrow HOBr + HBrO_3 \quad (R4)$$

$$H^+ + BrO_3^- + HBrO_2 \rightarrow 2\ BrO_2^{\cdot} + H_2O \quad (R5)$$

$$H^+ + BrO_2^{\cdot} + Ce^{3+} \rightarrow Ce^{4+} + HBrO_2 \quad (R6)$$

$$Ce^{4+} \rightarrow Ce^{3+} + Br^- \quad (R7)$$

The FKN scheme leaves one problem unsolved, namely the chemical mechanism of the bromide production. Instead, the scheme postulates that production of Br^- is proportional to reduction of Ce^{4+}. The scheme gives a good qualitative description of oscillations in the BZ reaction and remains at the core of more detailed schemes proposed later.

Prepared by Anatol Zhabotinsky

==

When did you complete your dissertation?

In 1965.

What was the topic?

Periodical chemical reactions in the liquid phase.

Was there any difficulty in defending your dissertation?

None.

What happened then?

I got a position as a junior research associate in the same Institute of Biological Physics. I worked in Pushchino, about 120 km from Moscow, where the institute had been relocated. The move was important for me for a personal reason. I was already married and we were expecting a child, and there was no hope for an apartment of our own in Moscow. In Pushchino, we got an apartment.

Did you continue working with oscillating reactions?

Yes. However, my main interest was not to study them by themselves but to use them for modeling analogous biological processes. The most attractive topic was propagation of excitation in the heart. Everyone knew that the normal operation of the heart is controlled by very long waves of excitation spreading from the sinus pacemaker. A theory existed that attributed the most dangerous cardiac arrhythmia to the emergence of short spiral waves of excitation in the myocardium. But very few believed in it. We started to study chemical waves in thin layers of solutions containing the oscillating reaction and found that they formed wonderful target and spiral patterns which were relevant to those in the myocardium. Our first paper on chemical waves in two-dimensional media appeared in *Nature* in 1971. Two years later, I returned to Moscow for personal reasons.

Where did you go to work upon your return to Moscow?

I couldn't find a position in the Academy of Sciences. Finally, I found a job in the new Institute of Biological Tests of Chemical Compounds, which belonged to the second-rank Ministry of Medical Industry. I was appointed Head of the Laboratory of Mathematical Modeling, which began with three people and gradually became a nice working group of about a dozen graduates of two biophysics departments, those at the Physical Faculty of MSU and at the Moscow Institute of Physics and Technology. We did interesting work on biochemical regulation in red blood cells. Then my interest shifted to attempts to employ my expertise in oscillations to the improvement of cancer chemotherapy. We discovered a resonance response of dividing cells to periodic administration of highly toxic anticancer drugs. This phenomenon is promising for the development of more efficient chemotherapy for cancer and probably for AIDS.

Despite very limited resources, things were going pretty well until 1985, when bureaucratic wars started to destroy the institute. After a couple of years, the situation became unbearable and I decided to leave. However, it was very difficult to find a new position. Only in 1989 was I able to join the National Scientific Center of Hematology, where I hoped to apply our results on resonance in dividing cells to the treatment of leukemia.

However, it turned out that I was involved in an endless reorganization, and finally I became the formal head of a large nonhomogeneous department, unable to perform real research. The overall situation in Soviet science deteriorated fast during the last years of existence of the Soviet Union. So I was happy when, in 1991, Irving Epstein offered me a one-year visiting position at Brandeis University.

What's your position at Brandeis?

I'm an Adjunct Professor of Chemistry, and my salary comes entirely from grants. But this is the only drawback — that I don't have a permanent position at my age. This is a very good place to work, and the people are also very nice.

Can you now devote all your time to do the things that interest you most?

I have returned to my original field, and now I am working with chemical waves and patterns. During the past four years, we have obtained a lot of interesting results. However, I would like to resume work on biological kinetics, and I believe that I will be able to do this in the rather near future.

Have you ever been treated as a celebrity?

No. You know, for instance, here in the Boston area, for the majority of people the local basketball or baseball team, which is not the best in the country, means much more than Harvard and MIT together.

Of course, I meant only among chemists.

Sometimes.

Did you get any official recognition for your work back in the Soviet Union?

I received the Lenin Prize.

You and who else?

I and three other members of the Institute of Biophysics.

Was Belousov included?

Yes. He became the fifth winner. His name was absent in the original nomination and was added during the selection process.

When was this?

In 1980.

But by then he had already been dead for seven years.

Yes, Belousov shared the prize with us posthumously.

Richard N. Zare, 1999 (photograph by I. Hargittai).

33

RICHARD N. ZARE

Richard N. Zare (b. 1939) is Marguerite Blake Wilbur Professor in Natural Science at Stanford University in Stanford, California. He got his B.A. in 1961 and his Ph.D. in 1964, both from Harvard University. Following appointments at the University of Colorado and the Massachusetts Institute of Technology, he was Professor of Chemistry at Columbia University between 1969 and 1977 and has been at Stanford University since 1977. He is a Member of the National Academy of Sciences of the U.S.A. and a Foreign Member of the Royal Society (London). He was Member (1992–1998) and Chair (1996–1998) of the National Science Board and Chair of the Commission on Physical Sciences, Mathematics, and Applications of the National Research Council (1992–1995) and served on the Council of the National Academy of Sciences from 1995 to 1998. His many awards include the Michael Polanyi Medal (1979, England), the National Medal of Science (1983), the Irving Langmuir Prize (1985), the Peter Debye Award (1991), the National Academy of Sciences Award in Chemical Sciences (1991), the NASA Exceptional Scientific Achievement Award (1997), and the Welch Award in Chemistry (1999). Our conversation was recorded at Stanford University on May 13, 1999.*

*This interview, somewhat abridged, was originally published in *Chemical Heritage* **2002**, *20*(2), 32–35.

Owing to your past and present administrative positions, you are one of the most visible personalities in chemistry in the United States. What started your interest in chemistry, and how did that cross over to an interest in public science policy?

I like to tackle problems that matter. It seems to be a well-kept secret but the amount of effort it takes to solve a really significant problem is not much more than an ordinary one. Let me tell you something about how I got the way I am.

My father was a failed chemist. He was studying to get a Ph.D. in organic chemistry at Ohio State University in the 1930s, working in overtone spectroscopy as a way to determine different kinds of C-H groups, long before NMR made such pursuits routine. He dropped out during the Depression and went from one job to another. I always wanted more love and recognition from my father, who seemed incapable of showing much emotion most of his life. He was a second-generation American. All my grandparents came from Russia and from Poland. Our family was poor and dysfunctional. My parents always fought with each other, and they separated after all the children left the nest. When I was about 10–12 years old and looked at some of my father's old chemistry books, my parents warned me that they would lead to nothing but unhappiness and urged me not to read them. This parental discouragement became its own driving force. So I took this forbidden literature to bed and read the chemistry books under the covers with a flashlight.

I did not make friends easily at school. I was antisocial and was treated as such. When I asked my parents to buy a chemistry set for me, they refused. Instead, I made friends with a pharmacist who supplied me with various ingredients for making explosives. By then, my parents understood that chemistry was what I wanted to do.

My mother never finished college and went through all her life feeling badly about it. Consequently, she read and read. She was reading books, trying to recapture something that she lost. She became more learned than most because she felt that she was never given the chance. Her love of learning affected me at an early age.

I was born in Cleveland, Ohio, but my family moved soon to Cleveland Heights, which was then a Jewish ghetto. Later, when I was about 8 years old, we moved to a farm area, Lyndhurst, Ohio. We were the only Jewish family there. Christmas was coming and the third-grade teacher asked why I was not singing any of the Christmas carols with the rest of the class.

I said to her that I did not know the words and that the words were false. I was sent to a corner of the classroom and punished by the teacher. Later, after school, I was beat up by my classmates. It was a traumatic experience for me, one that made me feel estranged from society.

Around the eighth grade, I was spending all my free time in the Cleveland Public Library. I got into a huge fight with my junior-high-school science teacher, who had just recently graduated from a school of education and did not know much science. He had been instructed, however, that he must be able to answer all questions because to do otherwise would mean a loss of classroom authority. I was at an age at which I was going to show him not only that there were a lot of things not known but that many of the things he said were not the way he said they were. The class loved this fight and the teacher told my parents in a parents-teacher conference, "I hate his guts." The school decided that the best thing for me was to leave the public school system. Fortunately, I won a scholarship and started tenth grade in the University School for Boys. It was an exclusive private school in Shaker Heights, Ohio. Part of the reason for my winning their scholarship was that it satisfied their Jewish quota, but I did not know that at the time.

In my senior year in high school, I built a binary computer. At its height, it was able to multiply 3 (binary 11) times 7 (binary 111) and get 21 (binary 10101), provided that the relays did not get stuck. It actually won a prize, which allowed me to go to Los Angeles to the U.S. National Science Fair with my high-school science teacher. On this trip, I met a distant cousin of mine, Dr. Abe Zarem, who had taken a second mortgage on his house so that he could start a scientific instrument business. He told me to come to see him at binary 9 Green Street in Pasadena. I had to figure out that the address was 1001 Green Street. He was a mentor and tormentor, and he had a huge impact on me. He posed a number of challenging problems for me to solve. He said that these problems would help me to judge whether I had the talent to pursue a career in science. At the time, the questions were quite beyond me. It would take through the end of four years of college for me to come up with approximate solutions. But he got me hooked onto scientific thinking.

When it was time to go to college, the headmaster at my high school urged me to go to Swarthmore College, but Swarthmore would not have me so I went to Harvard with a full scholarship. MIT, Caltech, and the University of Chicago accepted me, but Harvard was the one that paid the most, and my family and I needed as much financial aid as possible.

At Harvard, during my freshman year Bill Klemperer taught a course in quantitative analysis. He hated it and said that the class really ought to learn about atomic absorption spectroscopy rather than perform tedious gravimetric analyses. I told him that I wanted to learn about spectroscopy. He gave me a book by Herzberg, and I spent the next summer reading that book. I chose Klemperer as my undergraduate adviser and decided to take a double major in chemistry and physics. That summer, which was a turning point for me with respect to chemistry, I also built a homemade spectrograph that used a carbon arc filled with various chemical salts to record crude atomic spectra of the elements.

In my junior year, I got a summer job with a company later to be called Cleveland Crystal Corp. I published my first paper in *Nature* in 1959 on the absolute configuration of the cadmium selenide crystal, based on X-ray intensities. I was also interested in computers, but I understood that with the computer I would be working on problems for others whereas I wanted to work on my own problems. Being able to carry out a successful research project with my own hands gave me the courage and the conviction to pursue a career in experimental chemistry.

Before deciding on my graduate studies, I went to see Bill Klemperer, George Kistiakowsky, Bill Lipscomb, and E. B. Wilson, seeking their advice. Each one of them told me that there was not much to choose from outside of Harvard. But they all mentioned a former Harvard Junior Fellow, Dudley Herschbach, who was trying to set up crossed molecular-beam experiments at Berkeley. I went to Berkeley but ended up with a Harvard Ph.D. because within two years, Dudley was called back to be a full professor at Harvard. Herschbach proved to be the most influential person in my life, and I dedicated my book *Angular Momentum* to him.

With Herschbach I worked on photodissociation dynamics and fluorescence. When I finished my Ph.D., Dudley arranged a beginning faculty position for me with A. C. Cope who was then Department Head at MIT. It was a typical old-boy-network arrangement, which would be impossible today. However, I did not stay around and told Cope that I would take the faculty position in another year. I went out to Colorado to the Joint Institute for Laboratory Astrophysics (JILA), at the University of Colorado. I was a postdoc there with Gordon Dunn and Ed Condon. At JILA, I got my real first experience with experiments. I returned to MIT after a year, but I only spent nine months there. In the intervening time, Cope had been removed from being Head, and the MIT chemistry department was squabbling and in disarray. When I went to see the provost, Jerry

Wiesner, asking for help in getting access to a machine shop for making parts in stainless steel, he told me to be patient. And I told him that I did not come to MIT to be patient; I came to do the best science I could do. I also told him about an offer to return to JILA, and Wiesner told me, "No one leaves MIT to go to the University of Colorado. This is not a credible offer." That was 1965. When I put a resignation letter on his desk, I immediately got access to the machine shop. But I decided not to look back and I went off to Colorado anyway. At the age of 29, I became full professor at Columbia University, and then I came to Stanford in 1976 and have stayed here ever since. I work every day of the week, I am a workaholic, but it is not work for me, it is play. I marvel that people pay me for it.

Public policy has developed into my hobby. It is a way to give back something to the community that has given me so much. I was elected to the National Academy of Sciences and then got involved with the National Research Council (NRC). I proposed, with Jerry Berson from Yale, the idea that we should worry about prudent practices and safety in the chemistry labs. I got invited to chair a special study about how to establish national centers of science and technology. These are just two examples. Later, I also became Chairman of the Commission on Physical Sciences, Mathematics, and Applications of the NRC. That commission has boards in the fields of mathematics, physics, and chemistry, overlooks NIST and NASA, and gave me an insight into how the government works. NRC comprises the National Academy of Sciences, the National Academy of Engineering, and the Institute of Medicine. I had previous experience on the Scientific Advisory Board of IBM, where Paul Flory and I were the two chemistry members. Getting the experience of understanding how such a company works broadened my perspective of the world very much.

In 1992, President Bush appointed me to the National Science Board, which is the policy making body of the National Science Foundation (NSF). There has been a shift in the priorities of NSF. Concerning education, we must get away from an "ego" system to an "ecosystem." We cannot have so many institutions in this country that want to be "number one" in research. This is not good for the country. We want to have the strength of all the regions developed so that all the human talent in this country has a chance. We need community colleges, we need liberal arts colleges, we need research universities, and most of all we need coupling between them.

Could we now move to your research?

I am most excited about what we are doing in the lab for two basic reasons. One is discovery, and the other is working with people. It is my extended family. My research group consists of about 30 people, mostly graduate students. There are about 12 projects. My annual budget is over a million dollars, coming from NSF, the Air Force, the National Institutes of Health, the Department of Energy, the Army, and NASA, and also from various companies.

The main characteristic of my research is inventing ways of solving problems. I am a frustrated inventor; I like to make instruments and come up with new ways of doing things. I am very interested in how chemical reactions occur, collision by collision, in the gas phase. For some time, I have been in the detective business in looking at chemical reactions. The "clues" are provided by lasers, which not only measure how the molecules vibrate and rotate but also measure how fast they are traveling and how

Richard Zare and David W. Chandler in the early days (courtesy of Richard Zare).

they are oriented in space. When you have made enough measurements of different types, you can make correlations that lead back to what is happening at the molecular level.

In general, detailed studies of chemical reactions can be done in two different ways. There is the voyeuristic approach, just watch them happen. Unfortunately, they happen so fast that it is hard to get the timing right. Then there is the detective approach. You go to the scene of the traffic accident, you examine the wreck and measure the skid marks, and you imagine what it all was like before the accident occurred and what actually happened to cause the crash. I am much more involved in "before and after" research, trying to understand what happened in between than watching the chemical collisions take place in real time.

An example is the fundamental reaction of a hydrogen atom with a hydrogen molecule. To keep track of the players, we make it a hydrogen atom, H, with a D_2 molecule, forming HD + D. We prepare the hydrogen atoms with a laser beam, by photolyzing some precursor, like HBr, to produce fast H atoms. We detect the HD product, again with a laser, by making it absorb light at different wavelengths and turning it into an ion, which we can then count. We do all this with state preparation of the reactants and state resolution of the products and by measuring the laboratory speed distribution of the state-resolved products — enough information that we can also extract the angular distribution of the state-resolved products as well.

I have been involved in using lasers for detecting things, laser-induced fluorescence, resonance-enhanced multiphoton ionization, many schemes. Other investigators have been shooting beams at each other and then running around with a detector to determine the angular distribution. We have found a much simpler way that gives a larger signal when it can be applied.

Instead of shooting two beams at each other in the laboratory, we keep all the reagents together in one beam so that they are traveling at the same speed. We photodissociate one of the reagents to produce fast atoms. Because we know the energy of the light we put in, we know how fast the atoms recoil in the photolysis step. For example, we make HBr fall apart to an H atom and a Br atom. We know how fast that H atom is moving in the laboratory. We know if it collides with a D_2 molecule, with what speed it strikes the D_2. I am trying to be very careful in what I am saying. I am saying speed, that is, the magnitude of velocity.

HD is being formed in some vibrational and rotational state. Because we understand the energetics, the thermodynamics of this reaction, we also know how much energy is left over for translational motion, so we know what speed the HD molecule must have in some specific vibrational-rotational state. Then we measure a third speed in the laboratory, the speed of HD in that vibrational-rotational state moving in the laboratory. The speed of the center of mass plus the speed of HD in the center of mass makes the laboratory speed. Then, by knowing three sides of a triangle, we also know all the angles in the triangle, including the angle describing how the HD is scattered in the center-of-mass frame. By measuring the speed distribution in the laboratory, we are also measuring the angular distribution in the center of mass. We have found the way to get the differential cross section, to know what molecules appear at what angle, by never crossing molecular beams.

Let me give you another example. What happens when a laser beam is striking a molecular surface? I have found out that the laser can heat the surface at a very rapid rate of about 10^8–10^9 degrees per second. Of course, this does not last long; otherwise, we would have a thermonuclear reaction! But even for a millionth of a second, the temperature jump is huge, and the molecules hop off the surface into the gas phase, often without fragmenting. Then we come in with a second laser and we ionize those molecules that absorb a particular color of that second laser. Because we now have ions, we can do time-of-flight mass spectrometry. Laser-desorption/laser-ionization mass spectrometry is an invention of ours, and many other people have worked on this technique too.

With this technique, we can investigate sediments that you get by dredging a harbor. In the sediment we find organic molecules; some may be only on the outside of sediment particles, and others are permeating it. These questions have important implications for the environment. We need to distinguish between contamination and pollution, that is, whether the organics are sequestered rather permanently on the particle or are available for easy release into the biosphere. Using the same technique, we are also looking at particles that come to our planet from outside the Earth, so-called interplanetary dust particles.

Projects are initiated sometimes from the most unexpected directions. Once in the question-and-answer period following one of my talks at a national American Chemical Society meeting on the sensitivity of laser-induced fluorescence in detecting reaction products, a Dr. Larry Seitz, from the Manhattan, Kansas, test station of the U.S. Department of Agriculture, asked me whether I could detect aflatoxin. It is one of the most potent carcinogens and is found on all types of grains when they mold. Moldy grain leads to all types of cancers. Aflatoxin is labile, it is not easy to put it into the gas phase, but it does fluoresce. Aflatoxin is prevalent in moldy peanuts, and it can be deadly. Not only did I lose my appetite upon learning about this, I also saw the importance of detecting such materials. This led to a redirection of my research and helped me to get started in applying laser techniques to look at trace species. This was about 1975 at Columbia University.

We can now look at single molecules in room temperature solutions and see them by laser fluorescence spectroscopy. We can also capture them with light and move them about. The whole area of single-molecule detection is exciting because it breaks the averaging that you normally get when you look at ensembles. It allows you to see rare events that otherwise get lost in the average over observation of all the members of an ensemble.

An example of the power of this technique was our study of a single dye molecule in solution. We noticed that as we increased the laser power, we got more light out, more bursts of fluorescence. Indeed, the fluorescence intensity was at first directly proportional to the laser intensity. As the laser intensity was increased, however, we noticed that the molecule was getting "saturated" by the light, that is, the intensity of reemitted fluorescence was not keeping up with the intensity of the exciting light. For an ensemble of molecules in solution, complete saturation means that half of the molecules get into an excited state and the other half remain in the ground state. But for a single molecule it means that you are pumping energy into the molecule so fast that it does not have time to return to the ground state to absorb another photon. If this is true, the saturation behavior can be calculated because we know the lifetime of the excited state and we should be able to predict the saturation curve of how the fluorescence intensity varies with the excitation intensity. We made a prediction but it was off by a factor of 5. After a while we discovered that what people meant by the lifetime of the excited state was not the lifetime of an excited singlet state. Instead, the lifetime referred to the time that it took the molecule to return to the ground state on average. What we had failed to include was those rare times the molecule crossed from the excited singlet state to the lowest-lying triplet state. Once in the triplet state, the molecule essentially went to sleep. We had to wait a considerable time before the molecule woke up and crossed back into the ground state and came back into play. This example is just one vivid instance where you can learn secrets of Nature by following the time history of a single molecule, secrets that would remain hidden by looking instead only at snapshots of many, many molecules all at once. I could go on and on about other projects ...

I have one final question for you, about your role in analyzing the Martian meteorite found in 1984 in Antarctica. Do you really believe that your work shows that primitive life forms existed on Mars long ago?

What a painful episode in my life, one not totally of my own planning. My research group was working with Kathy Thomas-Keprta of NASA Johnson Space Center on interplanetary dust particles and cluster particles. She asked us as a favor to look at two rocks that were given the Walt Disney character code names of Mickey and Minnie to keep track of them and tell them apart. My graduate student Simon Clemett and I did not know what they were, but how could you possibly refuse such a request from a collaborator and friend. After all, looking at particles that you could only see under a microscope was so much more of a challenge, so the mass-spectrometric analysis of these samples should pose little problem. Simon found that the samples contained polycyclic aromatic hydrocarbons (PAHs), a class of organic molecules we had detected previously in interplanetary dust particles and in many different meteorites. What was so peculiar and special about these PAHs was that they were concentrated in the rims that enclosed carbonate globules in this basaltic, volcanic rock. When I told Kathy, she was quite excited and told us that Goofy would be sent to us in the mail at once and that we should stop everything and devote ourselves to looking at Goofy. I said I was doing nothing of the sort until I knew more about what we were actually working on. It was then that she told me that David McKay, Everett Gibson, herself, and others at NASA Johnson Space Center were secretly studying these samples from a Martian meteorite. Secrecy was necessary, she said, because this project had not been authorized by NASA and because of what they thought they might be finding. Simon and I were told that in the iron-rich rims that surrounded the carbonate globules, they were observing magnetite grains that looked like they came from biological activity and transmission electron microscope images that resembled fossilized bacteria-shaped objects. These findings, all located together, seemed to point to the possibility that some primitive form of life had been inside this meteorite. This meteorite had fallen to Earth in Antarctica about 13,000 years ago and had only been classified quite recently, largely based on isotope ratios, as one of the few, rare meteorites that came from Mars. What a wonderful sample we had. It would be the closest I imagine that I would come to Mars.

Our contribution to this project was to find and characterize these PAHs. In themselves, they could be biomarkers of fossilized organic matter of long ago, or they could have come from some nonliving source, such as metal-catalyzed reactions of organics on hot surfaces. If we could establish these PAHs as indigenous to the meteorite, then they would be the first observation of organic molecules found from Mars.

The rest is history, as they say. We wrote a paper titled "Search for Past Life on Mars: Possible Relic Biogenic Activity in Martian Meteorite ALH84001." This paper was ultimately published in *Science* after a lengthy and useful peer-review process, but not before the story was leaked to the news media and NASA decided to hold a press conference in Washington, DC. The reaction of many of my colleagues was to the press conference and to the hype surrounding it as opposed to the actual paper, which was to me quite disappointing. And the appetite for the press on this topic seemed insatiable. I refer to this period in my life as "Mars madness." It was exciting but also often unpleasant and stressful.

It was my intention in our *Science* paper to put forward a hypothesis — a hypothesis deserving serious examination, not a matter to be debated in the press but to be thoughtfully investigated. I believe that this hypothesis has not been refuted, but neither has it been confirmed. If anything, I think that the body of considered expert opinion is suggesting that the evidence we put forward is inconclusive, and many different parts of it may have different explanations. I have been gratified, though, that this work did play an important role in causing us to rethink the purpose of space exploration. Prior to this publication, speculation about life forms outside of Earth and how such could come about was almost exclusively the realm of science fiction. Today, it has become an issue that can be scientifically investigated and tested. I am proud to have played some small part in this transformation. Certainly, these issues involve chemistry among other disciplines. Progress will depend in no small part on the creation and perfection of new instruments that will allow us to explore with ever-increasing power the world of the very minute.

Paul J. Crutzen, 1997 (photograph by I. Hargittai).

34

PAUL J. CRUTZEN

Paul Josef Crutzen (b. 1933 in Amsterdam, Holland) is Director of the Division of Atmospheric Chemistry of the Max Planck Institute for Chemistry in Mainz, Germany. He has been associated with the Institute since 1980. He studied civil engineering in Amsterdam, earned equivalents of the M.Sc. and Ph.D. degrees at the University of Stockholm, the latter in meteorology in 1968. The title of his D.Sc. dissertation in Stockholm in 1973 was "On the photochemistry of ozone in the stratosphere and troposphere and pollution of the stratosphere by high-flying aircraft." Between 1974 and 1980, Dr. Crutzen did research at the National Center for Atmospheric Research in Boulder, Colorado. Dr. Crutzen has several honorary and part-time professorships, including those at the Scripps Institution of Oceanography in La Jolla, California, and at Utrecht University in Holland. Dr. Crutzen was co-recipient of the Nobel Prize in Chemistry for 1995, together with Drs. Mario J. Molina and F. Sherwood Rowland "for their work in atmospheric chemistry, particularly concerning the formation and decomposition of ozone." [An interview with F. Sherwood Rowland appeared in *Candid Science: Conversations with Famous Chemists.* Imperial College Press, London, 2000, pp. 448–465]. Paul Crutzen has many other awards and has been a member of many learned societies, including the Academia Europaea, the Royal Netherlands Academy of Science (Corresponding Member), the Royal Swedish Academy of Sciences, the U.S. National Academy of Sciences (Foreign Associate), and the Pontifical Academy of Sciences. Our conversation, whose timing coincided with the Kyoto Conference, was recorded in Dr. Crutzen's office in Mainz, Germany, on December 9, 1997.*

*This interview was originally published in *The Chemical Intelligencer* **1999**, *5*(1), 12–15 © 1999, Springer-Verlag, New York, Inc.

If you could have one wish fulfilled today about the environment, what would it be?

That we treat the environment very carefully, that we don't overload it. In fact, we're busily doing the latter. The Kyoto Conference is closing tomorrow. It seems that mainly the economic disadvantages of saving the environment are being considered there.

The first decisive steps should be taken by the industrialized nations, but the developing countries have an important task too. They should not be repeating the mistakes of the developed countries, and they should be helped to avoid them. This help should be in the form of clean energy and technology transfer for recycling and other processes. Eventually they should be able to take care of their own environment. They should act to diminish deforestation.

The impression at Kyoto is that the United States is unwilling to make sacrifices. There is a big gap between the proposals of the United States and the Western European countries. However, we should not make the United States a scapegoat. The promises made by the Europeans should be followed through. A little more than 10 years ago, the German government declared a reduction of greenhouse gas emissions by 25% by the year 2005. By now it's obvious that this promise will be impossible to fulfill. I was a member of the parliamentary commission, consisting of scientists and parliament members. When this goal was declared I was surprised to hear it because it sounded unrealistic, but how could I have gone against it?

Would you care to tell us about your research leading to the discovery of the importance of nitrogen oxides in the depletion of the ozone layer?

It was my first two papers. In 1970, I recognized that the ozone layer was controlled, to a large extent, by the catalytic actions of nitrogen oxides, NO_x, that is, NO and NO_2. This was something nobody had considered before in the stratosphere.

Already five years before my work it was becoming clear that the ozone production by solar radiation, a natural process, could not be balanced by the simple reactions between oxygen atoms and ozone, as proposed by Sydney Chapman in 1930. There had to be something else destroying ozone to reach the levels in the stratosphere we measured. Proposals had been made that OH and HO_2 molecules could also do the job. During my Ph.D. work in Stockholm I realized that these theories could not be right either. First of all, the rate constants — which had been introduced but never measured — gave us the wrong profile of ozone in the atmosphere. The right amount, roughly, but the wrong profile. Then other things were helping me think that there must be something else too. My work was mostly computer modeling, based on measurements by others in the atmosphere and on reaction rate constants obtained in the laboratory.

When I had my thoughts about the nitrogen oxides, I didn't publish them for about two years. There were no supportive measurements of nitrogen oxides in the stratosphere. Then in 1969–1970, some papers reported the presence of nitric acid in the stratosphere. I knew that wherever there is nitric acid, there must be also NO and NO_2. At that point I dared to publish my paper.

Did you talk about your results with your colleagues?

I did and a lot. Looking back, I appreciate the honesty of my colleagues, that they didn't take advantage of me. In the meantime we and others discovered a potential source of NO_x in the stratosphere. That was the oxidation of nitrous oxide, N_2O, by excited oxygen atoms. I put all this together in a simple model of the stratosphere and showed that nitrogen oxides dominate the destruction of ozone. The title of my paper was "The Influence of Nitrogen Oxides on the Atmospheric Ozone Content," *Journal of the Royal Meteorological Society* **1970**, *96*, 320–325. At that time I was a student of meteorology, I wasn't even a chemist. Then I read a report from a scientific meeting in the United States in which the issue of supersonic aircraft was discussed. Their potential role in the destruction of the ozone layer was denied, and they were the experts at that time. However, these conference proceedings gave me the first estimates of the amounts of nitrogen oxide emissions in the stratosphere. I could see the enormous danger for the ozone layer.

Did this realization cause you sleepless nights?

I didn't have sleepless nights about this, but I talked it over with my wife. I felt rather insecure being new in the field and not even being a chemist, my discovery was a little overwhelming for me. This was a nervous time. I felt I was right, but wondered also, why haven't the others seen this?

My discovery might not have had its proper impact had it not been for Professor Harold Johnston's work at the University of California at Berkeley. He, independently of me, called attention to the fact that nitrogen oxides would be added to the stratosphere in great amounts by supersonic aircraft. Harold didn't know about my paper, but others pointed it out to him and he recognized immediately that somebody had already done it. We've been extremely good colleagues ever since.

At that time there were plans to build enormous fleets of supersonic aircraft: 500 Boeing SSTs, Concordes, and in the Soviet Union the Tupolevs, nicknamed as Concordskys.

Your life has been very international, and you even named one of your daughters a Hungarian name, Ilona.

It's also a Finnish name, it means joy. My wife is Finnish. We speak Swedish at home. When our children were small we lived in Sweden. My wife is not a chemist, but she believed in me from the very beginning. We are natural environmentalists, not the campaigning kind, but we have always been very careful about waste in our household. We've never used spray cans. I bought two spray cans though when their ban was announced. They are on the shelf in this office along with some others, which I have received as souvenirs.

Do you think your Nobel Prize had anything to do with politics?

Our work has certainly had a strong societal impact, but this may in fact have delayed the Nobel Prize. Some may have thought that our research was too relevant, too applied. However, what is applied to some is fundamental to others. There are still people who doubt the validity of our claims in spite of all the hard evidence, but remember how many years it took Einstein's theory of relativity to get universally accepted. I am, of course, not comparing myself with Einstein.

Did the Nobel Prize facilitate your research?

It has not increased our funding, and the financial situation in Germany is not very rosy. Support has declined, but I don't complain because the Max Planck Society is supporting us quite well. We have had other sources of support and can't even take on more people because we're full. I'm director of a division of the Institute and this is the largest division with about 80 people of whom there are about 25 Ph.D.s. This Institute is one of about 70 Max Planck Institutes in Germany, but this was the very first. Originally it was the Kaiser Wilhelm Institute in Berlin from where it was moved here after World War II. Otto Hahn and Lise Meitner used to work in this Institute in Berlin.

Do you have any comment on the controversy about Lise Meitner and the Nobel Prize.

Only that it's very sad that she didn't get the Nobel Prize, whether sharing it with Otto Hahn in chemistry or in physics.

You had an unusual academic career and were a latecomer in science.

I started as a bridge builder, then I worked in house construction, then as a meteorologist, and finally, I switched to chemistry.

How much chemistry do you know?

[laughing heartily]: This is a mean question. Compared to real chemists, probably very little. Chemistry is an enormously broad field. Atmospheric chemistry is a part of it. I feel most at home in photochemistry, radical chemistry, spectroscopy, and quantum chemistry. Wherever I am weak we have others as excellent specialists and I support them. We need a lot of analytical chemistry, for example. Reaction simulation is also very important for atmospheric chemistry, and so is gas-phase kinetics, and more recently, surface reactions and atmospheric organic chemistry. We study the breakdown of hydrocarbons emitted by vegetation.

A very important part of my current research is tropical chemistry. The biosphere influences the chemistry of the atmosphere in a fascinating way. Furthermore, in less than 50 years, the most important impact on the atmosphere by human activities will come from the tropics and subtropics. However, for organic gases we must take into account the input by Nature. The global production of hydrocarbons from trees is a factor at least ten times higher than the emission of hydrocarbons from fossil fuels. There is a high potential to create photochemical ozone smog, especially in the tropics and subtropics. Another issue is the impact of burning of waste, dry grass, bushes, and forests in the tropics during the dry season. Of course, we can't tell people there what to do, but this practice deteriorates the environment and has an enormous influence on the chemistry of the atmosphere. I have been interested in these questions since the end of the 1970s. We carry out large-scale experiments in the region, involving a lot of aircraft of several countries. We're at the stage of collecting scientific information about the relationship between natural and anthropogenic processes in the atmosphere.

Will there be recommendations as a result of your studies?

Yes, but not very soon. We are in the initial stage of our work and I'm sure we're still in for very great surprises.

Do you have any message for our readers?

I'd just like to stress what you started with, the concern about the environment. Chemical research is very important for protecting the environment and especially in the recycling business. We should be very serious about the potential consequences of what we produce, from the beginning to the end product and beyond. The CFCs were once considered to be safe, the ideal, non-toxic substances, but the consequences weren't thought through to the end. In the future we should be more thorough. This is both to the advantage of mankind and to the advantage of the chemical industry.

Reiko Kuroda, 1999 (photograph by I. Hargittai).

35

Reiko Kuroda

Reiko Kuroda (b. 1947 in Japan) is at the Department of Life Sciences of Tokyo University. In fact, she is the first female full professor in the history of this most prestigious institution. My narrative is based on a conversation we recorded in May 1999 in Budapest.*

Some time in the early 1980s, I met Reiko Kuroda, for the first time, in London, where she was a postdoc. At that time she was facing an uncertain future. It was tough to get a permanent job in Britain, and she had been away from Japan for too long to hope to get a job there.

Reiko Kuroda graduated from the famous women's college Ochanomizu University in Tokyo in 1970 and got her M.Sc. and Ph.D. degrees from Tokyo University in 1972 and 1975, respectively. Her research was in X-ray crystallography, and her research supervisor was Professor Yoshihiko Saito, who did work on the determination of absolute configuration. For postdoctoral work, she decided to go to King's College in London. It wasn't simple, but her determination was strengthened by the difficulties of finding a job in Japan, especially for a woman with a high degree. At that time, the job market in Japan was not an open system. Rather, a professor would just mention to a colleague that he had a good student available. But Reiko Kuroda did not get this help either, since it was difficult to find jobs even for men. She was given the advice that the best thing for women was to get married, and she was offered help in finding a husband. She declined the offer. Kuroda was attracted by a postdoctoral position with Stephen Mason at the Department of Chemistry of King's

*This profile was originally published in *The Chemical Intelligencer* **2000**, *6*(2), 51–53

College in London [there is an interview with Stephen Mason in this Volume]. He was doing exciting things, and he needed someone to determine absolute configuration by X-ray crystallography.

Kuroda's English was not very good when she arrived in London but she was determined to improve it. First, she tried an adult education course but was not satisfied with its level. Next, she signed up for a course of allocution, which was not a course for foreigners, but she enjoyed it. They were learning things like how to sell umbrellas or holiday resorts. The course made her realize that becoming proficient in English would take more than learning the language. She learned English mainly from her colleagues and television. She got involved in teaching, and that helped too. Her efforts were truly recognized when Maurice Wilkins asked her to take over teaching his course on macromolecular assemblies.

Kuroda used to take X-ray diffraction photographs in the biophysics department. She also became more and more interested in the molecular basis of biology, and, in 1981, she moved from chemistry to biophysics. She learned base sequencing and other new techniques of molecular biology. She worked hard and joined in an investigation of the hypersensitive sites of immunoglobulin, which led to the publication of a paper in *Nature*. The same group also studied the molecular-recognition-type interactions between the DNA molecule and small molecules using standard techniques. She remembers with gratitude how much she learned from Mark Fisher and from Hannah Gould. Maurice Wilkins was the head of department, and his office was close to the X-ray lab. Kuroda found Wilkins quiet, thoughtful, hesitant, and certainly different from her image of a Nobel laureate. She had previously met Francis Crick, who radiated an inspirational flair.

Kuroda accompanied Wilkins on a trip to Japan. He was a little worried about the trip. He had been involved with the Manhattan Project during World War II as one of the British team working in California. Of course, his Japanese hosts knew about his work on the bomb. Wilkins asked Reiko, "What should I talk about?" He didn't want to upset anybody.

Kuroda thinks that Wilkins regrets his participation in the atomic bomb project and that this may have led to his interest in studying the social responsibility of scientists. She thinks that all Japanese condemn the dropping of the atomic bombs. She doesn't think that these bombs made much difference in ending the war because by then the situation in Japan had reached a point that would have made it impossible to continue the war. There was a shortage of food, a shortage of everything. I told her that I had often heard people in Japan condemn the dropping of the atomic bombs and the firebombing of Tokyo but never heard anybody mention the Japanese atrocities in China and Korea and elsewhere. Kuroda thinks

Reiko Kuroda in the company of four Nobel laureates in Tokyo, 1996: clockwise from her, Mrs. Watson, James Watson, Francis Crick, Mrs. Crick, Mrs. Jacob, François Jacob, and Susumu Tonegawa (courtesy of Reiko Kuroda).

that it is a matter of the way you put the question to the Japanese people. When they are asked about the bombs, they won't engage in discussing other things. She had never learned in school about what had happened during the war. She found out about many events later, from newspapers, and that was when she learned that other countries condemned the atrocities committed by the Japanese military during the war. It helped that she could read English. History class in her school started with the beginnings of civilization, and by the end of the academic year, they had not even reached the Meji period, which started in 1868. This was a recurring pattern. They would not have exam questions on the time after the Meji restoration. She remembers that they always had to finish history courses in a rush lest the school year end without their completing the curriculum.

Kuroda's research interests have always centered on chirality. She likes to go back to the famous story about Pasteur and Scacchi. Pasteur observed the spontaneous resolution of sodium ammonium tartrate whereas Scacchi obtained racemic samples when he repeated Pasteur's experiment. The difference was due to a phase transition at 27°C. Scacchi was working in Italy, so his lab was much warmer than Pasteur's.

About 20 years ago, Kuroda determined the crystal structures of Scacchi's and Pasteur's sodium ammonium tartrate at 35°C and at 20°C. The most interesting feature of these structures was the handedness of the adjacent molecules, that is, whether the next neighbor had the same handedness or opposite handedness. She has been interested in molecular packing in chiral systems. She also did calculations of chiral discrimination energies.

Considering just a pair of right- and left-handed molecules, the energy difference is very small. Extending these considerations to crystals, the difference is two or three orders of magnitude larger than in gaseous or liquid systems. The calculations must be kept simple because of the size of the system. The homochirality of the biological world is a fundamental question, and it is her primary interest.

Eventually Kuroda became Honorary Lecturer and finally a Research Fellow and won a permanent position in a tough competition. By then, she was in Steve Neidle's group, called the Biomolecular Structure Unit, which moved to the Institute of Cancer Research. She continued her teaching at King's College as Visiting Lecturer.

For the first several years in London, Kuroda was thinking of going back to Japan. She had been warned that if she stayed away for more than two years, she would lose an important feeling of harmony and would become too westernized, and then it would become impossible to get accepted again by Japanese society. After two years, Kuroda got a job offer from Japan to teach in a girls' private finishing school. The position was only for two years, and there was no chemistry department there; they just wanted her to teach some science and to teach English. She stayed in London and eventually got a permanent position.

She felt settled, but she was intrigued also to try her hand in the Japanese system and, after several years, she decided to submit an application for an opening at Tokyo University. She did this also to let them know about her and about her work. To her great surprise, she landed one of four associate professor positions from among 140 candidates. Thus, in 1986, she became the first female Associate Professor in natural sciences on her campus. When she was appointed full professor in 1992, she was the first female full professor in the history of Tokyo University. Some worried that she might not be able to understand the culture, the harmony, but there was no problem. When her colleagues go out in the evening for a drink, sometimes they ask her to join them. Normally, she is so involved with her experiments that she is not eager to go out anyway. When her colleagues have foreign visitors, they like her to join them because of her fluency in English.

Her good English may not be an unambiguous blessing though. She had been warned not to speak English in front of her colleagues. She had read articles in which students returning to Japan after having spent years abroad were advised not to use the proper pronunciation in an English class but adopt instead the Japanese pronunciation lest the English teacher become unhappy.

Kuroda's basic research support comes from the Ministry of Education, Monbusho. Everybody gets some money, regardless of performance. This is about the equivalent of $6000 per year. Thirty percent of this goes to the library. The rest is for chemicals, minor equipment, photocopying, and so on. To buy major equipment, she must obtain other grants. She was in a peculiar position because she was changing research topics. It is not common in Japan to change one's field of research. Kuroda got support to buy a circular dichroism instrument, two UV-visible spectrophotometers, an FTIR spectrometer, and electrophoresis equipment; she also has her own X-ray apparatus, a workstation, and other computers. However, even as a full professor, she has to do everything herself.

The focus of Kuroda's current research is the relationship between molecular chirality and macroscopic chirality. DNA helicity and solid-state chemistry are two important aspects of her work, while, on the biological side, there is the genetics of snail handedness. She is conducting complex investigations, and she was recently awarded an exceptionally large grant for five years. The amount is one and a half billion yen, which is roughly the equivalent of $12 million. She did not ask for this amount of money; the amount is not applied for but is set by the granting organization. Four such grants are awarded per year by the Science and Technology Agency of Japan. She has to find and rent lab space, outside the university campus, and has to employ 10 to 15 people, whose rank cannot be higher than postdoc. She is the sole professor in the project. She anticipates her hair growing white, and she can't afford to sleep more than four hours a day.

Reiko has coined a title for her project, "Chiral Morphology," and ascribes the success of her application to its adventurous nature. She was told that the most important thing expected to come out of her project was some new concept or some seeds of a new concept. The project leader is supposed to be in his or her mid-forties (so in her case they obviously made an exception) and is expected to employ postdocs. She has to advertise for them in *Science* and *Nature*. The enormity of the project is "killing" her.

Kuroda does not have a family background in natural sciences. Her father is a Professor of Japanese Literature, and books covered every wall in her family home. Her brother became a particle physicist.

When I asked Professor Kuroda if she could have one wish what would it be, she hesitated for a long time before answering. She wanted to make sure that the wish could be anything, however unrealistic. Finally, she told me that it would be nice to have a family and continue her work as well. Although she hardly has any free time, she likes to go to concerts. She also likes all kinds of housework, including cooking. She grows flowers on her tiny balcony. Fuchsias are her favorite. She looks at her fuchsias every day, and she has a photograph of Gregor Mendel holding fuchsias in his hands. Someone gave her this photograph after a TV appearance where she'd talked about her science and about her love of fuchsias.

Stephen Mason, 2000 (photograph by I. Hargittai).

36

Stephen Mason

Stephen Mason (b. 1923 in Leicester, England) studied chemistry at the University of Oxford, where he was awarded a B.A. with First Class Honours (1945), M.A. and D.Phil. (1947), and D.Sc. (1967). Since 1982, he has been a Fellow of the Royal Society of London. Following appointments at Oxford, in London, and Exeter, he was Professor of Chemistry at the University of East Anglia, Norwich (1964–70), and then at King's College London (1970–88). Since 1988, he and his fellow chemist-historian wife, Joan, have lived in Cambridge, where they are associated with the University Department of the History and Philosophy of Science. His first book was written before the age of 30, *A History of the Sciences: Main Currents of Scientific Thought* (Routledge & Kegan Paul, 1953). I asked Professor Mason about his career, and what lessons could be learnt from the history of science.

I was a student of chemistry during World War II and took up research on antimalarial drugs for the Army Medical Corps and then the Medical Research Council. We were asked to find a practical method of detecting whether Italian prisoners-of-war were taking their antimalarial pills. Malaria was endemic among the prisoners, employed as farm labourers from a camp north of Oxford, and it appeared that they passed on their antimalarial pills to local girls as contraceptives. Ultimately, a method was found, using the fluorescence of the antimalarial metabolites in the prisoners' urine, and I went on to study the physico-chemical factors possibly governing the relative efficacy of different antimalarial drugs.

The research was fairly routine, and I studied the history of chemistry, on which I wrote an essay for a Prize Fellowship in 1947. As a result I was offered a demonstratorship in the Oxford Museum of the History of Science to give lectures of general interest on the history of science. I held the post, in conjunction with a tutorship in chemistry for my College, Wadham, until 1953. The Warden of Wadham College, Sir Maurice Bowra (1898–1971), urged me to return to chemical research, since he foresaw no substantial place for science history at Oxford in my time, although he himself considered the subject a valuable link between the cultures of the arts and the sciences. In any event I missed the practical work of a chemistry laboratory, and the quest for new substances, mechanisms, and theories. Additionally I had acquired, from study of the history of chemistry, an appreciation of the factors making for success or failure in past scientific ventures, with abandoned enterprises sometimes becoming viable again through advances in technique or theoretical perspective.

In 1953, I joined the Department of Medical Chemistry of the Australian National University (ANU), then in London at the Wellcome Institute, pending the construction of buildings for the new university in Canberra. The

Stephen Mason, 1952 (courtesy of Stephen Mason).

Head of the Department, Adrien Albert (1907–89), was an international authority on biologically-active acridines, which included the antimalarial substances I had studied earlier, and now he was concerned with the purines and pyrimidines of the nucleic acids and other biochemically-important substances. My role was to extend the physico-chemical methods for studying organic structures and reactions from the electrochemical methods I had used in Oxford to the spectroscopic, using the infrared (IR) and ultraviolet (UV) spectrophotometers newly available commercially.

In 1955, I attended the first of the Summer Schools in Theoretical Chemistry, organised by Charles Coulson, then Rouse-Ball Professor of Mathematical Physics at Oxford, to introduce young chemists to the Molecular Orbital (MO) Theory, to which he made outstanding contributions. The MO theory gave me a wider perspective for the interpretation of molecular spectra and the study of molecular structures and reaction mechanisms.

My ANU research fellowship ended in 1956, when I moved to a lectureship in chemistry at the new University of Exeter, formerly the University College of the South West under the aegis of the University of London. Here I took up the spectroscopy of chiral molecules in connection with a general problem, which had come to concern me at Oxford, the use of synthetic racemic compounds in homochiral biochemical applications, medical and agricultural. Emil Fischer (1852–1919) had shown that the chemistry of living organisms is homochiral, involving mainly one of the two optical isomers of the organic compounds characterised, by Louis Pasteur (1822–95) in 1848, as handed (chiral, dissymmetric), with non-superposable mirror-image structural forms (enantiomers), and with equal and opposite rotations of the plane of linearly-polarized light in solution. Optical isomerism led Le Bel and van 't Hoff in 1874 to the insight that the four valencies of the carbon atom must be directed to the vertices of a tetrahedron, since models of four different groups bonded to such a carbon centre, as in known cases of optical isomerism, give two non-superposable mirror-image structures. The expectations of isomerism in chains of chiral carbon atoms were confirmed and used as a guide by Fischer in his investigations of the sugar series and then the amino-acid series in natural biochemical products. The chemical relationships he established led Fischer to conclude that virtually all structural and functional proteins were composed of L-amino acids, while the carbohydrates were composed of D-sugars, where L and D refer to a conventional absolute configuration of four different groups around a chiral carbon atom. In 1951, it was shown by X-ray diffraction methods that Fischer had made the correct choice in his convention for absolute sterochemical configuration.

Around 1950, most synthetic chiral pharmaceutical drugs were racemic mixtures of two enantiomers, although it was known to pharmacologists that the separated enantiomers had different biological activities. When I suggested to the Head of ICI Pharmaceuticals in the 1950s that racemic pharmaceuticals should be optically resolved and only the enantiomer with the selective activity be used medically, he responded that the expense of the resolution would be ruinous. This was before the thalidomide tragedy, when it turned out that only the laevorotatory (−)-enantiomer has teratogenic effects during embryonic development, while the other enantiomer is effective in chemotherapy. Afterwards drug regulations required the testing of the individual enantiomers of racemic pharmaceuticals, although racemic mixtures are still employed in biological applications. One enantiomer of warfarin, for example, retards blood-clotting more effectively than the other, by a factor of five in humans, and by a factor of nine in rats.

Stephen and Joan Mason in their Cambridge home, 2002 (photograph by M. Hargittai).

Fischer's L and D nomenclature referred to the spatial mirror-image structure of two enantiomers, but not to the sign of their optical rotation at a standard wavelength (such as the yellow sodium D-line). Studies of the variation of the optical rotation with wavelength had led Cotton in 1895 to discover an abrupt sigmoidal change in the rotation, "anomalous optical rotatory dispersion (ORD)," and the differential absorption of left- and right-circularly polarized light, "circular dichroism (CD)," in the wavelength range of an isotropic light-absorption band of an enantiomer. During the 1950s spectroscopic studies of these two aspects of the "Cotton effect" in series of similar chiral molecules led in 1961 to the Octant Rule, relating the stereochemical configuration of chiral ketones to the sign of the ORD or CD connected to the 290 nm carbonyl light absorption. At Exeter, we constructed a CD spectrophometer for the visible and quartz ultraviolet (UV) region, and later on in London, CD instruments for the vacuum-UV and the IR region, with the object of determining the absolute stereochemical configuration of an enantiomer from a comparison of the experimental CD with that calculated theoretically from model structures. At the same time we investigated new methods for resolving racemic mixtures into their individual enantiomers, and the stereoselective physico-chemical interactions between chiral systems. These studies were covered in my book, *Molecular Optical Activity & the Chiral Discriminations* (Cambridge University Press, 1982), which touches on the possible origin of the built-in handedness in the organic world, extended in the last chapter of my *Chemical Evolution: Origins of the Elements, Molecules and Living Systems* (Oxford University Press, 1991).

The Nobel Prize for Physics in 1979 was awarded for the unification of electromagnetism with the weak nuclear force, including the discovery of the universally-chiral neutral electroweak interaction. The choice of specifically L-amino acids and D-sugars, rather than their enantiomers, in the homochiral biochemistry of the organic world was generally regarded as a matter of chance, since specific mechanisms, such as the photochemical effects of cosmic circularly polarized radiation, might generate either enantiomer of the amino acids or of the sugars. The electroweak interaction prescribes a determinate helicity, which is left-handed for electrons (antiparallel spin and momentum vectors). By 1983, we had found, by incorporating the electroweak interaction as a perturbation into *ab initio* MO calculations, that the L-amino acids are inherently more stable, to a miniscule degree, than the corresponding D-enantiomers. My talented student George Tranter, as a postdoctoral research fellow, found that the D-sugars have a similar stability increment over their L-enantioners. The electroweak energy difference between enantiomers is minute, but subsequent improved calculations by others show that we underestimated the difference. Known reaction mechanisms demonstrate the evolution of a racemate to a determinate homochirality, starting from a very small energy difference between the enantiomers.

Did you have any hero when you were a student?

I was impressed and inspired by Charles Coulson (1910–74), the evangelical teacher of quantum chemistry. Quantum mechanics was not taught to chemistry students when I was an undergraduate, and Coulson's Summer Schools introduced some 35 to 60 chemistry graduate students, postdoctoral fellows and university lecturers to the subject each year from 1955 to 1973. His lectures were crystal clear, and his communication went beyond the power of his words, for one had to work hard at the problems he set, as part of the course, to really understand what he had said.

Coulson was a Methodist minister, and a polymath spanning mathematics and the physical sciences. My most treasured memory of Coulson was meeting him, clad in hiking gear with a rucksack, at Exeter railway station midweek in the late 1950s. Tell me, he asked, when am I due to lecture to the Chemical Society, the Physical Society, and the Mathematical Society of the University? My sermon of course will be delivered on Sunday, and could you come with me on the cliff walk from Exmouth to Budleigh Salterton on Saturday? His religion was an authentic part of his life, not conventional observance. He was a kindly man of singular probity and dedication over a range of scientific and social concerns. His criticisms were invariably constructive. Around 1970, I nominated Coulson for a Nobel Prize, citing his contributions to molecular orbital theory. The nomination was not successful, for supporters pointed out that Coulson, as a matter of principle, would not engage in the kind of lobbying in Stockholm that seemed to accompany successful Nobel awards.

I was inspired too by Linus Pauling (1901–94), another polymath with humanistic concerns. His *Nature of the Chemical Bond* (1939) brought a new perspective to theories of molecular structure, and refuted the implication of a popular examination question of the time, "Is inorganic chemistry a largely closed and finished subject?" Pauling's resonance theory, formally based on the quantum-mechanical valence-bond (VB) method for interpreting molecular properties, dominated chemistry during the 1940s, but the VB-procedure was rivalled by the MO (molecular orbital) method, which progressively gave deeper insights during the 1950s. By that time Pauling had moved on to become one of the founders of molecular biology, with his structures for the alpha-helix and beta-sheet conformations of the proteins (1951), based on model-building from established X-ray structures of simple peptide units. In addition, he originated the "molecular evolutionary clock" (1962), based on a comparison of the amino acid sequences of the haemoglobin proteins, from a variety of species, with the fossil record, calibrated to an average of one amino acid mutational change per polypeptide chain every seven million years (a unit termed *the pauling* by Kimura in 1969). It was ironic that such an all-American chemist as Pauling should have been termed "un-American" by the followers of Senator McCarthy, for opposing the testing of atomic weapons in the atmosphere, producing global radioactive pollution. His publicity of the ecological hazards of radioactive fallout and the addition of more carbon-14 to the atmosphere, followed by similar warnings by Andrei Sakharov (1921–89) in the USSR, led to international concern, resulting in the 1963 partial test-ban treaty. It was felt appropriate that the Nobel Peace Prize should be awarded to Pauling in 1963 and to Sakharov in 1975.

What are the highlights of the history of chemistry for you?

The two main aspects of the history of science that primarily concern me are the ways in which we have attained a progressively enlarged understanding of the natural world, and, through this development, how the sciences have been used to tackle social and technical problems.

The modern period of history begins around the time of the Protestant Reformation during the early 16th century, when it was perceived that what passed for the sciences needed reformation to bring about "the relief of man's estate," as Francis Bacon put it. The "Luther of medicine," Paracelsus (1493–1541) tried to tranform the wealth-seeking metallurgical alchemy of earlier times into a new iatrochemistry with more humanitarian medical aims, based on mineral as well as herbal remedies, and he secured a substantial following over the next century or so. The ancient notion that all substances were composed of a relatively passive body and an active spirit lived on, and the distillation and bottling of spirits containing the potent essences of the substrates flourished on an industrial scale. The iatrochemist van Helmont (1597–1644), under house arrest in Brussels by the Spanish

Inquisition for some 20 years, discovered that some spirits could not be condensed and bottled, yet they turned out to be powerful and individual derivatives of their substrates. He termed the non-condensible spirits "gases," opening up the era of pneumatic chemistry.

The Unitarian minister Joseph Priestley (1733–1804), made spectacular contributions to pneumatic chemistry, isolating and characterising more than a dozen different gases by their chemical reactions. Priestley was already concerned with the pollution of the atmosphere, during the early phase of the industrial revolution in Britain, and devised a method of measuring the "goodness of the air," from the changes in a volume of air after reaction with nitric oxide and condensation of the product in water. Subsequently he was delighted to discover that green plants in sunlight restore the 'goodness of the air' by producing the vital gas, oxygen. Priestley was a radical in politics and religion. His support for aspects of the French Revolution led to the torching of his manse, library and laboratory in Birmingham by a mob of "patriots" led by some of the county gentry, and he felt obliged to emigrate to the newly-independent U.S.A. in 1794. Yet Priestley remained enmeshed in the ancient alchemical body-spirit model at a time when it was being superseded during the Chemical Revolution led by Lavoisier (1743–94) and other French chemists, who interprested combustion as oxidation. Combustion, for Priestley, led to the *calx*, the "dead body," of the substrate, with the release of the spirituous *phlogiston*, which he came to identify with electricity. Electrochemistry was only just starting up in 1800, and another century elapsed before G. N. Lewis convincingly identified oxidation with electron-loss and reduction with electron-gain.

The Chemical Revolution, as it was termed in France, or the "new French chemistry" as it was referred to elsewhere, centred on the long-known "augmentation of the *calx*," the increase in weight of a substrate on combustion. The discovery of oxygen gas by Priestley enabled Lavoisier to ascribe the weight increase to the uptake of oxygen from the air, and the energy changes to the release of the imponderable matter of heat, *caloric*.

The nomenclature reform of the French chemists was of fundamental importance, replacing the old body-spirit terminology with new terms, based on oxygen. The *calx* was now termed the oxide, and the "spirit of vitriol" became sulfuric acid. The assumption that oxygen was the "acid generator," as its name implied, was flawed, but the systematic nomenclature based on the increase of acidity with increase of oxygen content lived on, e.g. the acidic component of the sulfides, sulfites, and sulfates. The definition of a chemical element as "the latest term whereat chemical analysis has arrived" was not really new, but now led to the proliferation of new inorganic compounds, hundreds from different elements taken two at a time from Lavoisier's list, thousands if taken three at a time. Some had remarkably useful properties, such as the new bleaching reagents, which saved England, during the vast expansion of the textile industry, from becoming a huge bleaching field, relying on sunlight and the oxygen of the atmosphere for a bleaching effect.

The theoretical rationalisations made 1800–1820 of the enormous body of chemical data now accumulating were largely ignored until the 1860s. The theory of John Dalton (1766–1844) postulated that the chemical elements fell into distinct species, with the identical atoms of one species differing in relative weight and combining propensities from the atoms of other species. His law of multiple proportions (1804) indicated that the weight ratios of two elements forming two or more compounds were simple integers, quantising atomic combining capacities, as in the oxides of nitrogen, N_2O, NO and NO_2. A similar simple integer ratio held for the volumes of combining gases, as Gay-Lussac (1778–1850) found in 1808: one volume of hydrogen and one volume of chlorine at the same temperature and presure combine to give two volumes of hydrogen chloride. Hence the gaseous elements must be composed of diatomic molecules, deduced Avogadro (1776–1856) in 1811, for each unit of hydrogen gas, as also each unit of gaseous chlorine, must split to produce two units of hydrigen chloride. Not necessarily so, argued Ampère (1775–1836) in 1814, for Gay-Lussac's law shows only that there are an even number of atoms in the molecule of an element. Molecules are the basic building blocks of three-dimensional crystals, and molecules should be three-dimensional too, composed at a minimum of four atoms in a tetrahedral array. The average practising inorganic chemist of the time maintained that none of the theoreticians could be taken seriously. In any event simple tables of equivalents, the combining weights of the elements relative to oxygen, or another standard element, sufficed for most practical purposes.

Such an approach led to confusion in the new field of organic chemistry emerging during the 1830s. So few elements entered the composition of organic compounds, basically carbon and hydrogen, then oxygen and nitrogen, and other elements only in more exotic cases. Gaseous acetylene and liquid benzene both analysed to $[CH]_n$, but a value for n could be obtained only from the molecular weights of the two molecules. Perceptive French organic chemists in the 1840s began to argue for a revival of Avogadro's hypothesis, according to which the molecular weight of a volatile organic substance, relative to the unit weight of a hydrogen atom, is given by twice the vapour density of the substance, relative to that of molecular hydrogen. German and British chemists joined the discussions in the 1850s, and the matter was largely solved after deliberations at a comprehensive international conference, the first of its kind, held at Karlsruhe in 1860, with 140 chemists attending. Here Cannizzaro (1826–1910) showed how a systematic application of the hypothesis of his fellow-Italian, Avogadro, to series of related molecules gave consistent sets of atomic and molecular weights, and the atomic combining numbers, their valencies.

Mendeléev (1834–1907) returned to Russia where he worked out his Periodic Classification of the 67 chemical elements then known, with gaps for missing elements, whose properties he predicted in detail (1869). A primary organiser of the conference, Kekulé (1829–96), worked out a testable flatland stereochemistry of aromatic molecules in 1865, based on the hexagonal ring structure for benzene. The structure rationalized the known features of aromatic chemistry, and its detailed expectations served as a guide for further explorations of the field, including the synthetic coal-tar dyes. Aliphatic organic chemistry was less well rationalised during the 1860s, for there remained the mysterious problem of optical isomers, two substances apparently identical in all chemical and physical properties, except for their equal and opposite rotation of the plane of polarized light in solution.

Pasteur in 1848 had provided a general morphological answer. A pair of optical isomers in solution must have overall dissymmetric forms, non-superposable mirror-image shapes, like their corresponding crystal forms in the solid state, whatever their internal structure might be. Pasteur then moved on, as a founder of microbiology, to study the diseases of wines and beers, of silkworms, dogs, cattle and humankind, surveying with a critical eye chemists who claimed to synthesise optically active molecules without the use of a chiral physical force which, he believed to the end of his days, was active throughout the cosmos. So it was left to Le Bel (1847–1930) and to van 't Hoff (1852–1911) to work out the internal structure of optical isomers, showing independently in 1874 that two non-superposable mirror-image forms are generated if four different atoms or groups in a tetrahedral array are bonded to a central carbon atom. The expectations of the tetrahedral model for the orientation of the four valencies of carbon were worked out and confirmed in detail, especially and spectacularly by Fischer for the sugar series (1884–1908) and then the natural amino acids and the peptides (1908–1919).

Fischer showed that Pasteur's chiral force of nature was not required to account for the homochirality of biomolecules, the prevalence of the D-series of sugars and the L-series of amino acids among natural products. The reactions of symmetric, non-chiral reagents adding a new potentially-chiral carbon atom to an enantiomer did not give the two expected products (diastereomers, containing two or more chiral carbon atoms) in equal quantity. Such reactions were stereoselective, and one of the diastereomeric products appeared in often substantial excess. With enzyme catalysis, such reactions became stereospecific, affording a single diastereomeric product. Fischer concluded that there is no need to postulate the continuous operation of a chiral force in biosynthesis, as Pasteur had supposed, "once a molecule is asymmetric, its extension proceeds also in an asymmetric sense." Given a primordial enantiomer, biochemical evolution necessarily gave rise to chiral homogeneity among the variety of natural products, through Fischer's "key and lock" mechanism, the survival of the best stereochemical fits in biomolecular reactions, ensuring an efficient and economic biosynthesis and metabolic turnover.

The chirality of the primal enantiomer appeared to be wholly a matter of chance, for left- and right-handed chiral forms of the classical forces and energy fields seemed to be equally abundant, and sum to zero over a time and space average, such as those involved in natural helical motions or circularly-polarized radiation. This viewpoint was enshrined in Wigner's (1927) principle of the conservation of parity, which implied that an organic world based on the D-amino acids and L-sugars would be wholly equivalent to the actual biomolecular world characterised by Fischer, based on the enantiomeric series. The weak interaction responsible for radioactive beta-decay was found to violate the principle of the conservation of parity in 1956. Subsequently the weak interaction was unified with electromagnetism, and the massive boson carriers of the integrated electroweak interaction were detected at CERN in 1983. It appeared that the neutral component of the electroweak interaction might provide an attenuated form of Pasteur's universal chiral force of nature, adequate enough to account for the minor prebiotic enantiomeric excess from which Fischer's "key and lock" biomolecular evolution started. A residual effect would be an inherent, if miniscule, greater stability of the L-amino acids and the D-sugars relative to the corresponding enantiomeric series, as was found to be the case in subsequent quantum mechanical estimations incorporating the neutral electroweak interaction.

Reading Lucretius today it is strikingly modern but we may be projecting our knowledge onto what we read. Did he and and the other Greek and Roman philosophers have an impact on the emergence of modern science?

Part of the Renaissance and Reformation of the early modern period lay in the revival of ancient philosophies, beliefs and styles, eclipsed during the middle ages. The Humanists revived the "pure" Latin vocabulary and style of Cicero and his contemporaries, to replace the "barbaric" medieval Latin with its Arabic and Germanic intrusions. The Protestant Reformers set out to revive the early authentic Christianity of the Church Fathers, as they interpreted it in individualistic and divergent ways. Philosophers looked to the precursors of Thomism, dominant in the thinking of the Roman Church since the 13th century, back to a "purified" Aristotelianism, freed from theological accretions, or to the Neoplatonism of the early centuries AD and other beliefs of early Imperial Rome associated with the official cult of the Divine Unconquerable Sun, such as Hermeticism. Philosophers of nature went back further to the Pre-Socratic Greeks, particularly the atomists, revived already by Lucretius (c. 95–55 BC) in the pre-imperial Republican Rome.

Descartes (1596–1650), for example, revived the vortex cosmology of Anaxagoras (c. 488–428 BC), who had been charged with impiety for his views, but saved by the intervention of Pericles, and Descartes withheld his mechanistic natural philosophy on hearing of the condemnation of Galileo in 1633. Descartes delayed full publication of his cosmology until he had worked out a theological buttressing for his views, but the censors were not satisfied, and the works of Descartes in 1663 were placed on the Roman Index of Prohibited Books (widely consulted by Protestant librarians for books to add to their libraries).

Copernicus (1473–1543) deleted a reference to the heliocentric cosmology of Aristarchus of Samos (c. 310–230 BC), declared impious in antiquity, from the manuscript of his own book, published in 1543, *On the Revolutions of the Heavenly Spheres*, reviving and elaborating the sun-centred astronomical system. The early use of the book was not problematic, for a Lutheran pastor, Osiander, who saw the book through the press in 1543, inserted an anonymous foreword declaring that the Copernican scheme was hypothetical, designed to save the astronomical appearances, not a physical cosmology, which it became in the hands of Kepler (1571–1630), the maverick Lutheran Imperial Mathematician to the Hapsburg Holy Roman

Emperors. But it was the qualitative telescopic discoveries of Galileo (1564–1642) that gave the Copernican system a widespread plausibility, the observation around 1610 of mountains and apparently seas on the moon, of spots on the sun, the phases of Venus, and the four moons of Jupiter. Copernicanism then became a doctrine banned by the Roman Church (1616) "until corrected," and Galileo was condemned for his support of the doctrine in 1633.

Classical atomism was especially abhorred during the early centuries AD and the middle ages as "atheistic." In fact, Epicurus and Lucretius held that the gods were indifferent to the autonomous workings of nature and the activities of humankind, a view akin to that of the deism widespread during the 18th century, holding that, after the Creation, God left the machine of the universe and its creatures to run themselves automatically. Such views jeopardized the claims of the Roman emperors and their historical "ghosts" (as Hobbes put it in his *Leviathan* of 1651), the Popes of the Roman Church, to temporal power based on a privileged direct communication with their cosmic representative, first the *Deus Sol Invictus* and then the Christian God. Lucretius appears to have been copied and read surreptiously during the medieval period. There are two 9th century copies of *On the Nature of Things* by Lucretius in the Leiden University Library. The title page of one copy had been torn out, and the name of the author had been erased and replaced by a pseudonym on the title page of the other.

Atomism was revived during the 16th and 17th centuries in the disguised form of corpuscularianism, a natural philosophy of particles in motion. Heat increased the motions of the particles of bodies, according to Francis Bacon. All of space was filled with very fine particles, Descartes supposed, and their motions set up the vortices that swept the earth and the planets in circular orbits around the sun, or of the sun and planets around the earth from the point of view of an observer beyond the sphere of fixed stars, he added, in a concession to the ban of the Roman Church on the Copernican scheme. Boyle supposed that corpuscules were polyatomic conglomerates, exchanging parts in chemical reactions. The corpuscles of the air might be stationary springs, with mutual repulsion as they approached one another. This mechanism would explain the inverse proportionality between pressure and volume (Boyle's law), as would a mechanism involving the air particles in motion, where the frequency of collisions between the particles and the walls of the container, and thus the pressure, would increase as the volume of the container were diminished.

Classical atomism was gradually made more theologically acceptable during the 17th century, through the postulate that God originally created atomic matter and set the atoms into such motions as produced the natural world and its creatures according to providential design. Atomism had an appeal for the mentality of assertive individualism, developed by Renaissance writers and radical Reformers, and prominent among the leaders of the great geographical explorations and their mercantile backers. Intellectual supporters were moved to give their own individual interpretations of the *Book of Nature* and the *Book of Holy Writ.* Interpretations and assertions as to the content of the *Book of Nature* mostly could be checked empirically, but rather less so in the case of those referring to *Scripture*, and a consequent progressive secularisation led to the Deists of the 18th century, who recapitulated much of the theological stance of the atomists of antiquity.

What can we learn about the relationship between science and authority in the history of science?

Broadly we learn that the natural sciences do not flourish in periods of zealous and effective authority, particularly in cases where socially-dominant institutions fear that scientific innovations may jeopardize their power, as in the case of the Roman Church during the 16th and 17th centuries. The Protestant Churches were more divided, and political power lay primarily in the hands of local princes, some pursuing scientific studies themselves, such as Landgrave Wilhelm IV of Hesse (1532–92), who built his own astronomical observatory at Kassel. The king of Lutheran Denmark, Frederick II, patronised the observational astronomy of one of his nobles, Tycho Brahe (1546–1601), whose death left his assistant Kepler with astronomical data, measured over 20 years, and the post of Imperial Mathematician to Emperor Rudolf II at Prague. In contrast, the observatory at Istanbul was destroyed, after only six years of operation (1575–80), by the command of Ottoman Sultan Murad III, despite his astrological interests, following the ruling (*fatwa*) of his Muslim juriconsults that astronomical prying into celestial affairs had caused the 1580 outbreak of bubonic plague in Turkey.

By the late 16th century, no religious authority in Europe could exercise such a control over science policy and, indeed, the Jesuits built observatories in order to compete with laymen in astronomy. Princes had become more and more resistant to the attempts of the Church to extend temporal power ever since the forgery in Rome of the Donation of Constantine during

Carolingian times, purporting to give the Bishop of Rome the temporal powers of the Western Roman Emperor, and to delegate those powers to a chosen nominee, crowned as the Holy Roman Emperor, beginning with Charlemagne (c. 742–814). These emperors had their own policies, and the Hapsburg emperor Rudolf II (1576–1612) collected to his Prague court notable scientists whatever their religion, appointing as Imperial Mathematician the Lutherans Tycho Brahe and then Kepler. During the 16th century, many princes strove to centralise their realms, taking over the Church within their territories as an instrument of state. During the 17th century, the princes and the learned among their professionals and gentry came to appreciate the potential usefulness of the natural sciences, as illustrated by the writings of Francis Bacon (1561–1626), and they founded national observatories and academies of science. While the science academies of Italy were ephermeral, and natural science was dormant in Italy for a century or so after the time of Galileo and his immediate disciples, the scientific societies of England and France were enduring, and they served as models for the 18th century science academies of North America, Russia, Germany, and elsewhere in Europe.

Thereafter the relationship between science and authority became more a question of political conformity, rather than religious dissent. Joseph Priestley was ostracised in London much more for his sympathy with the French Revolutionaries than for his Unitarianism, and John Dalton encountered disdain in London more for his association with industrial Manchester than for his Quaker beliefs. Prelates could be stirred to angry opposition by discoveries apparently at odds with their theological doctrines, such as, *On the Origin of Species by Means of Natural Selection, or the Preservation of Favoured Races in the Struggle for Life* (1859), by Charles Darwin (1809–1882), but theology was no longer widely regarded as the Queen of the Sciences, nor did prelates retain any more the social standing and power they enjoyed in Galileo's time.

During the 20th century, ideological factors were added to political conformity in the relationships between science and authority. In Germany, in 1933–1945, scientists, among others, with traceable Jewish antecedents, were expropriated and exiled or extinguished, with an even more efficient ruthlessness than the Spanish monarchs and their Inquisition during the 15th and 16th centuries. The wealth sequestered from the Jews expelled from Spain in 1492 financed the discovery of the New World by the expedition of Columbus in search of a westerly route to Asia. The Spanish

Inquisition pursued the *conversos*, descendants of Jewish converts to Christianity, and the *moriscos*, the corresponding Muslim converts, with vigour, and 45% of the cases recorded by the Court of the Inquisition of Toledo between 1575 and 1610 refer to these two categories, which included around 10% physicians and surgeons. But the Iberian monarchs and their grandees retained the services of *converso* physicians, reputed for their medical skills, just as Field Marshal Goering in Germany retained the services of the 1931 Nobel Laurate biochemist, Otto Warburg (1863–1970), despite his Jewish antecedents, since Warburg was reputed to have a cure for cancer, from which Goering feared he might suffer.

Historically such pogroms result in an enormous loss of talent, not only for the countries of the perpetrators, but also for humankind at large, and ultimately they fail in the quest for an extension of power over the natural and human world. The same holds for the lesser and more secular analogues after World War II, McCarthyism and Lysenkoism. The crusade of Lysenko (1898–76) in the USSR against Mendelian genetics over the years 1948–65 came to an end when it became apparent that his version of Lamarckism, the inheritance of acquired characters, failed both in agricultural practice and in laboratory test experiments. Lysenko's disciples whom he nominated as candidates for election to the USSR Academy of Sciences were rejected by the physicists on the grounds that these candidates, while excelling in the rhetoric of criticism, produced no authentic and tested innovations. The rise and fall of Lysenkoism provided an object lesson for those sociologists who regard all scientific theories as wholly socially-constructed ideologies, and who make no reference to the success or failure of the theories in the laboratory or the field.

Whom, among the chemists, did not get their proper recognition?

Scientists of a retiring disposition who make important discoveries, but publish them in journals of limited circulation, tend to be overlooked in their own day and receive appreciation only after they have died, when their innovations have been rediscovered. The work of Willard Gibbs (1839–1903) at Yale University on chemical thermodynamics and statistical mechanics, published in the *Transactions of the Connecticut Academy of Sciences*, was approved by Clerk Maxwell in 1875, but little known to others in Europe at the time. Yet the Gibbs Free Energy function and the Gibbs equations became standard in chemical thermodynamics student courses from the 1920s onwards. Chemists who work on non-standard topics receive little recognition in their own time, except from former students and a few appreciative colleagues.

G. N. Lewis (1875–1946), at MIT then Berkeley, University of California, appreciated the importance for chemistry of the early quantum theory and the discovery of the electron. He defined oxidation as electron-loss and reduction as electron-gain, and the "Lewis acids and bases" as respective acceptors and donors of an electron pair. In 1916, he postulated that the electrons in atoms of higher atomic number in the periodic table than helium were divided into an inner shell with the electronic structure of the preceding zero-valent noble gas, and an outer valence shell of electrons which determined the number and character of the bonds to other atoms, in which each atom attained the electronic structure of the preceding or the following noble gas. In the electron-pair covalent bond each atom shares an electron with that of its neighbour, as the case of the hydrogen molecule, where each hydrogen atom is surrounded by two electrons, as in the helium atom. In an ionic bond, one atom donates an electron to an acceptor atom, as in sodium fluoride, where both ions have the electronic structure of neon.

Lewis published these ideas in his 1923 book *Valence and the Structure of Atoms and Molecules*, and they were widely taken up and developed in the U.S.A. and Europe, for example, by N. V. Sidgwick at Oxford, whose *Electronic Theory of Valency* appeared in 1927. The Nobel Prize in Chemistry was left unfilled in 1919, 1924 and 1933 for "lack of candidates of suitable stature," and Lewis would have been an appropriate candidate for any of these years. In fact, he was nominated for a Nobel Prize by the inorganic chemist and historian of chemistry, J. R. Partington (1886-1965) at the University of London. For the first half-century after the award of the first Nobel Prize in Chemistry to van 't Hoff in 1901, the chemistry prize went to those who had discovered or characterised new chemical elements, new physico-chemical principles, new chemical reactions, or had elucidated the structure and accomplished the synthesis of natural products. The first award for "research into the nature of the chemical bond and its application to the elucidation of the structure of complex substances" went in 1954 to Linus Pauling at Caltech.

Another feature of the Nobel Prize in chemistry lies in its award on occasion, not to the original discoverer, but to a student or colleague who develops the discovery and makes known its scope and utility. The 1912 award in chemistry was divided: one half went to V. Grignard (1871–1935) at Nancy for "the discovery of the Grignard reagent," and the other half to P. Sabatier (1854–1941) of Toulouse for "his method of hydrogenating organic compounds in the presence of finely disintegrated metals," which is generally known as the Sabatier-Senderens process. Grignard was a research student at Lyons with F. A. P. Barbier (1848–1922), who gave him the problem of determining the scope of a reaction Barbier had discovered in 1899, that of an alkyl magnesium halide ("Grignard's reagent") with a ketone to form a tertiary alcohol. Grignard was eminently successful and widely publicised the vast scope of the reaction, making the field his own. Barbier carried on inventing new processes, and he is remembered for the Barbier-Wieland degradation of a carboxylic acid to its next lower homologue, discovered by Barbier in 1913 and developed by H. Wieland for his Nobel Prize "investigations of the constitution of the bile acids," awarded 1928. The Abbé J.-B. Senderens (1856–1936) collaborated for some 16 years with Sabatier at the University of Toulouse on the catalysis of gas-phase reactions by finely divided metals, and it is generally judged that their respective contributions to the Sabatier-Senderens process are not separable. Senderens was active in religious affairs, whereas Sabatier was more concerned with public affairs and became the first provincial (non-resident) member of the Paris Academy of Sciences. They parted company in 1911, when the award of half the Nobel Prize for 1912 to Sabatier alone became known.

The propensity to award a Nobel Prize to developers of a discovery, rather than its originator, was favoured by the regulation limiting the sharing of an award to no more than three persons, which tended to focus debate on the developers. The limitation to a trio became increasingly anachronistic during the late 20th century, particularly in experimental particle physics where the discovery of a new particle involved teams of several dozen scientists and engineers. An international team of 126 physicists from 11 institutions was involved in the detection of the neutral and two charged massive bosons mediating the unified electroweak interaction at CERN in 1983, but the physics Nobel Prize of 1984 for the achievement had to be limited by the regulation, and two of the leaders were selected for the award. Voices at the time urged that the institution of CERN itself should be honoured by the award. Such a change has been pioneered by the Nobel Peace Prize committee which made a recent award to the *Médecins sans frontières* as an institution, rather than to an individual, or duo or trio within it.

The Nobel regulation specifically restricting an award to no more than three persons appears to rest on ancient numerology. There is but a single mention of the concept and the term, "law of nature," in the entire

Aristotelian corpus (*On the Heavens*, 268a 12-16; 284b 24-25), where Aristotle tells us that, "as the Pythagoreans say, the world and all that is in it, are governed by the number three," and this is why extended bodies and space have three dimensions, and why we worship the gods in sets of three!

What message would you like to pass on from the history of science?

When I was a student, more than 50 years ago now, the textbooks usually outlined a brief history of each topic discussed, and chemical reviews were potted critical histories of the theme subject, indicating the apparent side-tracks and dead ends of some of the research lines. These days chemical reviews deal more with the current "state of the art," covering developments over the past few years or so, without much mention of the beginnings of the research field, or its early and often erratic progress. The practising scientist starting research half a century ago, particularly those who read up the salient history of his or her subject, usually had a broader perspective of the character and growth of science than those starting today, preoccupied with current and specialised concerns. The wider perspective of the past encouraged an interdisciplinary approach by spanning broad swathes of the methods and the notions of the natural world inherited from past generations, where ideas and methods from one field were often used analogically in another.

The significance of side-track and dead ends in historical research lines is illustrated by the apparently fanciful idea of Pasteur, which I first encountered in the 1940s, that there was a cosmic handed force of nature, omnipresent even in his crystallization dishes. Something of the notion remained with me, since Fischer's refutation of the idea, through his "key and lock" mechanism for the evolution of biomolecular homochirality, left open the problem of the origin and the particular chirality of the primordial enantiomer from which that evolution had begun. The demonstration of the non-conservation of parity of the weak nuclear interaction in 1956 and the subsequent unification of the force with electromagnetism, provided in the neutral electroweak interaction an attenuated version of Pasteur's postulate, a universal chiral force operational even in the ground state of atoms and molecules. Quantum mechanical computations including the electroweak interaction then showed that the series characterised by Fischer as dominant in the organic world, the L-amino acids and the D-sugars, have a slight energetic advantage over the corresponding enantiomeric series.

The universal nature of such a chiral force is evident from the analyses of the carbonaceous meterorites, where an excess of the L- over the D-enantiomer of the *non-biological* amino acids is found. The excess cannot be ascribed to biological contamination on the the surface of the earth after the fall of the meteorite, as was the corresponding excess of the L-enantiomer of the natural amino acids, detected earlier in meteoritic organic material. The non-biological and the natural amino acids were probably formed on the surface of a parent asteroid, by reactions which have been characterised in the laboratory, from precursor molecules in the giant molecular clouds, detected by microwave spectroscopy. There is no evidence as yet, however, for complex biological molecules, such as strands of DNA, free-floating in outer space or riding on cosmic dust, as advocated by the panspermists, who offer no account of the origin of such materials, but claim that cosmic DNA introduced life to the earth, despite the evidence suggesting the existence of an RNA organic world before DNA was evolved.

The history of science placed in its social context serves as a primary agent for the public understanding of science, informing all concerned of what became in the 19th century a major factor of historical change in many aspects of the life of humankind; agriculture, industry, medicine, war, world-view, and other features. Such changes were small and piecemeal before the 19th century, when they began to cohere and become effective. The sciences had humble beginnings in ancient Egypt and Mesopotamia with the mathematical arts of numerical and spatial reckoning, separate from mythological world-views. In Hellenic and Hellenistic times, cosmology became geometric and spatially-ordered, with a superior predictive capacity for calendrical and astrological applications (e.g., solar eclipses), compared with the Babylonian arithmetical cycle-schemes. But testing speculations by predictive success, or even conformity to known observations, remained a quite minor feature of rational natural philosophy from the time of the pre-Socratic Greeks to the 17th century Cartesian mechanists. Any observation contrary to hypothesis was no serious falsification, but a challenging anomaly, to be logically explained away, or avoided if potentially impious. Heresy was already a criterion for the rejection of hypotheses in natural and social philosophy during antiquity, and the criterion was strengthened during the medieval Christian and Islamic era.

The Renaissance and Reformation brought about a gradual secularisation of natural philosophy during the 16th and 17th century, with the decline of religious conformity as a serious criterion for scientific validity, and with the reorientation of logical coherence to the explanation of new observations and the results of the experimental probing of nature's secrets. Early modern proto-scientists were impressed by the new techniques which had come into late medieval Europe from China through the Arabs: gunpowder, the marine magnetic compass, paper and printing, which led them to value and respect empirical craft procedures. They were impressed too by the Hindu numerals which had come in with the Arabic algorithm and algebra, along with a refurbished alchemy and astronomy. Products coming in from the voyages of exploration indicated that many novelties remained to be discovered, in addition to the new lands.

A general mentality of exploration and discovery had developed by the 16th century, extending beyond the geographical surface of the terrestrial globe to the entire content of the cosmos. It was a thought-style leading to the foundation of several scientific academies, often ephemeral, to enquire into the secrets of nature. The social importance of science was stressed in a major way by Francis Bacon (1561–1626), a lawyer-philosopher and civil servant, who envisaged in his utopian *New Atlantis* (1627) a national academy of sciences, well equipped with workshops and support staff, dedicated to the discovery of "the knowledge of Causes, and secret motions of things; and the enlarging of the bounds of Human Empire, to the effecting of all things possible." This was to be achieved by means of Bacon's *New Instrument* (1620), his hypothetico-inductive method, applied to critical collections of information on natural phenomena, new and old, his *Histories*. Bacon's vision was influential in mid-17th century England, leading to the foundation of the Royal Society of London in 1660, in mid-18th century France among the *philosophes*, and in Britain again, among the founders of the British Association for the Advancement of Science (1831).

The principal scientific advances during the 17th century lay in the classical sciences of mechanics and astronomy, and they were made by the hypothetico-deductive mathematical method, culminating with the explanatory and predictive successes of Newton's *Principia* (1687). By 1700, there was some disenchantment with Bacon's vision among Fellows of the Royal Society, for there had been no important technological advances resulting from the applications of the new science over the past few decades. The vision had been narrowly interpreted to suppose that scientific discoveries would soon be rewarded with useful applications, if not immediately. In fact by 1712, the first new prime mover of the early industrial revolution, the atmospheric steam engine, had been set up in the Midlands, applying the 17th century discovery of the weight or pressure of the atmosphere at the surface of the earth, and its immense power, relative to the vacuum.

This power had been spectacularly demonstrated by the Mayor of Magdeburg, Otto von Guericke (1602–86), who showed in 1654 that two teams of eight horses were unable to pull apart the two halves of an evacuated hollow sphere.

Subsequent mechanical inventions showed a similar dependence on a general background knowledge of scientific principles and discoveries. In some of the new Baconian sciences of the 17th and 18th centuries applications were more direct, in magnetism, electricity, heat, light and, above all, chemistry. Modern studies of these sciences began in a Baconian fashion, with the experimental discovery and ordering of new effects, usually explained hypothetically in terms of the motions of particles or imponderable fluids, and new inventions came directly from the investigator, such as the lightning conductor by Benjamin Franklin (1706–90) from his kite experiment of 1752. Mathematical procedures and the hypothetico-deductive method entered studies of magnetism, electricity, heat and light, during the 19th century, and applications in these fields then began to depend on a wider background knowledge, as in the earlier case of mechanical inventions.

Chemistry and its off-shoots, biochemistry, geochemistry, molecular biology, and the like, have remained largely Baconian sciences, dependent upon the hypothetico-inductive method with mainly qualitative deductions, and a vast data base of compounds, ordered qualitatively into natural classes by properties and reactions, dependent upon the periodic table for inorganic materials, or the functional types (alcohols, acids and so forth) of the aromatic and aliphatic series for organic substances. The development of molecular stereochemistry during the 19th century was a wholly internal affair, quite independent of the kind of deductive mathematical superstructure involved in the unification of electromagnetism, embracing optics. Dirac in 1929 said of his quantum mechanics: "The underlying physical laws for for the mathematical theory of a large part of physics and the whole of chemistry are thus completely known." Over the 20th century, quantum mechanics, using radical approximations for the chemical field, largely told chemists what they already knew, although the new *post hoc* explanations were most enlightening and stimulated new lines of enquiry. Hitherto, quantum mechanics in chemistry (unlike the case for physics) has offered few new expectations or new predicted effects, except in the borderline of molecular physics.

The applications of chemistry in the early days were prompt and came from those associated with chemical discovery. The new bleaching agents of the 1790s for the textile industry originated with C.-L. Berthollet (1749–

1822), in the circle around Lavoisier, and the students of Kekulé were prominent in the development of synthetic coal-tar dyes from the 1860s on. By 1900, industrial laboratories had taken over most of the research and development on synthetic dyes, and then on synthetic pharmaceuticals. The last of the major independent chemical inventors was Alfred Nobel (1833–96). He had his own research laboratories and production plants for his new explosives in a number of countries, protected by some 355 patents across the world. His financial success was such that he was able to make provision for the Nobel Foundation (1900) to identify and award substantial prizes "to those who during the preceding year, shall have conferred the greatest benefit on mankind," in the fields of physics, chemistry, physiology or medicine, literature of an idealisatic tendency, and the promotion of world peace.

Critics asserted that the provision of the Foundation and Prizes was an attempt by Nobel to dispel the image, "Merchant of Death," acquired during his lifetime. Something of that image was attached to Fritz Haber (1868–1934), Nobel Laureate in Chemistry (1918/1919) for his synthesis of ammonia from its elements, when it became known that he had developed poison gases during World War I; even after he was obliged to leave Germany in 1933 on account of his Jewish antecedents. Nobel and Haber were cited as authentic exemplars of the mythology of Faustus and Frankenstein, subsequently augmented by the spendidly-preposterous Dr. Strangelove, all serving as scapegoats for perceived social ills and an apparent cosmic disarray. After World War II, scientists at large were targeted by maverick "Greens" as primarily responsible for industrial and agricultural environmental degradations, producing local and even global pollutions, despite the fact that, by that time, most scientists were placed in the position favoured by Winston Churchill; "Boffins should be on tap and not on top." Scientists could and should advertise to the public that technical solutions are accessible, where not already available, for most environmental problems, and that the real problem is social, the economic and political factors involved in the funding and implementation of scientifically-based solutions.

Name Index

Page numbers in bold refer to interviews.